10th INTERNATIONAL WORKSHOP

NONLINEAR EVOLUTION EQUATIONS & DYNAMICAL SYSTEMS

10th INTERNATIONAL WORKSHOP

NONLINEAR EVOLUTION EQUATIONS & DYNAMICAL SYSTEMS

Los Alamos, NM, USA 11–18 September '94

Edited by

Vladimir G. Makhankov, Alan R. Bishop & Darryl D. Holm

Los Alamos National Laboratory, Los Alamos, New Mexico, USA

World Scientific

Singapore • New Jersey • London • Hong Kong

Published by

World Scientific Publishing Co. Pte. Ltd.
5 Toh Tuck Link, Singapore 596224
USA office: 27 Warren Street, Suite 401-402, Hackensack, NJ 07601
UK office: 57 Shelton Street, Covent Garden, London WC2H 9HE

British Library Cataloguing-in-Publication Data
A catalogue record for this book is available from the British Library.

NONLINEAR EVOLUTION EQUATIONS AND DYNAMICAL SYSTEMS NEEDS '94

ISBN-13 978-981-02-2219-2
ISBN-10 981-02-2219-X

Dedicated
to the memory of
Irene Dorfman

ORGANIZING COMMITTEE

V. G. Makhankov (Charman, LANL, NM, USA)

F. Calogero (Co-Chairman, Rome)

B. Rhodes (Conference Secretary, LANL, NM, USA)

M. Ablowitz (Boulder, CO, USA)

A. R. Bishop (LANL, NM, USA)

G. D. Doolen (LANL, NM, USA)

D. D. Holm (LANL, NM, USA)

ACKNOWLEDGEMENTS

We would like to express our gratitude to those who helped in organizing the workshop. The support of Professor Don Cohen, at that time the Director of the Center for Nonlinear Studies (CNLS) at Los Alamos National Laboratory, is very much appreciated. The additional financial support provided by the Director of the Theoretical Division at Los Alamos National Laboratory, Richard Slansky, allowed us to invite and support a number of the participants, including those from Eastern Europe and the FSU.

Finally, we wish to thank all of the staff of CNLS involved in the organization of the conference, especially the Conference Secretaries Barbara Rhodes and Rose Vigil. We also benefitted from the expert assistance of the Visitor Support staff at the Los Alamos National Laboratory.

Index of Contributors

Alber M.
Biondini G.
Bouquet S.
Calogero F.
Caroll R.
De Lillo S.
Fordy A.
Gjaja I.
Guil F.
Heredero R.
Hietarinta J.
Iizuka T.
Kajiwara K
Kostelec P.
Lambert F.
Leznov A.
Luce B.
Manas M.
Marsden J.
Mattheus A.
Miller P.
Ohta Y.
Rakauskas R.
Robinson J.
Rybakov Y.
Sabatier P.
Sanyuk V.
Saxena A.
Springael J.
Svinolupov S.
Yukalova E.
Zagrodzinski J.

Baker S.
Blumenfeld R.
Brandt H.
Camassa R.
Deissler R.
Enolskii V.
Gabitov I.
Grammaticos B.
Habib S.
Herman R.
Holm D.
Jones D.
Kimura Y.
Kupershmidt B.
Lewis H.
Loris I.
Makhankov V.
Margolin L.
Martinez Alonso L.
Medina E.
Nimmo J.
Poje A.
Ramani A.
Rodin Y.
Ryne R.
Sadzius R.
Satsuma J.
Sokolov V.
Sulskus J.
Yukalov V.
Willox R.

PREFACE

The 10th Workshop on Nonlinear Evolution Equations and Dynamical Systems (NEEDS) was held in Los Alamos, New Mexico, USA, September 11-18, 1994. The previous nine workshops of this series have taken place in Crete (Greece) 1980, 1983, 1989; in Gallipoli (Italy) 1985, 1991, 1993; in Balaruc-les-Bains near Montpellier (France) 1987; and in Dubna near Moscow 1990, 1992. So the tradition was to hold workshops on the coast of European seas (natural or artificial). Decisively breaking with this tradition, the 10th Workshop "flew" over the Atlantic ocean, "climbed" the Rocky Mountains, and set up near what is probably the greatest volcano (now dormant, according to Los Alamos scientists!) ever known in the Earth's history. Hence, the NEEDS workshop series has acquired a new geographic and, we hope not just geographic, dimension.

The workshop was organized by the Center for Nonlinear Studies (CNLS) of the Los Alamos National Laboratory.

The workshop attracted about eighty participants from various nations: USA - 47, FSU (former Soviet Union) - 12, United Kingdom - 5, Italy and Japan - 4, Germany and Spain - 2, Belgium, Finland, France Israel and Poland - 1. We would like to note that again this time, many young scientists participated in the meeting from both the East and West, thanks to financial support provided by CNLS and the Theoretical Division of Los Alamos National Laboratory.

The program was interdisciplinary and highly intensive. More than fifty half-hour talks and twenty posters were presented.

The workshop followed the traditional pattern of including topics from practically all branches of nonlinear science, including integrable, near integrable and non-integrable dynamical systems (nonlinear ODE and PDE), their theory and possible applications in **classical and quantum physics** (condensed matter, fluids, plasmas, elementary particles, statistical mechanics, etc.) and others (oceanography, biophysics, etc.). The methods discussed ranged from pure mathematics and numerical computations to applied physics, including experiments.

The opening addresses were given by Professors Don Cohen and Alan Bishop and the workshop was closed by Professor Vladimir Makhankov.

The workshop participants were given an opportunity to visit one of the greatest calderas in the world, which reminded us of the gradiose volcano eruption held there a million years ago, as well as several other interesting and picturesque places related to the more recent history of the state of New Mexico, "Land of Enchantment."

On the last day of the workshop, a moment of silence was held in memory of Irene Dorfman who died last year. A letter expressing the sense of loss we felt due to her unexpected absence at the workshop and our gratitude for her beautiful contributions to our community was signed by the participants and sent to her mother. These Proceedings are dedicated to her memory.

This volume of NEEDS Proceedings includes papers from all of the talks received by the editors before the closing date for submissions, as well as several papers from the posters. For the reader's convenience, we have divided the presentations into two parts: "purely" theoretical papers and papers describing applications, although sometimes it was difficult to draw the line between them.

Los Alamos, December, 1994 V. G. Makhankov, A. R. Bishop; and D. D. Holm

CONTENTS

PART 1. THEORY

The Geometry of Weak Solutions of Certain Integrable Nonlinear PDE's

Mark S. Alber * Roberto Camassa †
Darryl D. Holm † and Jerrold E. Marsden ‡

Abstract

We investigate the geometry of new classes of soliton-like weak solutions for integrable nonlinear equations. One example is the class of peakons introduced by Camassa and Holm [1993] for their integrable shallow water equation. Alber, Camassa, Holm and Marsden [1994a] put this shallow water equation into the framework of complex integrable Hamiltonian systems on Riemann surfaces and use special limiting procedures to obtain new solutions such as quasiperiodic solutions, n-solitons, solitons with quasiperiodic background, billiard, and n-peakon solutions and complex angle representations for them. They also obtain explicit formulas for phase shifts of interacting soliton solutions using the method of asymptotic reduction of the corresponding angle representations. The method they use for the shallow water equation also leads to a link between one of the members of the Dym hierarchy and geodesic flow on N-dimensional quadrics. Amongst these geodesics, particularly interesting ones are the umbilic geodesics, which generate the class of umbilic soliton solutions. Umbilic solitons have the property that as the space variable x tends to infinity, the solution tends to a periodic wave, and as x tends to minus infinity, it tends to the same periodic wave with a phase shift. Elliptic billiards may be obtained from the problem of geodesics on quadrics by collapsing along the shortest semiaxis. The corresponding Hamiltonian billiard flows are associated to new classes of solutions of equations in the Dym hierarchy. Such billiard type solutions have discontinuous spatial derivative and, thus, are weak solutions for this class of PDE's.

*Department of Mathematics, University of Notre Dame, Notre Dame, IN 46556. Research partially supported by NSF grant DMS 9403861

†Center for Nonlinear Studies and Theoretical Division, Los Alamos National Laboratory, Los Alamos, NM 87545. Research supported in part by DOE CHAMMP and HPCC programs

‡Department of Mathematics, University of California, Berkeley, CA 94720. Research partially supported by the Department of Energy, the Office of Naval Research and the Fields Institute for Research in the Mathematical Sciences

1 Introduction

We are concerned with the geometry of new classes of soliton-like solutions. Examples of such solutions are the peaked soliton solutions (peakons) found by Camassa and Holm [1993] for the integrable equation

$$U_t - U_{xxt} + kU_x + 3UU_x = 2U_xU_{xx} + UU_{xxx} \tag{1.1}$$

which describes unidirectional propagation of waves at the free surface of a shallow layer of water at rest at infinity. The dependent variable $U(x,t)$ in (1.1) represents the fluid velocity in the x direction in appropriate nondimensional units, the constant k is related to the critical shallow water wave speed (see Camassa and Holm [1993]), and subscripts denote partial differentiation. Unlike the well known Korteweg-de Vries (KdV) and Benjamin, Bona, Mahoney (BBM) models, the derivation of equation (1.1) uses an asymptotic expansion that retains the Hamiltonian structure of an intermediate model, the Green-Naghdi system [1976], which in turn inherits its Hamiltonian structure from that of the original three dimensional Euler equations (as discussed, e.g., in Ge *et al.* [1993]).

The asymptotic expansion leading to equation (1.1) is carried out in terms of *only one* small parameter, *i.e.,* the typical amplitude of fluid velocities in a frame of reference moving at the critical shallow water wave speed. If one introduces an additional small parameter measuring the ratio between the water depth and a typical wavelength, or a long wave approximation, as in the derivation of the KdV and BBM models, then the right hand side of equation (1.1) would be of higher order with respect to the left hand side which has all the leading order terms. These leading order terms comprise the BBM equation. The higher order terms on the right are retained because the asymptotic expansion is performed on the level of the Hamiltonian structure itself rather than on the equations of motion. This procedure leads to retaining terms, among the higher order ones coming from the long wave asymptotics, that turn the BBM equation, perhaps surprisingly, into a completely integrable equation. In other words, the BBM equation can be seen as a perturbation of the shallow water equation (1.1), and so the latter provides an integrable structure behind the BBM model.

We remark that not all terms of the same order as the right hand side of (1.1) are retained by the Hamiltonian asymptotics. Hence, the shallow water equation (1.1) has the same asymptotic accuracy as the BBM or KdV equations. The derivation of (1.1) imposes conservation of the first two nontrivial terms of the asymptotic expansion of the Green-Naghdi Hamiltonian. The first term together with the Hamiltonian structure coming from Euler's equations yields the Hamiltonian flow for (1.1). The second term in the asymptotic expansion of the Green-Naghdi Hamiltonian is also conserved by (1.1) and leads to a *second* Hamiltonian structure for the equation. The two Hamiltonian structures define a Lax pair for (1.1) and show that the equation is completely integrable. See Camassa and Holm [1993] and Camassa, Holm and Hyman [1994] for details.

Camassa and Holm [1993] derive equation (1.1) and discuss its Lax pair and the dynamics of its peakon solutions in detail for the case $k = 0$. Alber, Camassa, Holm and Marsden

[1994a] investigate the phase geometry for solutions of this equation in the case that no restrictions are imposed on k. The main points of Alber, Camassa, Holm and Marsden [1994a] are as follows:

1. Finite dimensional Hamiltonian systems are found corresponding to peakon and billiard solutions of equation (1.1). These systems are complexified using the method of generating equations. This set up reveals finite dimensional integrable Hamiltonian systems whose solutions give new special solutions of the nonlinear equation (1.1) such as quasiperiodic and n-soliton solutions as well as solitons with quasiperiodic background.

2. Billiard solutions of these Hamiltonian systems are found. These solutions provide a possible route to the peakon solutions of equation (1.1) in the limit as k goes to zero.

3. A derivation of the shallow water equation hierarchy (the hierarchy of evolution equations that contains equation (1.1)) is given using the method of generating equations due to S. J. Alber [1979].

4. Phase shifts for soliton solutions are derived using the method of asymptotic reduction of complex angle representations.

5. A connection is established between equation (1.1), the problem of geodesics on quadrics and a nonlinear partial differential equation that belongs to the Dym equation hierarchy.

2 The geometry of the Dym equation

Equation (1.1) is expressible as

$$M_t = -(M\partial_x - \partial_x M)U, \tag{2.2}$$

with

$$M = U - U_{xx} + \frac{k}{2}.$$

This equation belongs to an associated hierarchy of equations determined from the recursion operator

$$R = J_2 J_1^{-1}, \quad \text{where} \quad J_1 = \partial_x - \partial_x^3 \quad \text{and} \quad J_2 = -M\partial_x - \partial_x M$$

are the first and second Hamiltonian operators. Applying R three times, starting from equation (1.1) gives the integrable equation (Camassa and Holm [1993]),

$$M_t = -(\partial_x - \partial_x^3)\frac{1}{\sqrt{M}}. \tag{2.3}$$

In the standard Dym equation (see Kruskal [1975] and Wadati *et al.* [1980]) the term ∂_x in (2.3) is dropped; thus, (2.3) is Dym-like but there are deep differences in its underlying complex geometry (see Alber, Camassa, Holm and Marsden [1994b]). The two Hamiltonian structures for the standard Dym equation are $\tilde{J}_1 = -\partial_x^3$ and J_2 above with

$$M = U_{xx} + \frac{k}{2}.$$

The equation located at the same level as (1.1) in the hierarchy of the standard Dym equation is

$$U_{xxt} = -2U_x U_{xx} - UU_{xxx} + kU_x, \tag{2.4}$$

which can be obtained by applying the recursion operator $\tilde{R} = \tilde{J}_1 J_2^{-1}$ three times to the Dym flow. Equation (2.4) is of special interest because it can be viewed as describing the dynamics of a stationary Hamiltonian system with a finite number of degrees of freedom associated with geodesic flow on n-dimensional quadrics. This connection leads to new interesting classes of solutions of equation (2.4). Every equation in the Dym hierarchy can be linked to the problem of geodesics on quadrics. However the corresponding time flows are different for each particular equation in the hierarchy.

Alber, Camassa, Holm and Marsden [1994b] describe a new class of soliton solutions, called umbilic solitons, for certain nonlinear integrable PDE's. These umbilic solitons have the property that as the space variable x tends to infinity, the solution tends to a periodic wave, and as x tends to minus infinity, it tends to a phase shifted periodic wave of the same period. The equations admitting solutions in this new class include the Dym equation and the equations in its hierarchy. The methods used to find and analyze these solutions are those of algebraic and complex geometry. Classes of solutions are sought by constructing associated finite dimensional integrable Hamiltonian systems on Riemann surfaces. For example, in this setting geodesics on n-dimensional quadrics provide the spatial, or x-flow, which, together with the commuting t-flow given by the equation itself, defines new classes of solutions. Among these geodesics, particularly interesting ones are the umbilic geodesics, which then generate the class of umbilic soliton solutions. This same setting allows the introduction of another class of solutions of "Dym-like" equations, which are related to elliptic and umbilic billiards.

Elliptic billiards may be obtained from the problem of geodesics on quadrics by collapsing along the shortest semiaxis; for details, see Alber [1986]. This process yields Hamiltonians and first integrals for the resulting billiard problem. The reflection conditions of the billiard can be expressed as a jump from one sheet of an associated Riemann surface to another. Alber, Camassa, Holm and Marsden [1994b] use the corresponding Hamiltonian billiard flows to construct new classes of solutions of equations in the Dym hierarchy. Such billiard type solutions have discontinuous spatial derivative and, thus, are weak solutions for the class of PDE's they consider. Particular weak solutions of equation (2.4) for the case $\kappa = 0$ are studied in connection with weakly nonlinear solutions of hyperbolic equations in Hunter and Zheng [1994]. (Alber, Camassa, Holm and Marsden [1994a] also characterize the peakon solutions of the shallow water equation (1.1) as solutions of billiard type).

This association of weak solutions of an equation in the Dym hierarchy with the dynamics of elliptic billiards raises the question of what types of algebraic perturbations of Hamiltonian billiard flows may be associated with quasiperiodic and umbilic-soliton solutions of Dym-like equations. For example, an interesting class of perturbations of the elliptic billiard flow is considered in Levallois and Tabanov [1993]. Elliptic billiards are investigated in Moser and Veselov [1991] in connection with the discrete C. Neumann problem, and in Deift, Li and Tomei [1992] in the more general context of loop groups. Also, Cewen [1990] discusses a link between stationary flows of the Dym equation and the problem of geodesics on an ellipsoid. A relation between stationary flows in the Dym hierarchy and certain mechanical problems is also discussed in Ragnisco and Rauch-Wojciechowski [1994].

Bibliography

M.S. Alber [1986], Hamiltonian formalism for finite-zone solutions of nonlinear integrable equations, *Proc. VIII Int.Cong.on Math. Physics*, Marseille, France (World Scientific) 447–462.

S.J. Alber [1979], Investigation of equations of Koteweg-de Vries type by the method of recurrence relations, *J. London Math. Soc. A* **19** 467-480.

M.S. Alber, R. Camassa, D.D. Holm and J.E. Marsden [1994a], The geometry of peaked solitons and billiard solutions of a class of integrable PDE's, *Lett. Math. Phys.* **32** (1994) 137–151.

M.S. Alber, R. Camassa, D.D. Holm and J.E. Marsden [1994b], On the link between umbilic geodesics and soliton solutions of nonlinear PDE's, *Proc. Roy. Soc.*, to appear.

R. Camassa and D.D. Holm[1993], An integrable shallow water equation with peaked solitons, *Phys. Rev. Lett*, **71**, 1661–1664.

R. Camassa, D.D. Holm and J.M. Hyman [1994], A new integrable shallow water equation, *Adv. in Appl. Mech.*, **31**, 1-33.

C. Cewen [1990], Stationary Harry-Dym's equation and its relation with geodesics on ellipsoid, *Acta Math. Sinica*, **6**, 35–41.

P. Deift, L.C. Li and C. Tomei [1991], Loop groups, discrete versions of some classical integrable systems, and rank 2 extensions, *Mem. Amer. Math. Soc.*, **479**.

Z. Ge, H.P. Kruse, J.E. Marsden and C. Scovel [1993] Poisson Brackets in the Shallow Water Approximation (*preprint*).

A.E. Green and P.M. Naghdi [1976] A derivation of equations for wave propagation in water of variable depth. *J. Fluid Mech.* **78**, 237-246.

J.K. Hunter and Y. Zheng [1994], On a Completely Integrable Nonlinear Hyperbolic Variational Equation *Physica D* (to appear).

M.D. Kruskal [1975] Nonlinear wave equations, *in* Dynamical Systems, Theory and Applications, J. Moser ed., Lecture Notes in Physics, **38**, Springer, New York.

P. Levallois and M.B. Tabanov [1993], Splitting of separatrices of elliptic billiard for a symmetric algebraic perturbation of the ellipse, *C.R. Acad. Sci. Paris*, **316**, Serie I, 589–592.

J. Moser and A.P. Veselov [1991], Discrete Versions of Some Classical Integrable Systems and Factorization of Matrix Polynomials, *Commun. Math. Phys.*, **139**, 217–243.

O. Ragnisco and S. Rauch-Wojciechowski [1994], Integrable mechanical systems related to the Harry-Dym hierarchy, *J. Math. Phys.*, **35**, 834–847.

M. Wadati, Y.H. Ichikawa and T. Shimizu [1980] Cusp soliton of a new integrable nonlinear evolution equation, *Prog. Theor. Phys.* **64**, 1959-1967.

Squared Eigenfunction Coordinates and Stationary Flows

S. Baker[a], V.Z. Enolskii[b] and A.P. Fordy[a]

November 29, 1994

[a] *Department of Applied Mathematical Studies and Centre for Nonlinear Studies, University of Leeds, Leeds LS2 9JT, UK*
[b] *Department of Theoretical Physics, Institute of Metal Physics, ul. Vernadsky 36, Kiev-680, 252142, Ukraine.*

Abstract

We present an alternative approach to establish the connection between the integrable cases of Hénon-Heiles and the stationary flows of some known integrable partial differential equations. We then extend this method to show a connection between some known integrable Hamiltonian systems with quartic potential and the stationary flows of some coupled KdV equations related to fourth and fifth order Lax operators. In particular, we give a connection between the Hirota-Satsuma coupled KdV system and the 1:6:1 integrable case quartic potential.

1 Introduction

A connection was shown in [1] between the known integrable cases of the Hénon-Heiles system and the stationary flows of some known integrable partial differential equations: the fifth order KdV, Sawada-Kotera and Kaup-Kupershmidt equations, which are associated with second and third order Lax operators. This connection was shown using ad hoc calculations which are difficult to extend to other systems.

Here we present an alternative method which can be used to establish this connection. In our approach we start with a skew or self adjoint Lax operator which, when factorised, gives rise to a Miura map. The Lax equation is used to generate our integrable pde, which we write in Hamiltonian form. We then look at the stationary case, which after some simple substitution can be integrated and then represented by some finite dimensional Hamiltonian system. We can determine the constants of motion as invariants of the Lax matrix, written in terms of canonical coordinates. We will illustrate our method by the Kaup-Kupershmidt and Sawada-Kotera cases. As these two equations are related through a Miura map we can determine a canonical transformation between the two Hamiltonians related to the stationary flows of these equations.

This approach can be extended to show a similar connection between some stationary flows associated with fourth and fifth order Lax operators and some generalisations of known integrable cases of the Hamiltonian system with quartic potential:

$$H = \frac{1}{2}(p_1^2 + p_2^2) + aq_1^4 + bq_1^2q_2^2 + cq_2^4. \tag{1}$$

This has the four nontrivial integrable cases:

1. $a : b : c = 1 : 2 : 1$,

2. $a : b : c = 1 : 12 : 16$,
3. $a : b : c = 1 : 6 : 1$,
4. $a : b : c = 1 : 6 : 8$.

2 Third Order Operators

We start with the skew adjoint third order operator written in the factorised form:

$$L_1 = (\partial + v)\partial(\partial - v). \tag{2}$$

With the Miura map

$$F_w[v] = w = -v_x - \frac{1}{2}v^2, \tag{3}$$

the operator becomes

$$L_1 = (\partial^3 + 2w\partial + w_x),$$

which, with the time evolution operator

$$M_1 = -(9\partial^5 + 30w\partial^3 + 45w_x\partial^2 + (20w^2 + 35w_{xx})\partial + 10w_{xx} + 20ww_x)$$

and the integrability condition

$$L_{it} = [M_i, L_i], \tag{4}$$

(with $i = 1$) gives the Kaup-Kupershmidt equation

$$w_t = (w_{xxxx} + 10ww_{xx} + \frac{15}{2}w_x^2 + \frac{20}{3}w^3)_x.$$

A rotation of the factors in (2) gives a second operator

$$L_2 = (\partial - v)(\partial + v)\partial. \tag{5}$$

This can be written as

$$L_2 = \partial^3 + u\partial,$$

where u is given by the Miura map:

$$F_u[v] = u = v_x - v^2. \tag{6}$$

The time evolution operator

$$M_2 = -(9\partial^5 + 15u\partial^3 + 15u_x\partial^2 + (5u^2 + 10u_{xx})\partial)$$

and the integrability condition (4) (with $i = 2$) give the Sawada-Kotera equation:

$$u_t = \left(u_{xxxx} + 5uu_{xx} + \frac{5}{3}u^3\right)_x.$$

The Kaup-Kupershmidt and Sawada-Kotera equations are both related through their Miura maps to the *same* modified equation [2]

$$v_t = (v_{xxxx} + 5v_xv_{xx} - 5v^2v_{xx} - 5vv_x^2 + v^5)_x.$$

This modified system has the Hamiltonian form

$$v_t = (\mathbf{B}_v)(\delta_v h),$$

with the Hamiltonian operator $\mathbf{B}_v = -\partial$ and the Hamiltonian $h = -\frac{1}{2}v_{xx}^2 - \frac{5}{2}v^2v_x^2 + \frac{5}{6}v_x^3 - \frac{1}{6}v^6$.

We now want to write the Kaup-Kupershmidt and Sawada-Kotera equations in their Hamiltonian forms:

$$w_t = \mathbf{B}_w\delta_w\mathcal{G}, \quad u_t = \mathbf{B}_u\delta_u\mathcal{H}.$$

The Hamiltonian operators are related to $\mathbf{B}_v$ by

$$\mathbf{B}_z = F_z'\mathbf{B}_v(F_z')^\dagger,$$

where F_z' denotes the Fréchet derivative of the Miura map and $\dagger$ denotes the adjoint. We find that

$$\begin{aligned}
\mathbf{B}_w &= (\partial^3 + 2w\partial + w_x), \quad \mathcal{G} = \frac{4}{3}w^3 - \frac{1}{2}w_x^2, \\
\mathbf{B}_u &= (\partial^3 + 4u\partial + 2u_x), \quad \mathcal{H} = \frac{1}{6}u^3 - \frac{1}{2}u_x^2.
\end{aligned}$$

2.1 Stationary Flows and Integrable Hénon-Heiles Systems

The stationary flows of the Kaup-Kupershmidt and Sawada-Kotera equations are respectively

$$\begin{aligned}
0 &= (\partial^3 + 2w\partial + w_x)\,\delta_w\left(\frac{4}{3}w^3 - \frac{1}{2}w_x^2\right), \\
0 &= (\partial^3 + 4u\partial + 2u_x)\,\delta_u\left(\frac{1}{6}u^3 - \frac{1}{2}u_x^2\right).
\end{aligned}$$

We can rewrite these equations as

$$\begin{aligned}
0 &= (\partial^3 + 2w\partial + w_x)A \text{ where } A = \delta_w(\frac{4}{3}w^3 - \frac{1}{2}w_x^2), \\
0 &= (\partial^3 + 4u\partial + 2u_x)B \text{ where } B = \delta_u(\frac{1}{6}u^3 - \frac{1}{2}u_x^2).
\end{aligned}$$

If we now set $A = \alpha y^2$ and $B = \beta z^2$ we find that

$$\begin{aligned}
&0 = 2yy_{xxx} + 6y_xy_{xx} + 4wyy_x + w_xy^2, \\
&0 = 2zz_{xxx} + 6z_xz_{xx} + 8uzz_x + 2u_xz^2.
\end{aligned}$$

These equations can easily be integrated, so we now have

$$y_{xx} + \frac{1}{2}wy = c_1y^{-3}, \quad \delta_w(\frac{4}{3}w^3 - \frac{1}{2}w_x^2) = \alpha y^2, \tag{7}$$

$$z_{xx} + uz = c_2z^{-3}, \quad \delta_u(\frac{1}{6}u^3 - \frac{1}{2}u_x^2) = \beta z^2, \tag{8}$$

where c_1 and c_2 are constants of integration. If we choose $\alpha = -\frac{1}{4}$ and $\beta = -\frac{1}{2}$ then the equations (7,8) can be derived respectively from the following Lagrangians:

$$\begin{aligned}
\mathcal{L}_{(kk)} &= \frac{1}{2}(w_x^2 + y_x^2) - \frac{4}{3}w^3 - \frac{1}{4}wy^2 - \frac{1}{2}c_1y^{-2}, \\
\mathcal{L}_{(sk)} &= \frac{1}{2}(u_x^2 + z_x^2) - \frac{1}{6}u^3 - \frac{1}{2}uz^2 - \frac{1}{2}c_2z^{-2}.
\end{aligned}$$

The standard Legendre transformation can be used for both cases. For the Kaup-Kupershmidt case choosing $q_1 = w$, $p_1 = w_x$, $q_2 = y$, $p_2 = y_x$ gives

$$h_{kk} = \frac{1}{2}(p_1^2 + p_2^2) + \frac{1}{4}q_1q_2^2 + \frac{4}{3}q_1^3 + \frac{1}{2}c_1q_2^{-2}, \tag{9}$$

and for the Sawada-Kotera case choosing $Q_1 = u$, $P_1 = u_x$, $Q_2 = z$, $P_2 = z_x$ we have

$$h_{sk} = \frac{1}{2}(P_1^2 + P_2^2) + \frac{1}{2}Q_1Q_2^2 + \frac{1}{6}Q_1^3 + \frac{1}{2}c_2Q_2^{-2}. \tag{10}$$

Both these Hamiltonians are generalisations of integrable cases of the Hénon-Heiles system.

Lax Representations

For our purposes of studying the stationary flows, the zero curvature representation is more convenient. We rewrite the equations:

$$L_1\phi = \lambda\phi, \quad M_1\phi = \phi_t,$$
$$L_2\psi = \lambda\psi, \quad M_2\psi = \psi_t,$$

in matrix form to get

$$\Phi_x = U_1\Phi, \quad U_1 = \begin{pmatrix} 0 & 1 & 0 \\ 0 & 0 & 1 \\ \lambda - w_x & -2w & 0 \end{pmatrix},$$

$$\Psi_x = U_2\Psi, \quad U_2 = \begin{pmatrix} 0 & 1 & 0 \\ 0 & 0 & 1 \\ \lambda & -u & 0 \end{pmatrix},$$

where $\Phi = (\phi, \phi_x, \phi_{xx})^T$, $\Psi = (\psi, \psi_x, \psi_{xx})^T$ and λ is the spectral parameter. The time evolution now takes the form:

$$\Phi_t = V_1\Phi, \equiv - \begin{pmatrix} a_{11} & -4w^2 - w_{xx} & 9\lambda \\ a_{21} & -6w\lambda & -4w^2 - w_{xx} \\ a_{31} & a_{32} & -8ww_x - w_{xxx} - 6w\lambda \end{pmatrix} \Phi,$$

$$\Psi_t = V_2\Psi \equiv - \begin{pmatrix} 6u\lambda & -u^2 + u_{xx} & 9\lambda - 3u_x \\ 3u_x\lambda + 9\lambda^2 & uu_x + u_{xxx} - 3u\lambda & -2u_{xx} - u^2 \\ \lambda(u_{xx} - u^2) & b_{32} & -u_{xxx} - uu_x - 3u\lambda \end{pmatrix} \Psi,$$

where

$$a_{11} = w_{xxx} + 8ww_x + 12w\lambda,$$
$$a_{21} = w_{xxxx} + 8w_x^2 + 8ww_{xx} + 3w_x\lambda + 9\lambda^2,$$
$$a_{31} = -2ww_{xxx} - 16w_xw^2 + 2\lambda(w_{xx} - 2w^2),$$
$$a_{32} = w_{xxxx} + 8w_x^2 + 10ww_{xx} - 3w_x\lambda + 9\lambda^2 + 8w^3,$$
$$b_{32} = u_{xxxx} + u_x^2 + 3uu_{xx} + 9\lambda^2 + u^3.$$

¿From the factorisation of the operators (2, 5) we have

$$(\partial + v)\partial\psi = \phi, \quad (\partial - v)\phi = \lambda\psi.$$

Using $\phi = \psi_{xx} + v\psi_x$ we get the following gauge transformation between the zero curvature representations:

$$\Phi = A\Psi, \quad A = \begin{pmatrix} 0 & v & 1 \\ \lambda & v^2 & v \\ \lambda v & \lambda + vv_x + v^3 & v^2 + v_x \end{pmatrix},$$

with

$$U_1 = A_x A^{-1} + AU_2A^{-1}, \quad \text{and} \quad V_1 = A_t A^{-1} + AV_2A^{-1}.$$

The stationary flows are given by the Lax representation $V_{ix} = [U_i, V_i]$, which can be written in terms of q_i, p_i, Q_i, P_i, and V_i are related through the similarity transformation $V_1 = AV_2A^{-1}$. The equations of motion generated by h_{kk} and h_{sk} are used to eliminate second and higher derivatives in V_i. We can use V_i to generate the constants of motion.

$$\begin{aligned} V_1 &= -\begin{pmatrix} 12q_1\lambda - \frac{1}{2}q_2p_2 & \frac{1}{4}q_2^2 & 9\lambda \\ a_{21} & -6q_1\lambda & \frac{1}{4}q_2^2 \\ q_1q_2p_2 - \frac{1}{2}\lambda q_2^2 - 12\lambda q_1^2 & a_{32} & \frac{1}{2}q_2p_2 - 6q_1\lambda \end{pmatrix}, \\ V_2 &= -\begin{pmatrix} 6\lambda Q_1 & -\frac{3}{2}Q_1^2 - \frac{1}{2}Q_2^2 & -3P_1 + 9\lambda \\ 3P_1\lambda + 9\lambda^2 & -Q_2P_2 - 3Q_1\lambda & Q_2^2 \\ -\frac{1}{2}\lambda Q_2^2 - \frac{3}{2}\lambda Q_1^2 & -P_2^2 + 9\lambda^2 - \frac{c_2}{Q_2^2} & Q_2P_2 - 3Q_1\lambda \end{pmatrix}, \end{aligned}$$

where

$$a_{21} = 9\lambda^2 + 3p_1\lambda - \frac{1}{2}p_2^2 + \frac{1}{4}q_2^2q_1 - \frac{1}{2}\frac{c_1}{q_2^2}, \quad a_{32} = 9\lambda^2 - 3p_1\lambda - \frac{1}{2}p_2^2 - \frac{1}{4}q_2^2q_1 - \frac{1}{2}\frac{c_1}{q_2^2}.$$

The determinants of V_i give

$$\begin{aligned} det(V_1) &= -729\lambda^5 + 162h_{kk}\lambda^3 + K_{kk}\lambda, \\ det(V_2) &= -729\lambda^5 + 162h_{sk}\lambda^3 + K_{sk}\lambda, \end{aligned}$$

where K_{kk} and K_{sk} are the corresponding second integrals of motion:

$$\begin{aligned} K_{kk} &= \frac{1}{32}q_2^6 + \frac{3}{16}q_1^2q_2^4 - \frac{9}{4}p_2^4 - \frac{9}{4}q_1p_2^2q_2^2 + \frac{3}{4}p_1p_2q_2^3 - \frac{3}{2}c_1\left(\frac{1}{2}q_1 + 3\frac{p_2^2}{q_2^2}\right) - \frac{9}{4}\frac{c_1^2}{q_2^4}. \\ K_{sk} &= -(3P_1P_2 + \frac{3}{2}Q_1^2Q_2 + \frac{1}{2}Q_2^3)^2 - 3c_2(\frac{3P_1^2}{Q_2^2} + 2Q_1). \end{aligned}$$

The Canonical Transformation

The Miura maps (3,6) and the equations of motion of the Hamiltonians (9,10) give

$$\begin{aligned} q_1 &= -v_x - \tfrac{1}{2}v^2 &= -2q_{2xx}q_2^{-1} + 2c_1q_2^{-4}, \\ Q_1 &= v_x - v^2 &= -Q_{2xx}Q_2^{-1} + c_2Q_2^{-4}. \end{aligned}$$

We can solve these for v to get

$$\begin{aligned} v &= 2\frac{q_{2x}}{q_2} + aq_2^{-2} \quad \text{where } a^2 = -4c_1, \\ v &= -\frac{Q_{2x}}{Q_2} + bQ_2^{-2} \quad \text{where } b^2 = -c_2. \end{aligned}$$

Therefore when $a = 4b$ we have a canonical transformation, which is *explicitly* invertible:

$$
\begin{aligned}
q_1 &= -\frac{3}{2}(-\frac{P_2}{Q_2} + bQ_2^{-2})^2 - Q_1, \\
p_1 &= 4\frac{P_2^3}{Q_2^3} + 4\frac{Q_1P_2}{Q_2} - P_1 - 4\frac{b}{Q_2^2}(Q_1 + 3\frac{P_2^2}{Q_2^2} - 3\frac{bP_2}{Q_2^3} + \frac{b^2}{Q_2^4}), \\
q_2 &= \frac{1}{Q_2^2}\left(-8P_1P_2Q_2^3 + 8P_2^2Q_1Q_2^2 - 4Q_1^2Q_2^4 - Q_2^6 + 8b[P_1Q_2^2 - 2P_2Q_1Q_2 + bQ_1]\right)^{1/2}, \\
p_2 &= \frac{1}{2Q_2^2}(-\frac{P_2}{Q_2} + bQ_2^{-2})(-8P_1P_2Q_2^3 + 8P_2^2Q_1Q_2^2 - 4Q_1^2Q_2^4 - Q_2^6 + 8b[P_1Q_2^2 \\
&\quad -2P_2Q_1Q_2 + bQ_1])^{1/2} + a\frac{1}{Q_2^2}(-8P_1P_2Q_2^3 + 8P_2^2Q_1Q_2^2 - 4Q_1^2Q_2^4 - Q_2^6 \\
&\quad +8b[P_1Q_2^2 - 2P_2Q_1Q_2 + bQ_1])^{-1/2}, \\
Q_1 &= -2(2\frac{p_2}{q_2} + aq_2^{-2})^2 - q_1, \\
P_1 &= 4\frac{p_2^3}{q_2^3} + 4\frac{q_1p_2}{q_2} - p_1 + 4\frac{a}{q_2^2}(q_1 + 3\frac{p_2^2}{q_2^2} + 3\frac{ap_2}{q_2^3} + \frac{a^2}{q_2^4}), \\
Q_2 &= \frac{1}{q_2^2}\left(-8p_1p_2q_2^3 + 8p_2^2q_1q_2^2 - 4q_1^2q_2^4 - q_2^6 - 8a[p_1q_2^2 - 2p_2q_1q_2 - aq_1]\right)^{1/2}, \\
P_2 &= -\frac{1}{q_2^2}(2\frac{p_2}{q_2} + aq_2^{-2})(-8p_1p_2q_2^3 + 8p_2^2q_1q_2^2 - 4q_1^2q_2^4 - q_2^6 - \\
&\quad 8a[p_1q_2^2 - 2p_2q_1q_2 - aq_1])^{1/2} + b(-8p_1p_2q_2^3 + 8p_2^2q_1q_2^2 - 4q_1^2q_2^4 - q_2^6 \\
&\quad -8a[p_1q_2^2 - 2p_2q_1q_2 - aq_1])^{-1/2}.
\end{aligned}
$$

3 Higher Order Operators

Here, we briefly describe the result of proceeding with higher order operators as we did in the above third order case. We start with the self adjoint fourth order factorised operator

$$L_1 = (\partial + v_1)(\partial + v_2)(\partial - v_2)(\partial - v_1), \tag{11}$$

which becomes

$$L_1 = (\partial^2 + f\partial + g + f_x)(\partial^2 - f\partial + g),$$

with the Miura map:

$$f = v_1 + v_2, \quad g = v_1v_2 - v_{1x}.$$

One of the isospectral flows is the coupled KdV system:

$$
\begin{aligned}
f_t &= -\frac{1}{2}(2f_{xxx} + 3ff_{xx} + 3f_x^2 - 3f^2f_x + 6fg_x + 6gf_x), \\
g_t &= \frac{1}{4}(2g_{xxx} + 12gg_x + 6fg_{xx} + 12gf_{xx} + 18f_xg_x - 6f^2g_x \\
&\quad +3f_{xxxx} + 3ff_{xxx} + 18f_xf_{xx} - 6f^2f_{xx} - 6ff_x^2).
\end{aligned}
$$

A rotation of factors of the fourth order factorised operator (11) leads us to

$$L_2 = (\partial - v_1)(\partial + v_1)(\partial + v_2)(\partial - v_2) = (\partial^2 + u + \phi)(\partial^2 + u - \phi),$$

where

$$u=\frac{1}{2}(v_{1x}-v_{2x}-v_1^2-v_2^2),\quad \phi=\frac{1}{2}(v_{1x}+v_{2x}-v_1^2+v_2^2).$$

One of the isospectral flows is the Hirota-Satsuma coupled KdV system:

$$\begin{aligned} u_t &= \frac{1}{2}u_{xxxx}+3uu_x-6\phi\phi_x,\\ \phi_t &= -\phi_{xxx}-3u\phi_x. \end{aligned}$$

Writing these coupled KdV equations in Hamiltonian form, we can proceed as before and integrate for the stationary flow cases after some simple substitutions. These integrated equations can then be written as generalisations of the integrable cases (4) and (3) respectively of the Hamiltonian (1). Furthermore since both the coupled KdV systems are related through a Miura map a canonical transformation can be found between the generalisations of the integrable cases (3) and (4).

The skew adjoint fifth order factorised operator is

$$L_1=(\partial+v_1)(\partial+v_2)(\partial)(\partial-v_2)(\partial-v_1).$$

After a rotation of factors we have the operator

$$L_2=(\partial-v_1)(\partial+v_1)(\partial+v_2)(\partial)(\partial-v_2).$$

¿From this operator, which becomes

$$L_2=(\partial^2+r)(\partial^3+2s\partial+s_x),$$

with the Miura map

$$r=v_{1x}-v_1^2,\quad s=-v_{2x}-\frac{1}{2}v_2^2,$$

we can show a connection between the equations

$$\begin{aligned} r_t &= -r_{xxx}+3s_{xxx}+6sr_x+12rs_x-6rr_x,\\ s_t &= -4s_{xxx}+3r_{xxx}-12ss_x+6sr_x+3rs_x, \end{aligned}$$

and a generalisation of the integrable case (2) of the Hamiltonian (1).

The details of these calculations will be presented in [3] and [4].

Acknowledgements

S. B. gratefully acknowledges the support of the E.P.S.R.C., both for the studentship and the travel grant which enabled her to attend the NEEDS meeting in Los Alamos.

References

[1] A.P. Fordy. The Hénon-Heiles system revisited. *Physica*, 52D:201–210, 1991.

[2] A. P. Fordy and J. Gibbons Factorisation of operators I : Miura transformations. *J.Math.Phys.*, 21:2508–10, 1980.

[3] S. Baker, V. Z. Enolskii, A. P. Fordy. Integrable quartic potentials and coupled KdV equations. Preprint, 1994.

[4] S. Baker. Integrable nonlinear evolution equations and related finite dimensional systems, Ph.D. Thesis. University of Leeds (in preparation), 1995.

The (2+1)-dimensional Eckhaus equation: display of some solutions

G. Biondini,[1,2] S. De Lillo[1,2] and F. Calogero[3,4]

[1] *Dipartimento di Fisica, Università di Perugia, 06100 - Perugia, Italy*
[2] *Istituto Nazionale di Fisica Nucleare, Sezione di Perugia*
[3] *Dipartimento di Fisica, Università di Roma "La Sapienza", 00185 - Roma, Italy**
[4] *Istituto Nazionale di Fisica Nucleare, Sezione di Roma*
* *On leave while serving as Secretary General, Pugwash Conferences on Science and World Affairs, Geneva, London, Rome*

Abstract

We display, analytically and graphically, solutions of the (2+1)-dimensional Eckhaus equation; they represent one and two solitons moving in a spherically symmetrical harmonic oscillator potential.

Physics and Astronomy Subject Classification: 02-30 jr, 03-40 kf, 03-65 nk

Introduction

In a recent paper [1] the "(N+1)-dimensional Eckhaus equation" was introduced, and reasons were given which motivate its study. Purpose and scope of the present paper is to display, not only analytically but also graphically, some representative solutions of this equation in two-dimensional space (N=2). In this case the (2+1)-dimensional Eckhaus equation reads as follows:

$$i\psi_t + \Delta\psi + V\psi + [|\psi|^4 + 2(|\psi|^2)_x]\psi + if\psi + 2g\psi_y + (g_y + g^2)\psi = 0\,, \qquad (1.1a)$$

$$g_x = (|\psi|^2)_y\,, \qquad (1.1b)$$

$$f_x + 2|\psi|^2 f = 2ig(\psi^*\psi_y - \psi\psi^*_y) + i(\psi^*\psi_{yy} - \psi\psi^*_{yy})\,. \qquad (1.1c)$$

Here the complex scalar dependent variable $\psi \equiv \psi(\mathbf{r}, t)$ is the main unknown field, and $f \equiv f(\mathbf{r}, t)$, $g \equiv g(\mathbf{r}, t)$ are two auxiliary real fields. The independent variable $\mathbf{r}$ is a vector in 2-dimensional space, $\mathbf{r} \equiv (x, y)$, and Δ is the corresponding Laplacian, $\Delta \equiv \partial^2/\partial x^2 + \partial^2/\partial y^2$. Subscripted Cartesian variables denote partial differentiation, f.i. $g_x \equiv \partial g/\partial x$, $(|\psi|^2)_y \equiv \partial(|\psi|^2)/\partial y$.

The real scalar $V(\mathbf{r})$ is a given "external potential"; in this paper we limit our consideration to the case of a rotationally symmetrical oscillator potential,

$$V(\mathbf{r}) = -r^2 \equiv -(x^2 + y^2)\,. \qquad (1.2)$$

Let us recall that, if the "second" space variable, y, were omitted, one could set $f = g = 0$ in (1.1), thereby recovering the standard one-dimensional Eckhaus equation in the quadratic potential [2].

Below we display some analytic solutions for $\psi(\mathbf{r},t)$; corresponding expressions for the two auxiliary fields $f(\mathbf{r},t)$ and $g(\mathbf{r},t)$ can be obtained by quadratures from the formulas

$$g(\mathbf{r},t) = \int_{-\infty}^{x} dx'(|\psi(x',y,t)|^2)_y\,, \tag{1.3a}$$

$$f(\mathbf{r},t) = -4\,\mathrm{Im}\Big\{ \int_{-\infty}^{x} dx' \psi^*(x',y,t)[2g(x',y,t)\psi_y(x',y,t) + \psi_{yy}(x',y,t)]\cdot$$
$$\cdot \exp\Big[2\int_{-x}^{x'} dx''(|\psi(x'',y,t)|^2)\Big]\Big\}\,, \tag{1.3b}$$

which follow from (1.1b,c) and the boundary conditions (assumed for simplicity)

$$g(-\infty,y,t) = f(-\infty,y,t) = 0\,. \tag{1.4}$$

We shall, accordingly, hereafter limit our consideration to solutions ψ that vanish as $x \to -\infty$, sufficiently fast to guarantee convergence of all integrations in the r.h.s. of (1.3).

The solutions $\psi(\mathbf{r},t)$ reported below depend on several parameters. The main contribution of this paper is to display graphically, for some representative values of these parameters, certain features of these solutions, including in particular various snapshots of $|\psi(\mathbf{r},t)|^2$ as it evolves over time. Of course a more striking representation would be provided by a movie; indeed the possibility to manifacture some such video is presently under consideration.

Let us finally note that, if all the nonlinear contributions are neglected, namely g, f and $|\psi|^2$ are vanishingly small, then the Eckhaus equation (1.1) becomes the linear Schrödinger equation

$$i\varphi_t + \Delta\varphi + V(\mathbf{r})\varphi = 0\,. \tag{1.5}$$

Solutions

Let us recall the simple prescription[1] to manifacture solutions of the Eckhaus equation (1.1): if $\varphi(\mathbf{r},t)$ is a solution (vanishingly sufficiently fast as $x \to -\infty$) of the linear Schrödinger equation (1.5), then

$$\psi(\mathbf{r},t) = \varphi(\mathbf{r},t)\Big[1 + 2\int_{-\infty}^{x} dx' |\varphi(x',y,t)|^2\Big]^{-1/2} \tag{2.1}$$

is a solution (clearly also vanishing as $x \to -\infty$) of the Eckhaus equation (1.1). Our consideration is in fact restricted to solutions ψ that vanish in all directions, as $\mathbf{r} \to \infty$; note however that this does not happen for the corresponding auxiliary fields

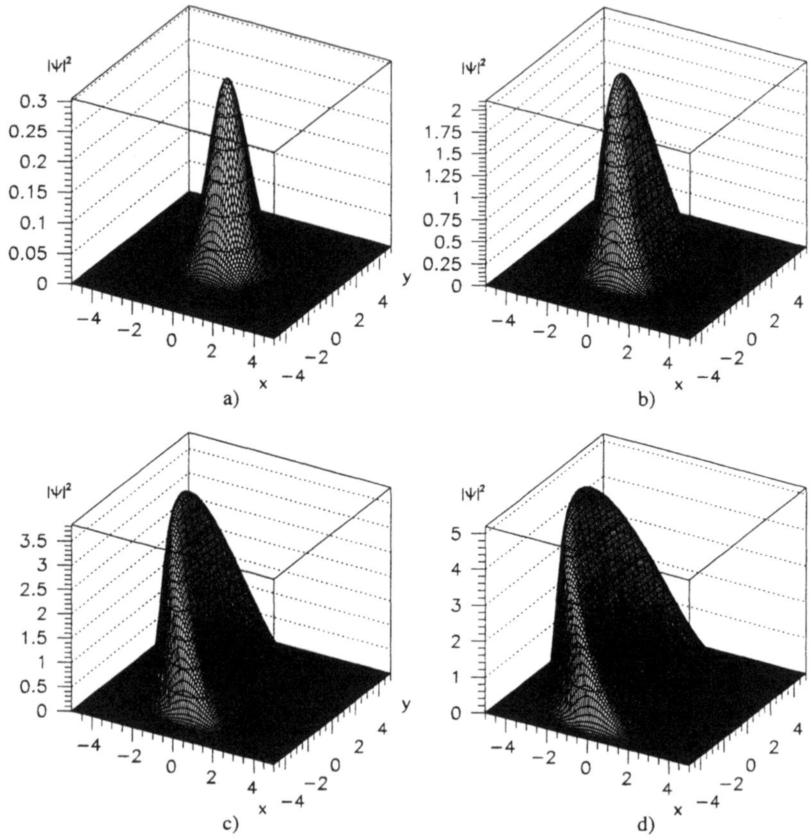

Figure 1: The shapes of the "single-soliton solution" (2.2–4) of the Eckhaus equation (1.1), at $t = 0$, for different values of the parameter a, see (2.4b). a) $c = 1$, b) $c = 10$, c) $c = 100$, d) $c = 1000$. The values of the other parameters are: $b^{(x)} = b^{(y)} = 0.5$, $p^{(x)} = p^{(y)} = q^{(x)} = q^{(y)} = t_0^{(x)} = t_0^{(y)} = 0$.

$g(\mathbf{r},t)$ and $f(\mathbf{r},t)$, which in some directions tend to finite values (see (1.3)). Of course (2.1) implies

$$|\psi(\mathbf{r},t)|^2 = |\varphi(\mathbf{r},t)|^2\Big[1 + 2\int_{-\infty}^{x} dx'|\varphi(x',y,t)|^2\Big]^{-1}. \tag{2.2}$$

The solution $\varphi(\mathbf{r},t)$ of the linear Schrödinger equation (1.5), representing a single particle moving in the harmonic oscillator potential (1.2), reads as follows:

$$\varphi(\mathbf{r},t) = c\,\exp[\alpha^{(x)}(t)x^2 + \alpha^{(y)}(t)y^2 + \beta^{(x)}(t)x + \beta^{(y)}(t)y + \gamma(t)]\,, \tag{2.3a}$$

$$\alpha^{(z)}(t) = -(i/2)\tan[2T^{(z)}(t)]\,, \qquad z = x, y\,, \tag{2.3b}$$

$$\beta^{(z)}(t) = s^{(z)}/\cos[2T^{(z)}(t)]\,, \qquad z = x, y\,, \tag{2.3c}$$

$$\gamma(t) = -\tfrac{1}{2}\ln\{\cos[2T^{(x)}(t)]\cos[2T^{(y)}(t)]\} + \\ + (i/2)\{[s^{(x)}]^2\tan[2T^{(x)}(t)] + [s^{(y)}]^2\tan[2T^{(y)}(t)]\}\,, \tag{2.3d}$$

$$T^{(z)}(t) = t - t_0^{(z)} - (i/2)b^{(z)}\,, \qquad z = x, y\,, \tag{2.3e}$$

$$s^{(z)} = p^{(z)} + iq^{(z)}\,, \qquad z = x, y\,, \tag{2.3f}$$

implying

$$|\varphi(\mathbf{r},t)|^2 = a[A^{(x)}(t)A^{(y)}(t)]^{1/2}\cdot \\ \cdot\exp\{-A^{(x)}(t)[x - \tilde{x}(t)]^2 - A^{(y)}(t)[y - \tilde{y}(t)]^2\}\,, \tag{2.4a}$$

$$a = 2|c|^2[\sinh 2b^{(x)}\sinh 2b^{(y)}]^{-1/2}\cdot\exp\{[p^{(x)}]^2\coth b^{(x)} + [p^{(y)}]^2\coth b^{(y)} \\ - [q^{(x)}]^2\tanh b^{(x)} - [q^{(y)}]^2\tanh b^{(y)}\}\,, \tag{2.4b}$$

$$A^{(z)}(t) = \frac{1}{2}\frac{\sinh 2b^{(z)}}{\cosh^2 b^{(z)} - \sin^2[2(t - t_0^{(z)})]}\,, \qquad z = x, y\,, \tag{2.4c}$$

$$\tilde{x}(t) = 2\frac{p^{(x)}\cos[2(t - t_0^{(x)})]\cosh b^{(x)} + q^{(x)}\sin[2(t - t_0^{(x)})]\sinh b^{(x)}}{\sinh 2b^{(x)}}\,, \tag{2.4d}$$

$$\tilde{y}(t) = 2\frac{p^{(y)}\cos[2(t - t_0^{(y)})]\cosh b^{(y)} + q^{(y)}\sin[2(t - t_0^{(y)})]\sinh b^{(y)}}{\sinh 2b^{(y)}}\,. \tag{2.4e}$$

As it is clear from (2.4), and of course was well known from the earliest days of quantum mechanics, this solution describes a localized entity, of gaussian shape, that moves around on an elliptical orbit (as indeed does a harmonic oscillator in the plane); as it moves, the amplitude and width of the gaussian function also change periodically, keeping constant the overall normalization,

$$\iint_{-\infty}^{+\infty} dx\,dy\,|\varphi(\mathbf{r},t)|^2 = \pi a\,. \tag{2.5}$$

The corresponding solution of the (2+1)-dimensional Eckhaus equation (1.1) with (1.2), as implied by (2.2) and (2.4), is described by the formula

$$|\psi(\mathbf{r},t)|^2 = \frac{a[A^{(x)}(t)A^{(y)}(t)]^{1/2}\,e^{-A^{(x)}[x-\tilde{x}(t)]^2 - A^{(y)}[y-\tilde{y}(t)]^2}}{1 + a[\pi A^{(y)}(t)]^{1/2}(1 + \mathrm{erf}\{\sqrt{A^{(x)}}[x - \tilde{x}(t)]\})\,e^{-A^{(y)}[y-\tilde{y}(t)]^2}}\,, \tag{2.6}$$

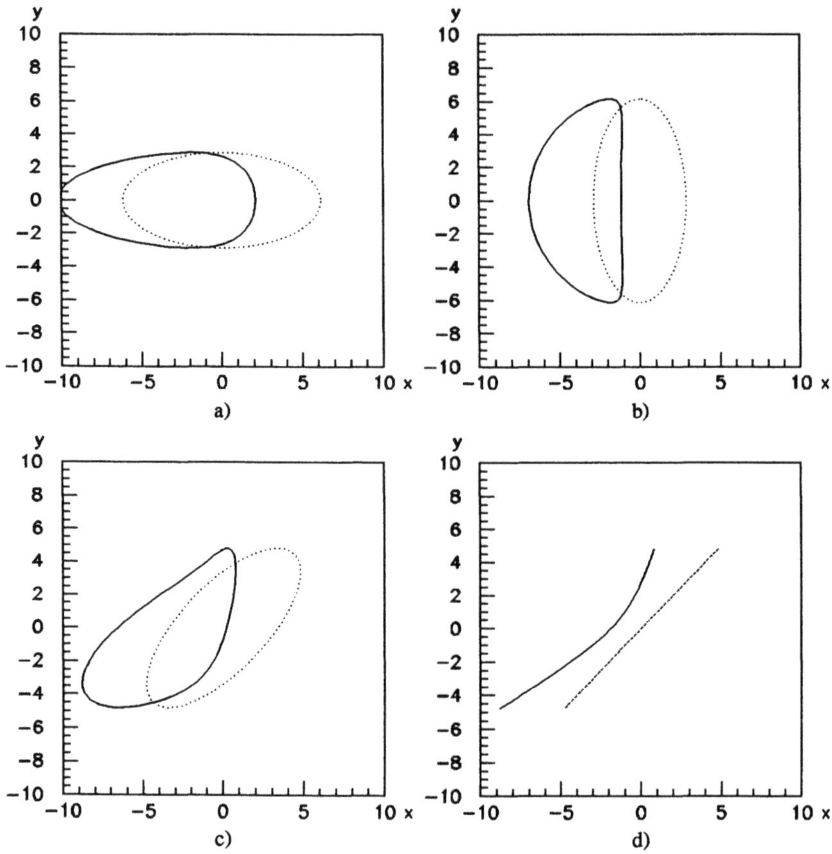

Figure 2: The elliptical trajectory (dotted line) of the "center of mass" of the solution of the linear Schrödinger equation (1.5) and the trajectory (continuous line) of the "center of mass" of the corresponding solution of the Eckhaus equation (1.1), for different choices of the parameters. a) $p^{(x)} = 3.2$, $p^{(y)} = 1.5$, $t_0^{(x)} = 0$, $t_0^{(y)} = \frac{1}{4}\pi$, $c = 1$; b) $p^{(x)} = 1.5$, $p^{(y)} = 3.2$, $t_0^{(x)} = \frac{1}{4}\pi$, $t_0^{(y)} = 0$, $c = 1$; c)$p^{(x)} = p^{(y)} = 2.5$, $t_0^{(x)} = \frac{1}{16}\pi$, $t_0^{(y)} = -\frac{1}{16}\pi$, $c = 1$; d) $p^{(x)} = p^{(y)} = 2.5$, $t_0^{(x)} = t_0^{(y)} = 0$, $c = 1$. In all four cases $b^{(x)} = b^{(y)} = 0.5$ and $q^{(x)} = q^{(y)} = 0$.

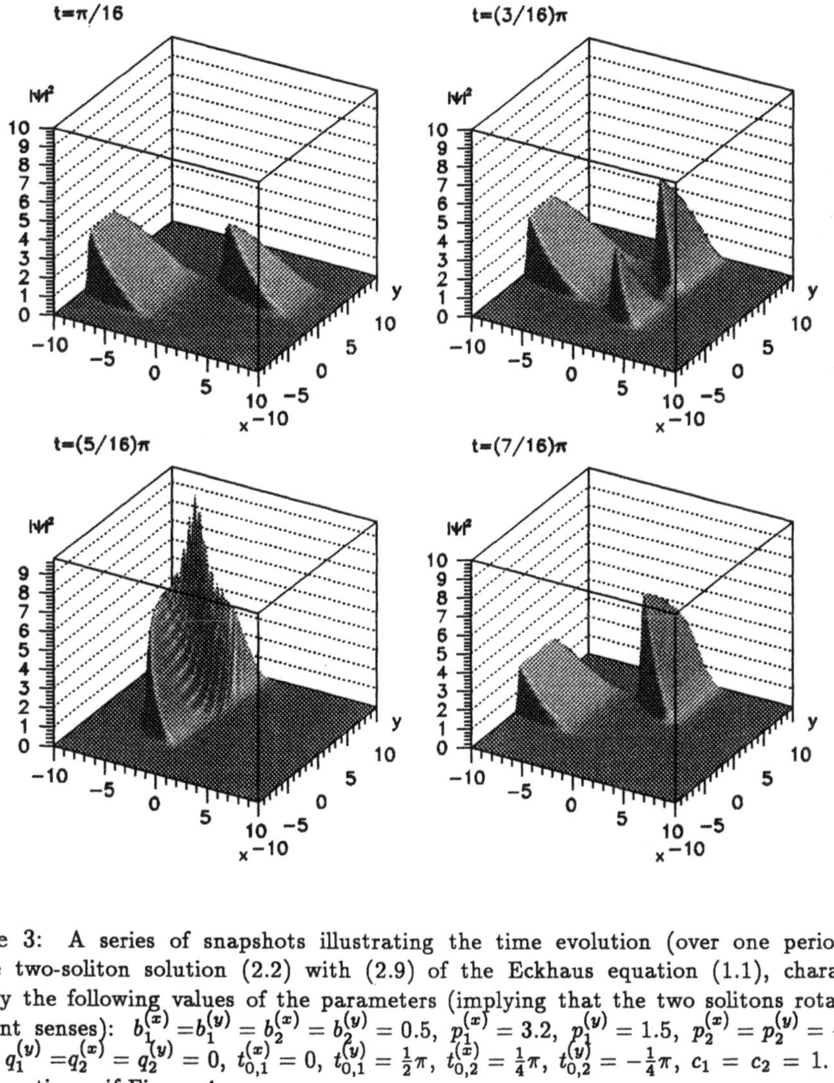

Figure 3: A series of snapshots illustrating the time evolution (over one period π) of the two-soliton solution (2.2) with (2.9) of the Eckhaus equation (1.1), characterized by the following values of the parameters (implying that the two solitons rotate in different senses): $b_1^{(x)} = b_1^{(y)} = b_2^{(x)} = b_2^{(y)} = 0.5$, $p_1^{(x)} = 3.2$, $p_1^{(y)} = 1.5$, $p_2^{(x)} = p_2^{(y)} = -2.5$, $q_1^{(x)} = q_1^{(y)} = q_2^{(x)} = q_2^{(y)} = 0$, $t_{0,1}^{(x)} = 0$, $t_{0,1}^{(y)} = \frac{1}{2}\pi$, $t_{0,2}^{(x)} = \frac{1}{4}\pi$, $t_{0,2}^{(y)} = -\frac{1}{4}\pi$, $c_1 = c_2 = 1$. The series continues if Figure 4.

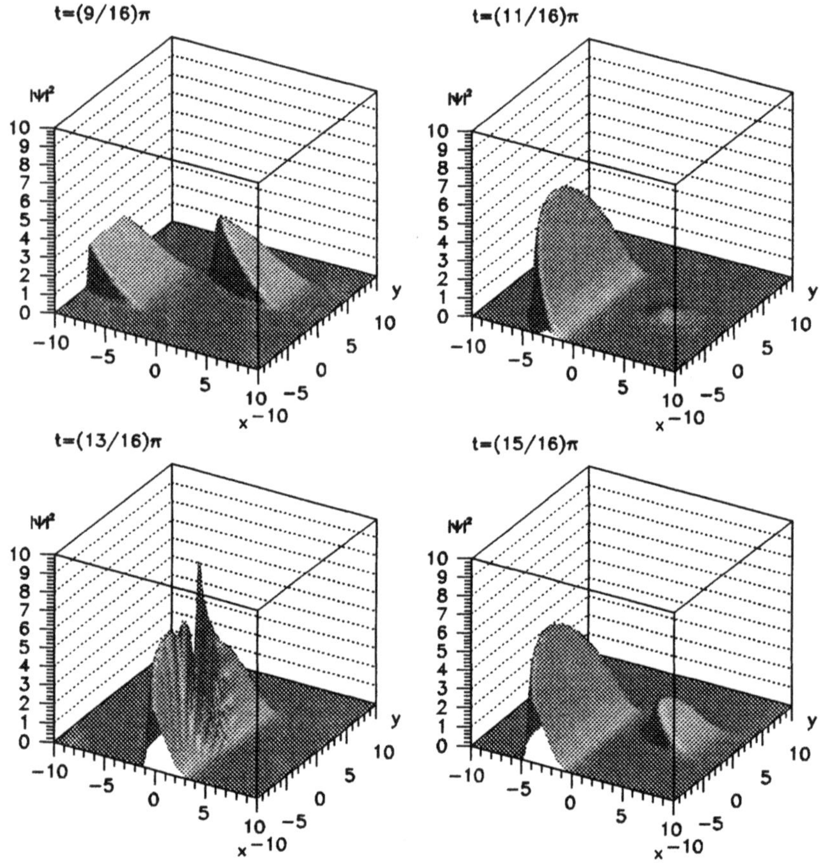

Figure 4: A series of snapshots illustrating the time evolution of the two-soliton solution of the Eckhaus equation. See Figure 3 for explanation.

where the error function $\mathrm{erf}(x)$ is defined as

$$\mathrm{erf}(x) = \frac{2}{\sqrt{\pi}} \int_0^x dz\, e^{-z^2} . \tag{2.7}$$

This is of course to be interpreted as the single-soliton solution of the (2+1)-dimensional Eckhaus equation (1.1) with (1.2); graphical "three-dimensional" representations of its shape (which is somewhat lopsided; recall the lack of circular symmetry of the Eckhaus equation) for different choices of the parameters that characterize its shape, are given in Fig. 1. Of course the positive parameter a provide some measure of the "nonlinearity" of this solution.

It is also of interest to compare the "motion" of this soliton to that of the correponding solution of the linear Schrödinger equation (1.5). This is displayed by the trajectories of the corresponding centers-of mass $\tilde{\mathbf{r}}(t) \equiv (\tilde{x}(t), \tilde{y}(t))$ and $\bar{\mathbf{r}}(t) \equiv (\bar{x}(t), \bar{y}(t))$, which are of course defined by the following formulas:

$$\tilde{\mathbf{r}}(t) = \iint_{-\infty}^{+\infty} dxdy\, \mathbf{r} |\varphi(\mathbf{r},t)|^2 \Big/ \iint_{-\infty}^{+\infty} dxdy |\varphi(\mathbf{r},t)|^2 , \tag{2.8a}$$

$$\bar{\mathbf{r}}(t) = \iint_{-\infty}^{+\infty} dxdy\, \mathbf{r} |\psi(\mathbf{r},t)|^2 \Big/ \iint_{-\infty}^{+\infty} dxdy |\psi(\mathbf{r},t)|^2 . \tag{2.8b}$$

Here of course the components of $\tilde{\mathbf{r}}(t)$ are explicitly given by (2.4d,e), as implied by (2.4a). A graphical representation of these trajectories is given in Fig. 2. Note that a clear picture of the time-evolution of the soliton along its trajectory can be evinced from (2.4e) and the identity of $\bar{y}(t)$ with $\tilde{y}(t)$, $\bar{y}(t) = \tilde{y}(t)$, as implied by (2.2) and (2.4a) and (2.8a,b).

We proceed next to consider 2-soliton solutions. In this case $|\psi(\mathbf{r},t)|^2$ is given by (2.2) with

$$\varphi(\mathbf{r},t) = \sum_{j=1}^{2} c_j \exp[\alpha_j^{(x)}(t)x^2 + \alpha_j^{(y)}(t)y^2 + \beta_j^{(x)}(t)x + \beta_j^{(y)}(t)y + \gamma_j(t)] , \tag{2.9}$$

with $\alpha_j^{(z)}(t)$, $\beta_j^{(z)}(t)$ for $z = x, y$, and $\gamma_j(t)$ given by formulas analogous to (2.3b–f), but with an appropriate index j appended to the parameters $t_0^{(z)}$, $b^{(z)}$ and $s^{(z)} = p^{(z)} + iq^{(z)}$. In this case little enlightment can be obtained from analytical calculations; it must be rather sought from numerical evaluations, as displayed, for one representative example, in Figs. 3–4.

References

1. F. Calogero, 1993 *"Universal C-integrable nonlinear partial differential equation in N+1 dimensions"*. J. Math. Phys. **34** 3197-3209

2. F. Calogero and S. De Lillo, 1992 *"The Eckhaus equation in an external potential"*. J. Phys. A **25** L287-290

ON DISPERSIONLESS HIROTA TYPE EQUATIONS

Robert Carroll
Mathematics Department
University of Illinois
Urbana, IL 61801

Abstract

Various connections between 2-D gravity and KdV, dKdV, inverse scattering, etc. are established. For KP we show how to extract from the dispersionless limit of the Fay differential identity of Takasaki-Takebe the collection of differential equations for $F = log(\tau^{dKP})$ which play the role of Hirota type equations in the dispersionless theory.

1. HIROTA EQUATIONS

In [7] we showed how second derivatives of $F = log(\tau^{dKP})$ survive in the dispersionless limit of the KP Hirota equations and in [31] it was shown how such second derivatives are contained in the dispersionless limit of the Fay differential identity (cf. [1]). We show here how to extract all such dispersionless Hirota type equations from the limiting Fay identity and indicate the pattern. Thus, referring to [1,6-9,31,32] for background in KP theory, we recall first some basic relations involving the Lax and gauge operator. Thus

$$L = \partial + \sum_1^\infty u_{n+1}\partial^{-n};\ W = 1 + \sum_1^\infty w_n\partial^{-n} \tag{1.1}$$

One has $L = W\partial W^{-1}$ and $\partial_n W W^{-1} = -L^n_-$ (Sato equation). Then define $\psi = Wexp(\xi) = w(x,\lambda)exp(\xi)$ $(w(x,\lambda) = 1 + \sum_1^\infty w_n\lambda^{-n})$ and $\psi^* = W^{*-1}exp(-\xi) = w^*(x,\lambda)exp(-\xi)$ $(w^*(x,\lambda) = 1 + \sum_1^\infty w_j^*\lambda^{-j})$. It follows that

$$B_n = L^n_+;\ \xi = \sum_1^\infty x_n\lambda^n\ (x = x_1);\ L\psi = \lambda\psi;\ \partial_n\psi = B_n\psi\ (\partial_n = \partial/\partial x_n); \tag{1.2}$$

$$L^*\psi^* = \lambda\psi^*;\ \partial_n\psi^* = -B_n^*\psi^*;\ \partial_n L = [B_n, L];\ \partial_n B_m - \partial_m B_n = [B_n, B_m]$$

One also has a tau function $\tau(x)$ and vertex operators X, X^* which satisfy

$$\psi(x,\lambda) = X(\lambda)\tau/\tau = e^{\xi}\tau_-/\tau = e^{\xi}G_-(\lambda)\tau/\tau = e^{\xi}\tau(x_j - 1/j\lambda^j)/\tau;\ G_{\pm}(\lambda) = \tag{1.3}$$

$$exp(\pm\xi(\tilde{\partial},\lambda^{-1}));\ \tilde{\partial} = (\partial_1, \frac{1}{2}\partial_2, ...);\ \psi^* = X^*(\lambda)\tau/\tau = e^{-\xi}G_+(\lambda)\tau/\tau = e^{-\xi}\tau_+/\tau$$

One writes also $exp(\xi) = exp(\sum_1^\infty x_n\lambda^n) = \sum_0^\infty p_j\lambda^j$ (Schur polynomials).

For dispersionless KP one takes $\epsilon \to \epsilon x = X$ and $t_n \to \epsilon t_n = T_n$ in say the KP equation $u_t = K(u) = u''' + 3uu' + \frac{3}{4}\partial^{-1}\partial_2^2 u$ with $\partial_n \to \epsilon\partial/\partial T_n$ and $u(x,t_n) \to \tilde{u}(X,T_n)$ to obtain $\partial_T\tilde{u} = 3\tilde{u}\partial_X\tilde{u} + \frac{3}{4}\partial^{-1}\partial^2\tilde{u}/(\partial T_2)^2$ when $\epsilon \to 0$ (see [10] for a more geometric approach). Thus, using (t_n) for (x_n), consider $L = \epsilon\partial + \sum_1^\infty u_{n+1}(\epsilon,T)(\epsilon\partial)^{-n}$. We think of replacing (t_n) by (T_n) now and assuming $u_n(\epsilon,T) = u_n(T) + O(\epsilon)$ etc. (in an obvious abuse of notation which seems to cause no confusion). Take then

$$\psi = (1 + O(1/\lambda))e^{(\sum_1^\infty T_n\lambda^n/\epsilon)} = e^{(\frac{1}{\epsilon}S(T,\lambda)+O(1))};\ \tau = e^{(\frac{1}{\epsilon^2}F(T)+O(\frac{1}{\epsilon}))} \tag{1.4}$$

and replace ∂_n by $\epsilon\partial_n$ where $\partial_n \sim \partial/\partial T_n$ now to get $L\psi = \lambda\psi$ with

$$\partial_n L = [B_n, L];\ \epsilon\partial_n\psi = B_n\psi;\ \psi = e^{(\sum_1^\infty \frac{1}{\epsilon}T_n\lambda^n)}\tau(\epsilon, T_n - \epsilon/n\lambda^n)/\tau(\epsilon,T) \tag{1.5}$$

The connection to standard semiclassical approximation ideas is clear. Now observe that for $P = \partial S$ ($\partial \sim \partial/\partial X$ now), $\epsilon^i\partial^i\psi \to P^i\psi$ as $\epsilon \to 0$ since $\epsilon\partial\psi = \partial S\psi = P\psi$, $\epsilon^2\partial^2\psi = (\epsilon\partial P)\psi + P^2\psi$, etc. Further $\epsilon\partial(\psi/P) = -((\epsilon\partial P)/P^2)\psi + \psi = \psi + O(\epsilon)\psi$ which implies $(\epsilon\partial^{-1})\psi \to P^{-1}\psi$ etc. Hence $L\psi = \lambda\psi$ becomes

$$\lambda = P + \sum_1^\infty u_{n+1}P^{-n};\ P = \lambda - \sum_1^\infty P_i\lambda^{-i} \tag{1.6}$$

(we write $u_{n+1}(T)$ or $u_{n+1}(t)$ depending on context). Also from $B_n\psi = \sum_0^n b_{mn}(\epsilon\partial)^m\psi$ one gets

$$B_n\psi = \sum_0^n b_{mn}P^m\psi + O(\epsilon)\psi = \mathcal{B}_n(P)\psi + O(\epsilon)\psi;\ \epsilon\partial_n\psi = \partial_n S\psi = \mathcal{B}_n(P)\psi \tag{1.7}$$

$$+O(\epsilon)\psi \Rightarrow \partial_n S = \mathcal{B}_n(P) \Rightarrow \partial_n P = \hat{\partial}\mathcal{B}_n(P)\ (\hat{\partial}\mathcal{B}_n = \partial_X\mathcal{B}_n + \partial_P\mathcal{B}_n(\partial P/\partial X))$$

We write $B_n = L_+^n \to \mathcal{B}_n = \lambda_+^n$ (where + referes to powers of P now) and we note that $(L,\lambda) \to (\lambda, P)$ (the notation is different in [31,32] for example where in particular k is used for P). One has a conventional notation for $A = \sum a_i\partial^i \sim \sum a_i\xi^i$, etc., namely

$[A,B] = A\circ B - B\circ A$; $A\circ B = \sum_0^\infty \partial_\xi^n A\partial^n B/n! = A(x,\xi)exp(\overleftarrow{\partial_\xi}\overrightarrow{\partial_X})B(x,\xi)$. Putting $\xi = \epsilon\partial$ the term of first order in ϵ is $\{A,B\} = \partial_\xi A\partial B - \partial A\partial_\xi B = \sum\sum(ia_i b_{jX} - jb_j a_{iX})\xi^{i+j+1}$. Thinking of action on ψ one replaces ξ by P and obtains ($\mathcal{B} \sim \lambda_+^n$) : $\partial_n\lambda = \{\mathcal{B}_n, \lambda\}$. Further, putting in ϵ at appropriate places one obtains from [31,32] ($S = S(T_n,\lambda)$, $T_1 = X$, $S_j \to S_j(T_n)$)

$$\partial_n\mathcal{B}_m - \partial_m\mathcal{B}_n + \{\mathcal{B}_m,\mathcal{B}_n\} = 0;\ \partial_n S = \mathcal{B}_n;\ S = \sum_1^\infty T_n\lambda^n + \sum_1^\infty S_{i+1}\lambda^{-i} \tag{1.8}$$

and we note explicitly $\mathcal{B}_1 = \lambda_+ = P$. Further $\mathcal{B}_n = \lambda^n + \sum_1^\infty \partial_n S_{i+1}\lambda^{-i}$ and one shows that $\partial S_{j+1} = -P_j$. Finally the tau function for dKP is simply defined via $\partial_n log(\tau^{dKP}) = -nS_{n+1}$ ($n \geq 1$) where $log(\tau^{dKP}) \sim F$ and hence $\partial S_{n+1} \sim -P_n \sim \partial\partial_n F/n$.

Now the Hirota bilinear identity for KP is $\oint_\infty \psi(x,\lambda)\psi^*(y,\lambda)d\lambda = 0$ and this contains all of the Hirota bilinear equations for KP (cf. [6,7]). Such equations correspond to Plucker relations and in fact a subset of such relations embodied in the Fay differential identity is also equivalent to the KP hierarchy (cf. [31]). The Fay identity can be written (cf. [1] - $[s_i] = (s_i, \frac{1}{2}s_i^2, \frac{1}{3}s_i^3, ...)$ and c.p. $\sim$ cyclic permutations of $1,2,3$)

$$\sum_{c.p.}(s_0 - s_1)(s_2 - s_3)\tau(t + [s_0] + [s_1])\tau(t + [s_2] + [s_3]) = 0 \tag{1.9}$$

Differentiating this in s_0 and manipulating variables suitably leads to the differential Fay identity (with ϵ included for dKP limits - cf. [1,31])

$$\epsilon\partial_X\tau(t - \epsilon[\mu^{-1}])\tau(t - \epsilon[\lambda^{-1}]) - \tau(t - \epsilon[\mu^{-1}])\epsilon\partial_X\tau(t - \epsilon[\lambda^{-1}]) \tag{1.10}$$
$$-(\lambda - \mu)\tau(t - \epsilon[\mu^{-1}])\tau(t - \epsilon[\lambda^{-1}]) + (\lambda - \mu)\tau(t)\tau(t - \epsilon[\mu^{-1}] - \epsilon[\lambda^{-1}]) = 0$$

($X \sim T_1$). Then in [31] it is shown that this has a dispersionless limit ($\epsilon \to 0, F = log(\tau^{dKP})$)

$$\sum_{m,n=1}^\infty \mu^{-m}\lambda^{-n}\frac{\partial_m\partial_n F}{mn} = log[1 + \sum_1^\infty \frac{\lambda^{-n} - \mu^{-n}}{\mu - \lambda}\frac{\partial\partial_n F}{n}] \tag{1.11}$$

Hence the partial differential equations for F arising from (1.11) should characterize the dKP system. Now, although (1.11) provides a collection of partial differential equations for F in principle, their determination is not immediate and we provide here an effective way of determining these equations. This rests upon an observation made in [10]. Thus in [10] we used (1.6) ($P = \lambda - \sum_1^\infty P_i\lambda^{-i}$) and $\partial\partial_n F = nP_n$ to write

$$\sum_1^\infty \lambda^{-n}\frac{\partial\partial_n F}{n} = \sum_1^\infty \lambda^{-n}P_n = \lambda - P(\lambda) \tag{1.12}$$

Hence the right side of (1.11) becomes $log(\frac{P(\mu)-P(\lambda)}{\mu-\lambda})$. Thus for $\mu \to \lambda$ and $\dot{P} \sim \partial_\lambda P$

$$log\dot{P}(\lambda) = \sum_{m,n=1}^{\infty} \lambda^{-m-n}\frac{\partial_m\partial_n F}{mn} \tag{1.13}$$

But we already know via (1.6) (since $P_n = \partial\partial_n F/n$)

$$\dot{P}(\lambda) = 1 + \sum_1^{\infty} jP_j\lambda^{-j-1} = 1 + \sum_1^{\infty} \partial\partial_j F\lambda^{-j-1} \tag{1.14}$$

Looking at the first terms in the expansion of $log(\dot{P}(\lambda))$ from (1.14) one has ($log(1+x) = x - \frac{1}{2}x^2 + \frac{1}{3}x^3 + ...$)

$$log(\dot{P}) \sim \partial^2 F\lambda^{-2} + \partial\partial_2 F\lambda^{-3} + \partial\partial_3 F\lambda^{-4} + \partial\partial_4 F\lambda^{-5} + ... \tag{1.15}$$

$$-\frac{1}{2}[(\partial^2 F)^2\lambda^{-4} + 2\partial^2 F\partial\partial_2 F\lambda^{-5} + ...] + \frac{1}{3}[(\partial^2 F)^3\lambda^{-6} + ...] + ...$$

Hence the consistency equations with (1.13) begin with $\partial^2 F = \partial^2 F$, and $\partial\partial_2 F = \partial\partial_2 F$, which are tautological, and thereafter we obtain in particular

$$\frac{1}{2}(\partial^2 F)^2 - \frac{1}{3}\partial\partial_3 F + \frac{1}{4}(\partial_2^2 F) = 0;\ \partial^2 F\partial\partial_2 F - \frac{1}{2}\partial\partial_4 F + \frac{1}{3}\partial_2\partial_3 F = 0; \tag{1.16}$$

$$-\frac{1}{3}(\partial^2 F)^3 + \partial^2 F\partial\partial_3 F + \frac{1}{2}(\partial\partial_2 F)^2 - \frac{3}{5}\partial\partial_5 F + \frac{1}{9}\partial_3^2 F + \frac{1}{4}\partial_2\partial_4 F = 0; ...$$

THEOREM 1.1 The relations (1.16) obtained from (1.13)-(1.14) are a system of partial differential equations for $F = log(\tau^{dKP})$ which should characterize F and thus play the role of dispersionless Hirota equations.

REMARK 1.2 It might be interesting to utilize the equations (1.16) in the study of the WDVV equations of [17] which classify certain 2-D topological field theories. Mr. Seung Hwan Son has shown how the equations (1.16) can be generated via a Mathematica program. We note also that the quantity $P(\lambda) - P(\mu)$ is an important ingredient in the Hamilton-Jacobi theory of Kodama-Gibbons (cf. [7,8,19,29]).

2. KdV, mKdV, dKdV and GRAVITY.

We indicate here a few observations which will be greatly expanded in [11]. First, referring to [7,8] one recalls that a dispersionless Jevicki-Yoneya action principle was derived for the KdV-Hermitian matrix model (HMM) situation based on an action (cf. [7,8,23,34])

$$\tilde{A} = \int Res_P(-\hat{P}Q - S + \xi)dX \tag{2.1}$$

and in [10] it was shown that in fact the term $S - \xi$ arises as a dispersionless version of the Sato equation $\partial_n WW^{-1} = -L^n_-$ since for $W \sim exp(-\frac{1}{\epsilon}H)$, $-H \sim S - \xi$, and $\epsilon\partial_n WW^{-1} \to \partial_n(S - \xi)$ so $\partial_n(S - \xi) = -\lambda^n_-$. It could be interesting to examine further the meaning here of this dispersionless Sato equation as an action term. We note also that e.g. the conformal flows pass to a dispersionless form via Sato type equations. Thus (cf. [1,9,16,21,26]) one defines an Orlov operator M via

$$\partial_\lambda\psi = M\psi = W(\sum_1^\infty kx_k\partial^{k-1})W^{-1}\psi = (G + \sum_2^\infty kx_k\lambda^{k-1})\psi; G = WxW^{-1} \tag{2.2}$$

and the conformal flows are $\partial_{m+1,1}u = \partial Res(ML^{m+1})$ with $\partial_{m+1,1}W = -(ML^{m+1})_-W$. One knows (cf. [7,8,31,32]) that $M \to \mathcal{M} = S_\lambda$ and $\epsilon\partial_{m+1,1}W_\epsilon W_\epsilon^{-1} \to -(\mathcal{M}\lambda^{m+1})_-$ leading to $\partial_{m+1,1}(S-\xi) = -(\mathcal{M}\lambda^{m+1})_-$ with similar results for other additional symmetry flows. In this context we recall also that $(L, G) \to (\lambda, \hat{\xi})$ where $(\lambda, -\hat{\xi})$ are action angle variables in the Hamilton-Jacobi theory of [19,29].

Now based on a less sophisticated action idea for HMM situations (cf. [7,8,20]) $(\delta/\delta u)\int R_{k+1}(u)dx = -(k+\frac{1}{2})R_k(u)$, where $R_k \sim$ Gelfand-Dickey resolvant terms, one develops a unitary matrix model (UMM) action via (cf. [2-4,14,15])

$$I = \int \sum_1^\infty [4t_{2k+1}R_{k+1} - v^2x]dx \tag{2.3}$$

where $u = v^2 - v', \delta u = 2v\delta v - \partial\delta v, \hat{D} = \partial + 2v$, so that $(\delta I/\delta v) = -\sum_1^\infty(2k+1)t_{2k+1}\hat{D}R_k - vx = 0$. Since $R_k \to \frac{1}{2}r_{k-1}(p)$ $(p = -\frac{1}{2}u)$ in the dispersionless limit (cf. [7,8]) with $r_{n-1}(p) = (n-\frac{1}{2})...\frac{1}{2}/n!(2p)^n$ and $\epsilon\partial R_k \to 0$, upon shifting to T' variables $(\epsilon t_n = T_n,\ nT_n = T_n')$, the UMM yields a limiting string equation

$$\sum_1^\infty(2k+1)T'_{2k+1}r_{k-1}(p) = -X \tag{2.4}$$

which is in fact the Landau-Ginsburg (LG) equation of HMM. Thus

PROPOSITION 2.1 In the dispersionless limit the UMM, based on mKdV, corresponds to the HMM, based on KdV, via the LG equation.

This is meant to be heuristic since there has been little physical input; one argues in in [15] for example that KdV and mKdV are descriptions of an open-closed string theory involving gravity. In particular, the unique real nonsingular solutions provided by UMM map to open string generalizations of the nonsingular solutions of the KdV hierarchy found in [14]. Those solutions in turn arise by first assuming that the KdV flows are preserved nonperturbatively, which leads uniquely to the HMM string equations for closed strings, and to a unique pole free string susceptibility $\sim u$. For such closed (resp. open) strings one is dealing with HMM and Painleve 1 (resp. UMM and Painleve 2). The details here are complicated and we will say more about this in [11]. Further gravity topics in [11] will include material from [25] on a direct passage from KdV to gravity and from [22,27-30,33] on Liouville- Beltrami gravity and connections to KdV, mKdV, and Ur-KdV.

3. ON INVERSE SCATTERING AND DISPERSIONLESS LIMITS

In [7,8] we outlined some connections between inverse scattering for KdV and mKdV, dKdV. This is developed in more detail and generality in [11,13] (cf. also [5]). Thus in dKdV with $\epsilon x = X$, $x \to \pm\infty \sim \epsilon \to 0$, so there should be a natural relation between some inverse scattering quantities and some corresponding dispersionless quantities. Here the analysis of [18,19,35] is particularly relevant. We recall that $u_t = u''' - 6uu', u = v^2 - v', v_t = v''' - 6v^2v'$ are related and via a Galilean transformation $u \to u-\lambda, t \to t, x \to x-6\lambda t$, KdV is unaltered while $mKdV \to v_t = 6\lambda v' + v''' - 6v^2 v'$. Then for $u + \lambda = v^2 - v'$ with $v = -\psi'/\psi$ one has $\psi'' - u\psi = \lambda\psi = -k^2\psi$ with Jost solutions $\psi_\pm(k,x) \sim exp(\pm ikx)$ as $x \to \pm\infty$. We recall that the transmission coefficient T and reflection coefficients R, R_L arise in formulas $T(k)\psi_- = R(k)\psi_+ + \psi_+(-k,x)$; $T(k)\psi_+ = R_L(k)\psi_- + \psi_-(-k,x)$ and we set $\psi_- = exp(-ikx + \phi(k,x))$ with $\phi(k,-\infty) = 0$. This leads to (*) $\phi'' - 2ik\phi' + \phi'^2 = u$, so $-\psi'_-/\psi_- = -(-ik + \phi') = v$ and one expects a solution of (*) in the form $\phi' = \sum_1^\infty \phi_n/(ik)^n$ compatible with $u + \lambda = v^2 - v'$, i.e. $v \sim ik + \sum_1^\infty v_n/(ik)^n$. This leads to $\phi_n = v_n$.

Next recall (cf. [6]) that for suitable u with say $\int_{-\infty}^\infty (1 + |x|^2)|u|dx < \infty$ one has available all of the inverse scattering machinery and for convenience one could even assume $u \in \mathcal{S}$ (Schwartz space). One expects T(k) to be meromorphic for $Im(k) > 0$ with (possibly) a finite number of simple poles at $\beta_j = ik_j$. Further (cf. [6]) $1/T = (1/2ik)W(\psi_+,\psi_-) = (1/2ik)(\psi'_+\psi_- - \psi_+\psi'_-)$. Given $|R(k)|$ small for large $|k|$ real (which will be true for $u \in \mathcal{S}$ say) one expands $1/(\xi - k) = -\sum_0^\infty (\xi^n/k^{n+1})$ and writes ($\bar{R}(k) = R(-k)$ for k real)

$$log(T) \sim \sum_0^\infty \frac{c_{2n+1}}{k^{2n+1}};\, c_{2n+1} = \frac{2}{2n+1}\sum_1^N (i\beta_j)^{2n+1} - \frac{1}{2\pi i}\int_{-\infty}^\infty \xi^{2n} log(1 - |R|^2)d\xi \quad (3.1)$$

Now for $x \to \infty$, $\psi_- e^{ikx} \sim (R/T)e^{2ikx} + (1/T) \to 1/T$ when $Im(k) > 0$. Taking logarithms we get $log(\psi_-) + ikx = \phi(k,x) \to -log(T)$ as $x \to \infty$ so $\phi(k,\infty) = -log(T)$. Hence $\int_{-\infty}^{\infty} \phi_{2m}dx = 0$ and

$$-log(T) = \sum_1^\infty \frac{1}{(ik)^n}\int_{-\infty}^{\infty}\phi_n dx = -\sum_0^\infty \frac{c_{2n+1}}{k^{2n+1}};\ i^{2m}c_{2m+1} = \int_{-\infty}^{\infty}\phi_{2m+1}dx \qquad (3.2)$$

THEOREM 3.1. The scattering information or x-asymptotic information, given via R,T is related to the k or λ asymptotics of the wave function ψ_-, given via ϕ_n, through the formulas (3.2), and $\phi_n = v_n$ establishes a further connection to the mKdV expansion terms.

We remark that there are arguments based on a Gardner type transform $u = \alpha\epsilon^2v^2 + v + \beta\epsilon v'$ for example which lead to an identification of the $c_{2n+1}(u)$ with Hamiltonians H_n. This tells us that the ϕ_{2n+1} are nontrivial conserved densities and hence so are the corresponding v_n. Now from the dispersionless theory of Sections 1,2 we have (cf. also [7,8,32]) $-u = q = 2p$, $\lambda = -k^2$, $W\partial^2W^{-1} \to \hat{P} = P^2 + q = (ik)^2$, and $ik = P(1+\sum_1^\infty \binom{\frac{1}{2}}{m}q^mP^{-2m}$. This corresponds to (1.6) with $\lambda = ik$ and there will be a corresponding inversion $P = ik - \sum_1^\infty P_j(ik)^{-j}$. Consider $\psi \sim \psi_+ = exp(\frac{1}{\epsilon}S)$ with $\psi^* \sim \psi_- = exp(-\frac{1}{2}S)$ (cf. [10] for a proof of the last statement). We note explicitly that $\epsilon\partial\psi_-/\psi_- = -P = -S_X$ and $\epsilon^2\partial^2\psi_-/\psi_- = P^2 + \epsilon P' \to P^2$ leading to $(ik)^2 = P^2 + q$. But $\psi_- = exp(-ikx + \phi) \sim \psi_- = exp(-\frac{1}{\epsilon}S)$ for $|x|$ large or ϵ small which suggests an identification $-ikX/\epsilon + \phi(X/\epsilon, k) = -\frac{1}{\epsilon}S(X,k)$ with $-S' = -ik + \phi'$ where ϕ' is the derivative in the x argument. Here in the same notation where e.g. $u_n(T,\epsilon) = u_n(T) + O(\epsilon)$ we assume it is permissible to write $\phi'(X/\epsilon,k) = \phi'(X,k,\epsilon) = \phi'(X,k) + O(\epsilon)$. Then one can deduce that $-ik + \phi' = -S_X = -P; v \sim P$. Consequently,

THEOREM 3.2 Given the above relations betwen variables x and X, the dispersionless KdV theory interacts with scattering data ϕ_n and mKdV data v_n via $\phi_n \sim v_n \sim -P_n$.

Now consider the use of $cklog|T| = \alpha(k)$; $\beta(k) = \hat{c}arg(R_L/T)$; $Q_n = C_n\beta_n^2$; $\hat{P}_n = \hat{C}_n log|b_n|$ as action-angle variables in a Hamiltonian theory for KdV (cf. [6] - the c, $\hat{c}$, C_n, $\hat{C}_n$ are constants which depend on notation and the b_n are normalization constants). Under standard Poisson brackets $\{\alpha(k),\beta(k')\} = \delta(k-k')$ and $\{Q_n,\hat{P}_n\} = \delta_{mn}$ with other brackets vanishing (note such Gardner type brackets do not go over to our canonical phase space brackets as in (1.8)). Now recall that $1/T = (1/2ik)W(\psi_+,\psi_-)$ and consider $\psi_\pm \sim exp(\pm\frac{1}{\epsilon}S)$ as above. Then $W(\psi_+,\psi_-) \sim \epsilon\partial e^{\frac{1}{\epsilon}S}e^{-\frac{1}{\epsilon}S} - e^{\frac{1}{\epsilon}S}\epsilon\partial e^{-\frac{1}{\epsilon}S} = 2S' = 2P$; $|T| = |k/P|$. Note here $\epsilon \to 0 \sim x \to \infty$ so there is no contradiction to W being independent of x while P is just a phase space variable on an (X,P) phase space. Hence $\alpha(k) \sim cklog(|k/P|)$. On the other hand from [6] $R_L = W(\psi_-(-k,x),\psi_+)/W(\psi_+,\psi_-)$ and $\psi_-(-k,x) \sim exp(-\frac{1}{\epsilon}S(X,k))$ so the exponentials do not cancel. However some analysis based on [18] (details in [12]) allows us to say that $ImS = 0$ and $\beta(k) \to \hat{c}[\pi \text{ or } 2\pi]$. Thus

THEOREM 3.3. The action variable $\alpha(k) = cklog|T|$ in KdV passes to a phase space variable $cklog|k/P|$ in dKdV and locally $\beta(k) \to$ constant.

References

[1] M. Adler and P. vanMoerbeke, *Comm. Math. Phys.*, 147 (1992), 25-56

[2] K. Anagnostopoulos, M. Bowick, and N. Ishibashi, *Mod. Phys. Lett.*, 6A (1991), 2727-2739

[3] K. Anagnostopoulos, M. Bowick, and A. Schwartz, *Comm. Math. Phys.*, 148 (1992), 469-485

[4] K. Anagnostopoulos and M. Bowick, SU-4238-504

[5] R. Carroll, *On the philosophy of the spectral variable, Proc. First World Congress Nonlinear Analysts*, deGruyter, 1994, to appear

[6] R. Carroll, *Topics in soliton theory*, North Holland, 1991

[7] R. Carroll, *Remarks on dispersionless KP, KdV, and 2-D gravity, Jour. Nonlin. Sci.*, 1994, to appear

[8] R. Carroll, *On canonical coordinates in soliton hierarchies, Teor. i Mat. Fizika*, 1994, to appear

[9] R. Carroll, *Applicable Anal.*, 49 (1993), 1-31; *Symmetries, Sato theory, and tau functions, Applicable Anal.*, to appear

[10] R. Carroll, *Remarks on phase space variables and quasiclassical soliton theory*, to appear

[11] R. Carroll, *A sketch of KdV-gravity connections*, in preparation

[12] R. Carroll, *On inverse scattering and dispersionless limits*, in preparation

[13] R. Carroll and Y. Kodama, in preparation

[14] S. Dalley, C. Johnson, and T. Morris, *Nucl. Phys.*, 368B (1992), 625-654 and 655-670; *Nucl. Phys. B Supplement*, 25A (1992), 87-91

[15] S. Dalley, C. Johnson, T. Morris, and A. Watterstam, *Mod. Phys. Lett.*, 7A (1992), 2753-2762

[16] L. Dickey, *Soliton equations and Hamiltonian systems*, World Scientific, 1991

[17] B. Dubrovin, *Integrable systems, Verdier Memorial Conf.*,Birkhauser, 1993, pp. 313-359

[18] V. Geogdzaev, *Singular limits of dispersive waves*, Plenum, 1994, pp. 53-59

[19] J. Gibbons and Y. Kodama, *Singular limits of dispersive waves*, Plenum, 1994, pp. 61-66

[20] P. Ginsparg, M. Goulian, M. Plesser, and J. Zinn-Justin, *Nucl. Phys.* 342B (1990), 539-563

[21] P. Grinevich, *Singular limits of dispersive waves*, Plenum, 1994, pp. 67-88

[22] H. Guo, Z. Wang, and K. Wu, *Phys. Lett.* 264B (1991), 277-282 and 283-291; *Comm. Theor. Phys.*, 14 (1990), 99-106 (with S. Wang)

[23] A. Jevicki and T. Yoneya, *Mod. Phys. Lett.*, 5A (1990), 1615-1621

[24] Y. Kodama and J. Gibbons, *Proc. Inter. Workshop Nonlin. and Turb. Proc. Physics*, Kiev, 1989, pp. 166-180

[25] D. Montano and G. Rivlis, *Nucl. Phys.* 360B (1991), 1613-1651

[26] A. Orlov and E. Schulman, *Lett. Math. Phys.*, 12 (1986), 171-179

[27] J. Shen, Z. Sheng, and K. Li, *Phys. Lett.* 325B (1994), 377-382

[28] J. Shen, Z. Sheng, and Z. Wang, *Phys. Lett.* 291B (1992), 53-62

[29] J. Shen and Z. Sheng, *Phys. Lett.* 279B (1992), 285-290

[30] J. Schiff, hep-th 9205105

[31] K. Takasaki and T. Takebe, hep-th 9405096

[32] K. Takasaki and T. Takebe, *Inter. Jour. Mod. Phys. A, Supp.*, 1992, pp. 889-922

[33] Z. Wang, *Phys. Lett.* 264B (1991), 39-44

[34] T. Yoneya, *Comm. Math. Phys.*, 144 (1992), 623-639; *Inter. Jour. Mod. Phys.* 7A (1992), 4015-4038

[35] V. Zakharov, *Singular limits of dispersive waves*, Plenum, 1994, pp. 165-174

Stationary Reductions of Nonlinear Evolution Equations : Hamiltonian Structures and Canonical Transformations

Allan P. Fordy,
Department of Applied Mathematical Studies and
Centre for Nonlinear Studies,
University of Leeds,
Leeds LS2 9JT, UK.

February 8, 1995

Abstract

We consider the restriction of isospectral flows, Miura maps and Darboux transformations to stationary manifolds. Specifically, we give a brief presentation of:

1. A systematic construction of Hamiltonian structures on stationary manifolds.
2. The reduction of Darboux transformations to stationary manifolds. Using the relationship between the factorisation of the Schrödinger operator and the MKdV hierarchy, we derive interesting canonical transformations which preserve **both** Hamiltonian structures and **all** commuting Hamiltonians for the stationary flows.

Finally, we mention some interesting connections between stationary flows of coupled KdV systems and integrable, natural Hamiltonian systems with quartic potentials.

1 Introduction

In this paper we consider the restriction of an integrable nonlinear evolution equation to its stationary manifold. We start, in section 2, by constructing the Poisson brackets and Hamiltonians associated with the isospectral flows of a given spectral problem. By choosing the correct coordinates (and reversing the roles of x and t) these can be adapted to the stationary manifolds. These ideas are illustrated through the stationary KdV equation.

The Darboux transformation is related to reversing the factorisation of the Schrödinger operator. In section 2 we consider the restriction of the Darboux transformation to the stationary manifolds of the MKdV hierarchy. This gives rise to a remarkable, explicit (and explicitly invertible) transformation, which preserves *both* Poisson brackets and *all* the commuting Hamiltonians in the hierarchy.

2 Stationary Manifolds : Poisson Brackets

In this section we explain how to construct the Poisson brackets for stationary flows of integrable nonlinear evolution equations. These are illustrated through the KdV hierarchy. The mth stationary flow in the KdV hierarchy is defined by

$$u_{t_m} = \left(\frac{1}{2}\partial^3 + 2u\partial + u_x\right)\delta_u\mathcal{H}_m = \frac{1}{2}\partial\delta_u\mathcal{H}_{m+1} = 0, \quad m \geq 1,$$

which is in the form of a generalised Lagrangian equation:

$$\delta_u \mathcal{L}_{m+1} = 0, \quad \mathcal{L}_{m+1} = \mathcal{H}_{m+1} - c_m u. \tag{1}$$

It is possible to write these equations as an m-degrees of freedom canonical Hamiltonian system [12]. If we take c_m as a dynamical variable then the Poisson bracket is degenerate, with c_m as Casimir:

$$\{f,g\} = \frac{\partial f}{\partial x_i} J_{ij} \frac{\partial g}{\partial x_j} = (\nabla f)^T J(\nabla g), \quad J = \begin{pmatrix} 0 & I_m & 0 \\ -I_m & 0 & 0 \\ 0 & 0 & 0 \end{pmatrix}, \tag{2}$$

where I_m is the $m \times m$ unit matrix. Bogoyavlenskii and Novikov [4] showed that this Hamiltonian system is completely integrable with first integrals given by the fluxes $\mathcal{F}_{km}$ of the Hamiltonians $\mathcal{H}_1, \cdots, \mathcal{H}_m$. The equation $\mathcal{F}_{0m} = c_m$ is an ODE of order $2m$. This is used to eliminate higher derivatives from the fluxes $\mathcal{F}_{km}$, which can then be written as functions of the variables (q_i, p_i, c_m), these being labelled h_k.

Example 2.1 (The stationary KdV and MKdV equations)

The stationary KdV equation (which defines c_1) is:

$$2u_{xx} + 6u^2 = c_1, \tag{3}$$

with Lagrangian $\mathcal{L}$ given by $\mathcal{L} = 2u^3 - u_x^2 - c_1 u$ and canonical co-ordinates: $q = u, \quad p = -2u_x, \quad c_1$ (the Casimir of the degenerate canonical bracket). The Legendre transformation gives $h_{\mathcal{L}} = c_1 q - 2q^3 - \frac{1}{4}p^2$, which is the flux of $\mathcal{H}_1 = u^2$.

The Miura map induces the diffeomorphism M given by:

$$\begin{aligned} q &= \frac{1}{4}\lambda - \frac{1}{2}\tilde{p} - \tilde{q}^2, \\ p &= -\tilde{c}_1 + 4\tilde{q}^3 + 2\tilde{q}\tilde{p} - 3\lambda\tilde{q}, \\ c_1 &= \frac{1}{2}\tilde{p}^2 - 2\tilde{q}^4 + 2\tilde{q}\tilde{c}_1 + 3\lambda\tilde{q}^2 + \frac{3}{8}\lambda^2, \end{aligned}$$

relating (3) to the stationary MKdV equation:

$$-2v_{xx} + 4v^3 - 3\lambda v = \tilde{c}_1,$$

which has Lagrangian $\mathcal{L} = v_x^2 + v^4 - \tilde{c}_1 v - \frac{3}{2}\lambda v^2$ and canonical variables: $\tilde{q} = v, \quad \tilde{p} = 2v_x, \quad \tilde{c}_1$. The canonical Poisson bracket between the modified variables:

$$\{\tilde{q}, \tilde{p}\} = 1, \quad \{\tilde{q}, \tilde{c}_1\} = \{\tilde{p}, \tilde{c}_1\} = 0$$

induces :

$$\{q,p\} = -2q - \lambda, \quad \{q, c_1\} = -p, \quad \{p, c_1\} = -2c_1 + 12q^2. \tag{4}$$

This is a one parameter family of Poisson brackets: the coefficient of λ is just the canonical bracket, while the remaining part is the *second* bracket [2]. The two Poisson brackets are thus compatible.

A similar calculation for system (1) gives rise to a $(2m+1) \times (2m+1)$ non-canonical Poisson bracket, which is compatible with (2).

2.1 Reversing the role of x and t

Each member of the KdV hierarchy is bi-Hamiltonian, in both the PDE and stationary flow cases. However, the latter case required a Legendre transformation and a Miura map, so showed no clear relationship between the Hamiltonian structures of the PDE and those of the stationary flow. However, if we take t as the 'spatial' variable and x as the evolution parameter, then the KdV equation becomes a system of three equations in x, with a 3×3 matrix Poisson bracket, *which reduces to the stationary case when the potential functions are independent of t* [1, 9, 11]. The Poisson brackets can be systematically constructed by considering the corresponding spectral problem and its isospectral flows.

In order to consider the t_m stationary flow we just reverse the roles of U (the original '$x-$part' of the zero-curvature representation) and $V_{(m)}$ (the original 't_m-part'). The latter now plays the role of the spectral problem. We write the zero curvature equations as:

$$V_{(m)x} = U_\xi - [V_{(m)}, U],$$

where $\xi = t_m$. It should be noted that $V_{(m)}$ depends upon $u, u_x, \cdots, u_{x\cdots x}$, where the last term has $2m$ derivatives, so can be written in terms of any set of co-ordinates on the stationary manifold. For comparison of results we choose the canonical co-ordinates.

Example 2.2 (The KdV equation) The spectral problem is

$$\Psi_\xi = \mathcal{U}\Psi, \quad \text{where} \quad \mathcal{U} \equiv V_{(1)} = \begin{pmatrix} \frac{1}{2}p & \lambda + 2q \\ \frac{1}{4}\lambda^2 - \frac{1}{2}\lambda q + q^2 - \frac{1}{2}c_1 & -\frac{1}{2}p \end{pmatrix}. \tag{5}$$

We are interested in the hierarchy rather than just the KdV flow, so consider the time evolution:

$$\Psi_\tau = \mathcal{V}\Psi \equiv \begin{pmatrix} A & B \\ C & -A \end{pmatrix} \begin{pmatrix} \psi_1 \\ \psi_2 \end{pmatrix},$$

where A, B and C are functions of q, p, c_1, their $\xi-$derivatives and λ.

In terms of $\mathcal{U}$ and $\mathcal{V}$ the integrability conditions are just:

$$\mathcal{U}_\tau = \mathcal{V}_\xi - [\mathcal{U}, \mathcal{V}].$$

We are interested in the equation for $\mathbf{q} = (q, p, c_1)^T$ and have:

$$\mathcal{U}_\tau = \frac{D\mathcal{U}}{D\mathbf{q}} \cdot \mathbf{q}_\tau,$$

where $\frac{D\mathcal{U}}{D\mathbf{q}}$ is the Fréchet derivative. If we represent $\mathcal{U} = \begin{pmatrix} a & b \\ c & -a \end{pmatrix}$ by the vector $\mathbf{b} = (b, c, a)^T$, and $\mathcal{J}_\varphi$ is the Fréchet derivative of the map $\varphi : \mathbf{q} \mapsto \mathbf{b}$, then

$$\mathbf{b}_\tau = \mathcal{J}_\varphi \, \mathbf{q}_\tau, \quad \text{where} \quad \mathcal{J}_\varphi = \begin{pmatrix} 2 & 0 & 0 \\ 2q - \frac{1}{2}\lambda & 0 & -\frac{1}{2} \\ 0 & \frac{1}{2} & 0 \end{pmatrix},$$

whilst from the standard Poisson bracket associated with the general 2×2 matrix spectral problem, we have:

$$\mathbf{b}_\tau \equiv \begin{pmatrix} b \\ c \\ a \end{pmatrix}_\tau = \begin{pmatrix} 0 & -\partial + 2a & -b \\ -\partial - 2a & 0 & c \\ b & -c & -\frac{1}{2}\partial \end{pmatrix} \begin{pmatrix} C \\ B \\ 2A \end{pmatrix} \equiv \tilde{\mathbf{B}}\delta_b \tilde{h}, \tag{6}$$

where $\partial \equiv \frac{\partial}{\partial \xi}$ and $(C, B, 2A) = -\left(\frac{\delta \tilde{h}}{\delta b}, \frac{\delta \tilde{h}}{\delta c}, \frac{\delta \tilde{h}}{\delta a}\right)$, $h(q, p, c_1) = \tilde{h} \circ \varphi(\mathbf{q})$. We pull this gradient back to the $\mathbf{q}$ space:

$$\delta_q h = \mathcal{J}_\varphi^\dagger \cdot \delta_b \tilde{h} \quad \Rightarrow \quad \begin{pmatrix} \delta_q h \\ \delta_p h \\ \delta_{c_1} h \end{pmatrix} = \begin{pmatrix} -2 & \frac{1}{2}\lambda - 2q & 0 \\ 0 & 0 & -\frac{1}{2} \\ 0 & \frac{1}{2} & 0 \end{pmatrix} \begin{pmatrix} C \\ B \\ 2A \end{pmatrix}.$$

In this case it is possible to invert $\mathcal{J}_\varphi$ to obtain:

$$\mathbf{q}_\tau = \mathcal{J}_\varphi^{-1} \tilde{\mathbf{B}} \left(\mathcal{J}_\varphi^{-1}\right)^\dagger \delta_q h = \mathbf{B}\, \delta_q h, \tag{7}$$

where

$$\mathbf{B} = \begin{pmatrix} 0 & -2q - \lambda & -p + \partial \\ 2q + \lambda & -2\partial & 12q^2 - 2c_1 \\ p + \partial & -12q^2 + 2c_1 & (4q - \lambda)\,\partial + \partial\,(4q - \lambda) \end{pmatrix}. \tag{8}$$

This is related to the simple canonical structure of $x - t$ reversed MKdV equation through the Miura map. This is an alternative way of proving that $\mathbf{B}$ is indeed Hamiltonian.

The isospectral flows of (5) are thus bi-Hamiltonian:

$$\mathbf{q}_{\tau_r} = (\mathbf{B}_1 - \lambda \mathbf{B}_0)\delta_q h_{(r)} = \mathbf{B}_1 \delta_q h_r, \quad h_{(r)} = \lambda^{r+1} h_{-1} + \cdots + h_r,$$

where h_i are easily calculated from the fluxes of the *known* series of Hamiltonians in the usual representation. The first few are given by:

$$h_{-1} = \frac{1}{2}c_1, \quad h_0 = -\frac{1}{4}p^2 + qc_1 - 2q^3, \quad h_1 = \frac{1}{4}c_1^2 + q_\xi p,$$

giving:

$$\begin{pmatrix} q \\ p \\ c_1 \end{pmatrix}_{\tau_{-1}} = \begin{pmatrix} -\frac{1}{2}p \\ 6q^2 - c_1 \\ 2q_\xi \end{pmatrix}, \quad \begin{pmatrix} q \\ p \\ c_1 \end{pmatrix}_{\tau_0} = \begin{pmatrix} q \\ p \\ c_1 \end{pmatrix}_\xi,$$

$$\begin{pmatrix} q \\ p \\ c_1 \end{pmatrix}_{\tau_1} = \begin{pmatrix} \frac{1}{2}c_{1\xi} - 2qq_\xi - \frac{1}{2}pc_1 \\ -2q_{\xi\xi} - 2qp_\xi + 6q^2 c_1 - c_1^2 \\ (4c_1 q - 4q^3 - \frac{1}{2}p^2 - p_\xi)_\xi \end{pmatrix}.$$

The time evolution Ψ_{τ_r} is given in terms of $\delta_q h_{(r)}$:

$$A = -\frac{\delta h_{(r)}}{\delta p}, \quad B = 2\frac{\delta h_{(r)}}{\delta c_1}, \quad C = -\frac{1}{2}\frac{\delta h_{(r)}}{\delta q} + (\frac{1}{2}\lambda - 2q)\frac{\delta h_{(r)}}{\delta c_1}.$$

Thus h_{-1} generates the KdV equation and h_0 the 'translational flow'. We can use the τ_{-1} flow to write p, c_1 and q_ξ in terms of q and its τ_{-1} derivatives after which:

$$q_{\tau_1} = q_{xxxxx} + 10qq_{xxx} + 20q_x q_{xx} + 30q^2 q_x,$$

where we have written $\tau_{-1} = x$.

Thus, the hierarchy is just the KdV hierarchy in disguise. However, without the τ_{-1} flow, this hierarchy shows no hint of being reducible from the (q, p, c_1) space to the q space. In these co-ordinates the reduction to the stationary flow $q_\xi = 0$ is very natural, with (8) reducing to (4) and the bi-Hamiltonian ladder becoming finite.

The KdV equation itself is deceptively simple, since the map $\varphi : \mathbf{q} \mapsto \mathbf{b}$ is invertible, so the systems (6) and (7) are equivalent. However, when $m \geq 2$, the stationary manifold of the mth member of the KdV hierarchy is of dimension $2m+1$. Thus, whilst $\mathbf{b}$ is a 3−vector, $\mathbf{q}$ is a $(2m+1)$−vector and $\mathcal{J}_\varphi$ *projects* $\mathbf{q}_\tau$ onto $\mathbf{b}_\tau$: $\mathbf{b}_\tau = \mathcal{J}_\varphi \mathbf{q}_\tau$, since $\mathcal{J}_\varphi$ is a $3 \times (2m+1)$ matrix. Thus, the simple formula (7) is no longer possible. Nevertheless, it *is* possible to modify the procedure to find $\mathbf{B}_0$ and $\mathbf{B}_1$ so that:

$$\mathbf{q}_{\tau_r} = (\mathbf{B}_1 - \lambda \mathbf{B}_0)\, \delta_q h_{(r)}.$$

The details of this and other examples can be found in [7, 10].

3 The Darboux Transformation

We now consider the classical Darboux transformation $L \mapsto \tilde{L}$, with:

$$\begin{aligned} L &= -\partial^2 + u = -(\partial + f)(\partial - f) \\ \tilde{L} &= -\partial^2 + v = -(\partial - f)(\partial + f) = -(\partial + g)(\partial - g), \end{aligned}$$

where $\partial = \frac{d}{dx}$. The KdV variable v is related to two *different* MKdV variables f and g:

$$v = -f_x + f^2 = g_x + g^2.$$

The transformation $L \mapsto \tilde{L}$, given above, can be represented as a gauge transformation between the corresponding 'modified' spectral problems. If we define $U(g)$ and $T(g)$ by:

$$U(g) = \begin{pmatrix} g & 1 \\ \lambda & -g \end{pmatrix}, \quad T(g) = \begin{pmatrix} 1 & 0 \\ g & 1 \end{pmatrix},$$

then, under the gauge transformation:

$$U \mapsto TUT^{-1} + T_x T^{-1},$$

we have:

$$U(g) \mapsto \begin{pmatrix} 0 & 1 \\ \lambda + v & 0 \end{pmatrix}, \quad \text{where} \quad v = g_x + g^2.$$

Thus, combining the two transformations:

$$U(-f) \mapsto \begin{pmatrix} 0 & 1 \\ \lambda + v & 0 \end{pmatrix} \mapsto U(g),$$

we have:

$$T(-f-g) : U(-f) \mapsto U(g),$$

where:

$$-f_x + f^2 = g_x + g^2.$$

This gauge transformation reduces to the identity when $f + g = 0$.

Any isospectral flow of $U \equiv U(g)$ satisfies a zero curvature condition:

$$U_t - V_x + [U, V] = 0,$$

which is invariant under the gauge transformation if V transforms as:

$$V \mapsto TVT^{-1} + T_t T^{-1}.$$

For stationary flows, this reduces to a similarity transformation.

For the stationary MKdV equation, we use the equation to eliminate second order derivatives to obtain V in canonical co-ordinates:

$$\begin{aligned} V^q &= \begin{pmatrix} -b-4kq & 4k+2(p-q^2) \\ 4k^2-2k(p+q^2) & b+4kq \end{pmatrix}, \\ V^Q &= \begin{pmatrix} b+4kQ & 4k-2(P+Q^2) \\ 4k^2+2k(P-Q^2) & -b-4kQ \end{pmatrix}. \end{aligned}$$

The similarity transformation:

$$TV^qT^{-1} = V^Q, \quad \text{where} \quad T = \begin{pmatrix} 1 & 0 \\ -q-Q & 1 \end{pmatrix},$$

gives the formulae:

$$P+p = q^2-Q^2 \quad \text{and} \quad (q+Q)(p-q^2) = b,$$

which can be solved to give:

$$Q = -q + \frac{b}{p-q^2} \quad , \quad P = -p+q^2-Q^2.$$

Remark 3.1 *Since $p-q^2 = -P-Q^2$, we can similarly write:*

$$q = -Q - \frac{b}{P+Q^2}.$$

The first integrals are defined by the determinant:

$$-\det V^q = 16k^3 - 4k(p^2-q^4-2bq) + b^2,$$

which is preserved by this similarity transformation.

For the stationary 5th order MKdV equation we use the equation to eliminate fourth order derivatives to obtain V in canonical co-ordinates:

$$\begin{aligned} V^q &= \begin{pmatrix} -16q_1\lambda^2+4\lambda(2q_1^3-p_2)+c & 16\lambda^2+8\lambda(q_2-q_1^2)-2d_+(q) \\ 16\lambda^3-8\lambda^2(q_2+q_1^2)+2\lambda d_-(q) & 16q_1\lambda^2+4\lambda(p_2-2q_1^3)-c \end{pmatrix}, \\ V^Q &= \begin{pmatrix} 16Q_1\lambda^2-4\lambda(2Q_1^3-P_2)-c & 16\lambda^2-8\lambda(Q_2+Q_1^2)+2d_-(Q) \\ 16\lambda^3+8\lambda^2(Q_2-Q_1^2)-2\lambda d_+(Q) & -16Q_1\lambda^2-4\lambda(P_2-2Q_1^3)+c \end{pmatrix}, \end{aligned}$$

where $d_\pm(q) = p_1 \pm 2q_1p_2 \mp 3q_1^4 - 4q_1^2q_2 \mp q_2^2$.

Under the trivial gauge transformation $f \mapsto -g$, $c \mapsto -c$, we have: $V^q \mapsto V^Q$. The similarity transformation:

$$V^Q = TV^qT^{-1}, \quad T = \begin{pmatrix} 1 & 0 \\ -q_1-Q_1 & 1 \end{pmatrix}$$

leads to the set of equations:

$$\begin{aligned} (q_1+Q_1)d_+(q) &= c, \\ q_2+Q_2 &= q_1^2-Q_1^2, \\ p_2+P_2 &= 2(q_1+Q_1)(q_2+Q_1^2-q_1Q_1), \\ d_+(q) &= -d_-(Q), \end{aligned}$$

which can be solved to give

$$\begin{aligned}
Q_1 &= -q_1 + \frac{c}{d}, \\
Q_2 &= -q_2 + \frac{2\,c\,q_1}{d} - \frac{c^2}{d^2}, \\
P_2 &= -p_2 + \frac{2\,c}{d}\left(2\,q_1^2 + q_2\right) - \frac{6\,c^2\,q_1}{d^2} + \frac{2\,c^3}{d^3}, \\
P_1 &= -p_1 - \frac{2\,c}{d}\left(p_2 - 6\,q_1^3 - 4\,q_1\,q_2\right) \\
&\quad - \frac{2c^2}{d^2}\left(11\,q_1^2 + q_2\right) + \frac{16\,c^3\,q_1}{d^3} - \frac{4\,c^4}{d^4},
\end{aligned}$$

with $d = d_+(q)$. Since we have $d_+(q) = -d_-(Q)$, these formulae can be *explicitly* inverted.

The first integrals of the fifth order stationary MKdV equation are coefficients in:

$$det\, V^q = -256\lambda^5 - 32\lambda^2 h_1 + 8\lambda h_2 - c^2,$$

where

$$\begin{aligned}
h_1 &= \frac{1}{2}p_2^2 + q_2 p_1 - 5q_1^2 q_2^2 - q_1^6 - cq_1, \\
h_2 &= \frac{1}{2}p_1^2 - 2q_1^2 p_2^2 + 6q_1^5 p_2 - \frac{9}{2}q_1^8 - 4q_1^2 q_2 p_1 + 2q_1 q_2^2 p_2 + 5q_1^4 q_2^2 - \frac{1}{2}q_2^4 + c(p_2 - 2q_1^3) \\
&= \frac{1}{2}d_+(q)d_-(q) + c(p_2 - 2q_1^3).
\end{aligned}$$

These are invariant under the similarity transformation.

Remark 3.2 *It is interesting that the denominator $d = d_+(q)$ is a factor of the integral h_2 when $c = 0$.*

Remark 3.3 *In both cases, the transformation is not just canonical, but preserves* both *Poisson brackets and* all *commuting Hamiltonians.*

4 Conclusions

We have been interested in various aspects of the stationary flows of integrable NLEEs. In this paper we presented a systematic way of constructing Poisson brackets for these stationary flows and presented a restriction of the Darboux transformation to stationary manifolds.

The usual canonical coordinates of the stationary KdV hierarchy are associated with the first Hamiltonian structure $\mathbf{B}_0 = \partial$. If we use the 'squared eigenfunction' representation of the *second* Hamiltonian structure, then we obtain a different set of canonical coordinates, which in the case of the stationary 5th order KdV equation, transforms the equation onto an integrable case of the Hénon-Heiles system. This connection (using a reverse procedure) was first presented in [6]. This squared eigenfunction procedure can be used in conjunction with other non-canonical Poisson brackets to obtain interesting coordinate systems for related stationary flows. In particular, the Hirota-Satsuma equation can be related to (a generalisation of) an interesting integrable quartic potential.

The Lax representation of the PDE gives rise to a Lax representation for the related stationary system. Gauge transformations on the PDE level give rise to canonical transformations between the Hamiltonian systems on the reduced, stationary level. This is discussed more fully in the contribution of Baker, et al in this volume. The details will be presented elsewhere [3].

Acknowledgements

I thank V.G.Makhankov and the other organisers of the NEEDS Workshop for inviting me to present this material in Los Alamos in September, 1994.

References

[1] M. Antonowicz and M. Blaszak. On a non-standard Hamiltonian description of NLEE. In S. Carillo and O. Ragnisco, editors, *Nonlinear Evolution Equations and Dynamical Systems*, pages 152–156. Springer, Berlin, 1990.

[2] M. Antonowicz, A.P. Fordy, and S. Wojciechowski. Integrable stationary flows : Miura maps and bi-Hamiltonian structures. *Phys.Letts.A*, 124:143–50, 1987.

[3] S. Baker, V.Z. Enolskii, and A.P. Fordy. Integrable quartic potentials and coupled KdV equations. Preprint, 1994.

[4] O.I. Bogoyavlenskii and S.P. Novikov. The relationship between Hamiltonian formalisms of stationary and nonstationary problems. *Func.Anal & Apps.*, 10:8–11, 1976.

[5] R.K. Dodd and A.P. Fordy. On the integrability of a system of coupled KdV equations. *Phys.Letts. A*, 89:168–70, 1982.

[6] A.P. Fordy. The Hénon-Heiles system revisited. *Physica*, 52D:201–210, 1991.

[7] A.P. Fordy and S.D. Harris. Hamiltonian flows and stationary manifolds. Preprint, 1994.

[8] A.P. Fordy, A.B. Shabat, and A.P. Veselov. Factorisation and Poisson correspondences. Preprint, 1994.

[9] B. Fuchssteiner and W. Oevel. New hierarchies of nonlinear completely integrable systems related to a change of variables for evolution parameters. *Physica*, 145A:67–95, 1987.

[10] S.D. Harris. Integrable nonlinear evolution equations and their stationary flows. Ph.D. Thesis, University of Leeds, 1994.

[11] O.I. Mokhov. On the Hamiltonian property of an arbitrary evolution system on the set of stationary points of its integral. *Math.USSR Izvestiya*, 31:657–664, 1988.

[12] E.T. Whittaker. *A Treatise on the Analytical Dynamics of Particles and Rigid Bodies.* CUP, Cambridge, UK, 1988.

WHY ARE THERE SO MANY INTEGRABLE EVOLUTION EQUATIONS OF THIRD ORDER?

R. HERNÁNDEZ HEREDERO, V. V. SOKOLOV AND S. I. SVINOLUPOV

November 29, 1994

ABSTRACT. Every integrable evolution equation which has a nontrivial algebra of evolutionary classical symmetries, generates a family of integrable equations. The equations of this family are related to the original one through Cole-Hopf type differential substitutions. The number of generated equations is given by the number of inequivalent subalgebras of the algebra of classical symmetries of the original equation. We describe here the family generated by the integrable equation $u_t = (u_{xxx})^{-1/2}$, which possesses a maximal algebra of classical symmetries.

1 INTRODUCTION

The problem of classifying integrable third order scalar evolution equations of the form

$$u_t = F(t, x, u, u_x, u_{xx}, u_{xxx}) \tag{1.1}$$

has been undertaken in several references [1,2]. There are many integrable equations of such type, inequivalent up to classical invertible transformations (i.e. point and contact transformations). This is why the classification must be made up to more general transformations, such as the Miura transformation $v = -u_x + u^2$, which links the modified KdV equation $u_t = u_{xxx} - 6u^2u_x$ with the KdV equation $v_t = v_{xxx} + 6vv_x$. We shall use Miura-type differential substitutions called differential substitutions, which have the form:

$$y = \phi(x, u, u_x, \ldots, u_N), \qquad u_i = \frac{\partial^i u}{\partial x^i}, \tag{1.2}$$

$$v = \psi(x, u, u_x, \ldots, u_N). \tag{1.3}$$

Key words and phrases. Classical symmetries, differential substitutions, factorization procedure, integrable equations of KdV-type.

Some kind of differential substitutions [3] between a pair of equations

$$u_t = f(x, u, u_x, \ldots, u_k), \tag{1.4}$$

$$v_t = g(y, v, v_y, \ldots, v_k), \tag{1.5}$$

is related with the classical symmetries [4] of (1.4). A symmetry group can provide two lowest order invariants ϕ and ψ which serve as the functions in (1.4) and (1.5). Then (1.2)–(1.3) is called a *group substitution* and (1.5) the *factor equation* of (1.4) with respect to the group. For example, the heat equation $u_t = u_{xx}$ admits the scaling symmetry $u \to \lambda u$. The two independent invariants x and u_x/u, provide the Cole-Hopf transformation $y = x$, $v = \frac{u_x}{u}$, which thus is a group substitution relating the heat equation with the Burgers equation $v_t = v_{yy} + 2vv_y$. In fact, every finite-dimensional subgroup of the heat's equation symmetry group leads through factorization to a linearizable equation of the Burgers-type [4]. Thus, the heat equation generates a whole family of nonlinear integrable equations.

The previous ideas suggest that equations of type (1.1) posessing many symmetries produce many integrable equations through factorizations, and they should be studied. It is known [3] that $k+3$ is an upper bound on the dimension of the group of classical t-independent symmetries of an equation (1.4). For $k = 3$, there are only two equations with a group of dimension 6, namely

$$u_t = (u_{xxx})^{-1/2}. \tag{1.6}$$

$$u_t = u_x \left(\frac{u_{xxx}}{u_x} - \frac{3}{2} \frac{u_{xx}^2}{u_x^2} \right)^{-1/2}.$$

Remarkably, both equations turn out to be completely integrable, in the sense that they can be related to the well-known KdV equation by differential substitutions. In this paper we describe the family of evolution equations generated by (1.6). It is to be noted that Harry-Dym equation $v_t = v^3 v_{xxx}$ is a factor equation of (1.6), using the substitution by the differential substitution $y = x$, $v = \sqrt[3]{\frac{15}{16} u_{xxx}^{-1/2}}$, which is linked to the symmetry group of (1.6) characterized by $u \to u + \lambda_1 + \lambda_2 x + \lambda_3 x^2$.

We will make now a rigorous settlement of the classifying problem we are dealing with. The concept of factorization of differential equations appears to be a very general one [4]. We have given above a restricted definition of factorization, specialized for the case of evolution equations (1.4) and groups of t-independent point symmetries. We can restrict ourselves to such kind of symmetries because they are the only ones admitted by the equation under study (1.6).

Our definition of factorization is based on the following theorem [4,7].

Theorem 1. *For any N-parameter group of point transformations acting on a space of one independent variable x and one dependent variable u, there exist strictly two functionally independent differential invariants ϕ, ψ of order not greater than N.*

Let us consider an evolution equation (1.4) which possesses a continuous group G of t-independent point symmetries. These symmetries are the ones for which the generators of the associated vector fields have the form $a(x,u)u_x + b(x,u)$. We can regard G as the group of Theorem 1, obtaining two invariants ϕ, ψ that will be taken as the right hand sides (1.2)–(1.3). It can be proved theoretically [3] that there exists some evolution equation (1.5) linked to (1.4) through such differential substitution. We will call this differential substitution a G-substitution.

The degree of arbitrarieness in the choice of ϕ, ψ, produced by using a different pair of independent functions $\phi'(\phi,\psi)$, $\psi'(\phi,\psi)$, leads to an equation (1.5) related with the previous one through a point transformation. Thus, the factor equation (1.5) is defined up to an arbitrary point transformation.

It is possible that different subgroups of the symmetry group G of (1.4) give rise to the same factor equation. Let G_1 be a subgroup of G and let us denote by σ_1 a G_1-substitution between (1.4) and (1.5), and by ρ a point transformation that preserves equation (1.4). It is clear that the composition $\sigma_2 = \sigma_1 \circ \rho$ is also a differential substitution (1.2)–(1.3) between the same equations (1.4)–(1.5). It is easy to see that it is the G_2-substitution, where

$$G_2 = \rho^{-1} G_1 \rho. \tag{1.11}$$

It must be noted that ρ need not be only a continuous symmetry from G, but it may be a discrete point symmetry, not belonging to G.

As it is usual, any subgroups G_1 , G_2 related through (1.11) with $\rho \in G$, will be considered as *equivalent.*

As a consequence of the fact that two equivalent subgroups of symmetries of (1.6) lead, as we have seen, to the same factor equation, the first task is to find all the inequivalent subgroups of the symmetry group of (1.6). This will be accomplished in Section 1, with essentially the habitual technique used in the research of self-similar solutions of differential equations [6]. But in our case we must extend its application to the classification of multi-parameter subgroups, contrasting with the standard search for one-parameter subgroups only that is performed when looking for self-similar solutions.

In Section 2 we display the invariants ϕ, ψ calculated for all the inequivalent subgroups found in the previous section, thus finding all the inequivalent factoring substitutions for (1.6).

In Section 3 we present the explicit form of the factor equations, giving the whole family generated by equation (1.6) through the substitutions found in Section 2. It consists of quasilinear equations of the form

$$u_t = a(x, u, u_x, u_{xx})u_{xxx} + b(x, u, u_x, u_{xx}).$$

The list of factor equations shows how are the dependences of a and b in their argument.

It turns out that some pairs of inequivalent subgroups of symmetries of (1.6) lead to the same factor equation. The explanation of this fact will be given in Section 4.

2. Optimal system of subalgebras

The Lie group of point symmetries of (1.6) is generated by the following one-parameter subgroups:

$$\begin{aligned}
&x \to x, \quad u \to u + \tau_1, && x \to x, \quad u \to u + 2\tau_2 x,\\
&x \to x, \quad u \to u + \tau_3 x^2, && x \to x + t_1, \quad u \to u, \\
&x \to 2\exp(t_2)x, \quad u \to 2\exp(t_2)u, && x \to \frac{x}{1 - t_3 x}, \quad u \to \frac{u}{(1 - t_3 x)^2},
\end{aligned} \tag{2.1}$$

The corresponding vector fields have the form

$$R_1 = \frac{\partial}{\partial u}, \qquad R_2 = 2x\frac{\partial}{\partial u}, \qquad R_3 = x^2\frac{\partial}{\partial u}, \tag{2.2}$$

$$S_1 = \frac{\partial}{\partial x}, \qquad S_2 = 2x\frac{\partial}{\partial x} + 2u\frac{\partial}{\partial u}, \qquad S_3 = x^2\frac{\partial}{\partial x} + 2xu\frac{\partial}{\partial u}. \tag{2.3}$$

These vector fields form a basis of the Lie algebra $\mathfrak{g}$, whose commutation relations are described in teh following table.

Table 1.

Commutators						
	R_1	R_2	R_3	S_1	S_2	S_3
R_1	0	0	0	0	$2R_1$	R_2
R_2	0	0	0	$-2R_1$	0	$2R_3$
R_3	0	0	0	$-R_2$	$-2R_3$	0
S_1	0	$2R_1$	R_2	0	$2S_1$	S_2
S_2	$-2R_1$	0	$2R_3$	$-2S_1$	0	$2S_3$
S_3	$-R_2$	$-2R_3$	0	$-S_2$	$-2S_3$	0

¿From Table 1 it follows that the algebra $\mathfrak{g}$ is the semidirect sum of the commutative radical $\mathfrak{r}$ generated by R_i, $i = 1, \ldots, 3$ and the simple subalgebra $\mathfrak{s}$ generated by S_i, $i = 1, \ldots, 3$, which is isomorphic to $sl(2)$. Any element $g \in \mathfrak{g}$ can be expressed as a sum $g = g_{\mathfrak{r}} + g_{\mathfrak{s}}$, where $g_{\mathfrak{r}}$, $g_{\mathfrak{s}}$ denote the projections of g onto $\mathfrak{r}$ and $\mathfrak{s}$, respectively.

We are going to classify subalgebras of $\mathfrak{g}$ up to automorphisms (1.11). Any such automorphism is a composition of automorphisms

$$\mathbf{Ad}\,(\exp(\tau_i R_i)), \qquad \mathbf{Ad}\,(\exp(t_i S_i)), \qquad i = 1, \ldots, 3, \qquad \tau_i, t_i \in \mathbb{C}, \tag{2.4}$$

corresponding to the one-parameter groups (2.1). To calculate the action of automorphisms (2.4) on $\mathfrak{g}$, one can just apply the transformation groups (2.1) to the vector fields (2.2), (2.3). For example,

$$\mathbf{Ad}\,(\exp(\tau_1 R_1))\,(S_3) = x^2\frac{\partial}{\partial x} + 2x(u + \tau_1)\frac{\partial}{\partial u} = S_3 + \tau_1 R_2.$$

The result of such type of calculations is described by the following table.

Table 2.

Adjoint actions						
	R_1	R_2	R_3	S_1	S_2	S_3
$\exp(tR_1)$	R_1	R_2	R_3	S_1	S_2+2tR_1	S_3+tR_2
$\exp(tR_2)$	R_1	R_2	R_3	S_1-2tR_1	S_2	S_3+2tR_3
$\exp(tR_3)$	R_1	R_2	R_3	S_1-tR_2	S_2-2tR_3	S_3
$\exp(tS_1)$	R_1	R_2+2tR_1	$R_3+tR_2+t^2R_1$	S_1	S_2+2tS_1	$S_3+tS_2+t^2S_1$
$\exp(tS_2)$	$e^{-2t}R_1$	R_2	$e^{2t}R_3$	$e^{-2t}S_1$	S_2	$e^{2t}S_3$
$\exp(tS_3)$	$R_1-tR_2+t^2R_3$	R_2-2tR_3	R_3	$S_1-tS_2+t^2S_3$	S_2-2tS_3	S_3

It can be checked that the projection of any subalgebra $\mathfrak{a}$ of $\mathfrak{g}$ onto $\mathfrak{s}$ is a subalgebra of $\mathfrak{s}$, which we shall denote by $\mathfrak{a}_{\mathfrak{s}}$. Furthermore, the fact that $\mathfrak{r}$ is an ideal allows to restrict the action of any authomorfism $\mathbf{Ad}\,(\exp(Y))$ with $Y \in \mathfrak{g}$ to the space $\mathfrak{s}$, according to the formula

$$[\mathbf{Ad}\,(\exp(Y))\,(X)]_{\mathfrak{s}} = \mathbf{Ad}\,(\exp(Y_{\mathfrak{s}}))\,(X_{\mathfrak{s}}). \tag{2.5}$$

This implies that the classification of inequivalent subalgebras can be started in $\mathfrak{s}$, classifying inequivalent subalgebras of this algebra. As $\mathfrak{s}$ is isomorphic to $sl(2)$, it is well-known that an optimal set of its subalgebras is formed by $\{0\}$, $\{S_1\}$, $\{S_2\}$, $\{S_1, S_2\}$ and $\{S_1, S_2, S_3\}$.

We have reached the conclusion that for the projection onto $\mathfrak{s}$ of an optimal subalgebra $\mathfrak{a}$ of $\mathfrak{g}$ we can restrict ourselves to the next five possibilities:

$$i.\ \mathfrak{a}_s = \{0\}; \quad ii.\ \mathfrak{a}_s = \{S_1\}; \quad iii.\ \mathfrak{a}_s = \{S_2\}; \quad iv.\ \mathfrak{a}_s = \{S_1, S_2\};$$

$$v.\ \mathfrak{a}_s = \{S_1, S_2, S_3\}.$$

Formula (2.5) guarantees that two subalgebras having inequivalent projections onto $\mathfrak{s}$ are inequivalent.

We shall classify now one of the previous five groups of equations as an example, and afterwards we shall display the whole classification of all five cases. Let us consider the subalgebras of type *iv*, i.e. the ones with projection $\{S_1, S_2\}$. Let $\mathfrak{a}$ be such a subalgebra. Then it must contain two elements of the form

$$e_1 = S_1 + \alpha_1 R_1 + \alpha_2 R_2 + \alpha_3 R_3, \qquad e_2 = S_2 + \beta_1 R_1 + \beta_2 R_2 + \beta_3 R_3.$$

If the dimension of $\mathfrak{a}$ is greater than two, $\mathfrak{a}$ must contain elements of the form

$$e_3 = \gamma_1 R_1 + \gamma_2 R_2 + \gamma_3 R_3. \tag{2.6}$$

Lemma 1. *If $\mathfrak{a}$ contains an element e_3 of the form (2.6) then the vectors*

$$\gamma_1 R_1,\ \gamma_2 R_1,\ \gamma_2 R_2,\ \gamma_3 R_1,\ \gamma_3 R_2,\ \gamma_3 R_3$$

belong to $\mathfrak{a}$.

Proof. Commuting e_3 with e_2 we get $[e_3, e_2] = 2\gamma_1 R_1 - 2\gamma_3 R_3$, so $\gamma_1 R_1 - \gamma_3 R_3 \in \mathfrak{a}$. Commuting the last element with e_2 we see that $\gamma_1 R_1 + \gamma_3 R_3 \in \mathfrak{a}$, so $\gamma_1 R_1$, $\gamma_2 R_2$, $\gamma_3 R_3 \in \mathfrak{a}$. Commuting these elements with e_1 it can be concluded that the lemma holds.

¿From Lemma 1 several results about the dimension of the subalgebras can be readily obtained. If some subalgebra $\mathfrak{a}$ contains an element (2.6) with $\gamma_3 \neq 0$, then Lemma 1 implies that it must be the maximal subalgebra $\{S_1, S_2, R_1, R_2, R_3\}$, of dimension 5. Then any subalgebra of dimension less than 5 can only have elements (2.6) with $\gamma_3 = 0$. In this case, if $\gamma_2 \neq 0$ the subalgebra must be four dimensional and of the form $\{e_1, e_2, R_1, R_2\}$. The lemma shows analogously that three dimensional subalgebras are of the form $\{e_1, e_2, R_1\}$. Obviously, any two dimensional subalgebra must be of the form $\{e_1, e_2\}$.

The use of the adjoint actions of table 2, with proper choices of the parameter, allows to simplify the form of e_1 and e_2. The action $\mathbf{Ad}(\exp(\tau_3 R_3))$ allows to put $\beta_3 = 0$. Now using $\mathbf{Ad}(\exp(\tau_1 R_1))$, $\mathbf{Ad}(\exp(\tau_2 R_2))$ we can take $\beta_1 = 0$,

$\alpha_1 = 0$ respectively. So we have $e_1 = S_1 + \alpha_2 R_2 + \alpha_3 R_3$, $e_2 = S_2 + \beta_2 R_2$. It is easy to check that $1/2[e_1, e_2] - e_1 = \beta_2 R_1 - \alpha_2 R_2 - 2\alpha_3 R_3$. As $\mathfrak{a}$ is a subalgebra, we must have

$$\beta_2 R_1 - \alpha_2 R_2 - 2\alpha_3 R_3 \in \mathfrak{a}. \tag{2.7}$$

For the four dimensional case we knew that α_3 in (2.7) must be 0, and then it is clear that the subalgebra coincides with $\{S_1, S_2, R_1, R_2\}$. In the three dimensional case, $\alpha_3 = \alpha_2 = 0$, and the subalgebra has the form $\{S_1, S_2 + \beta R_2, R_1\}$. If $\beta \neq 0$ then through the scaling $\mathbf{Ad}(\exp(\tau S_2))$ we can make $\beta = 1$, obtaining the subalgebra $\{S_1, S_2 + R_2, R_1\}$. If $\beta = 0$ the subalgebra has the form $\{S_1, S_2, R_1\}$. In the two dimensional case, (2.7) implies that $\alpha_2, \alpha_3, \beta_2 = 0$ and we have only the subalgebra $\{S_1, S_2\}$.

Proceeding in an analogous way to the case *iv*, we can obtain the optimal set of subalgebras for all the remaining cases. It is given in the tables of Section 3.

3. Calculation of the factor equations

In this section we shall explain how to derive the factor equation associated with a symmetry subalgebra. As we did before, we will study one particular case as an example, the other cases being solved in an analogous way.

Let's consider the subalgebra $\mathfrak{a} = \{R_1, R_2, S_1, S_2\}$. According to the explanations given in the Introduction, the first step is to calculate two functionally independent invariants of order ≤ 4 of $\mathfrak{a}$. The existence of such invariants is assured by Theorem 1.

The problem of the calculation of differential invariants is a classic one in the theory of Lie symmetries [7]. In our case, the vector fields corresponding to R_1, R_2, S_1, and S_2 are written in (2.2)–(2.3). We must consider their 4-prolongation

$$R_1^{(4)} = \frac{\partial}{\partial u}, \qquad R_2^{(4)} = 2x\frac{\partial}{\partial u} + 2\frac{\partial}{\partial u_1}.$$

$$S_1^{(4)} = \frac{\partial}{\partial x}, \qquad S_2^{(4)} = 2x\frac{\partial}{\partial x} + 2u\frac{\partial}{\partial u} - 2u_2\frac{\partial}{\partial u_2} - 4u_3\frac{\partial}{\partial u_3} - 6u_4\frac{\partial}{\partial u_4},$$

The invariants of $\mathfrak{a}$ of order ≤ 4 are the solutions $I(x, u, u_1, u_2, u_3, u_4)$ of the system of partial differential equations

$$R_1^{(4)} I = 0, \qquad R_2^{(4)} I = 0, \qquad S_1^{(4)} I = 0, \qquad S_2^{(4)} I = 0.$$

¿From the first three equations it follows that I doesn't depend on x, u and u_1. The fourth equation is written

$$u_2\frac{\partial I}{\partial u_2} + 2u_3\frac{\partial I}{\partial u_3} + 3u_4\frac{\partial I}{\partial u_4} = 0.$$

The general solution of this equation can then be written as

$$I = F(\frac{u_3}{u_2^2}, \frac{u_4}{u_2^3}),$$

where F is an arbitrary function.

In order to calculate the factor equation, we have to choose two functionally independent differential invariants that will play the role of ϕ and ψ in the formulas (1.2) and (1.3). We have some freedom in the choice of these invariants, so we choose the ones that lead to a factor equation as simple as possible. In the case we are considering now, it turns out that the differential substitution

$$y = -\log\left(\frac{u_3}{u_2^2}\right), \qquad v = -u_3^{-3/2}u_4 + 2\frac{u_3^{1/2}}{u_2}$$

leads to the simplest equation

$$v_t = v^3 v_{yyy} + 3v^2 v_y v_{yy} - \frac{1}{4}v^3 v_y + \frac{3}{2}\exp(-y)\, v^2, \tag{3.1}$$

through the following procedure. It is known [3] that given an evolution equation (1.4) and a differential substitution (1.2)–(1.3), the transformed equation (1.5) is given by the formula

$$v_t = \left(\psi_* - \frac{D\psi}{D\phi}\phi_*\right)\psi, \tag{3.2}$$

being $D = \sum_i u_{i+1}\frac{\partial}{\partial u_i}$ the total derivative with respect to x and $\rho_* = \sum_i \frac{\partial \rho}{\partial u_i}D^i$ the "linearization" of ρ operator. Formula (3.2) is given in terms of x, u, $u_1, \ldots, u_i, \ldots$, and to really get (1.5) we must rewrite it in terms of $y = \phi$, $v = \psi$, $v_1 = \frac{D\psi}{D\phi}$, $\ldots$, $v_i = \left(\frac{D}{D\phi}\right)^i \psi$, $\ldots$. The fields u_i, $i \geq 3$ can be written in terms of x, u, u_1, u_2, ϕ, ψ, $\frac{D\psi}{D\phi}$, $\frac{D}{D\phi}\left(\frac{D\psi}{D\phi}\right)$, $\frac{D}{D\phi}\left(\frac{D}{D\phi}\frac{D\psi}{D\phi}\right)$. We have

$$u_3 = u_2^2 e^{-\phi}, \qquad u_4 = 2u_2^3 e^{-2\phi} - u_2^3 e^{-3/2\phi},$$

$$u_5 = \frac{6u_2^4}{e^{3\phi}} - \frac{7\psi u_2^4}{e^{5\phi/2}} + \frac{3\psi^2 u_2^4}{2e^{2\phi}} - \frac{\psi\frac{D\psi}{D\phi}u_2^4}{e^{2\phi}},$$

an so on. Substituting these expresions in (3.2), we obtain an equation that, if the differential substitution is admitted, does not depend on x, u, u_1 and u_2. It is the sought evolution equation (1.5).

In the following tables, we give in the first column the symmetry subalgebra, in the second one the two invariants ϕ, ψ and in the third the resulting factor equation. We use the notation $v_i = \frac{\partial^i v}{\partial y^i}$.

Table 3.

Projection $a_s = \{0\}$		
$\{R_1\}$	$x, \quad u_1$	$v_t = v_2^{-3/2} v_3$
$\{R_2\}$	$x, \quad u - xu_1$	$v_t = y^{3/4} v_2^{-3/2}\left(v_3 + \frac{3}{2} v_2/y\right)$
$\{R_1,R_2\}$	$x, \quad u_2$	$v_t = v_1^{-3/2}\left(v_3 - \frac{3}{2} v_2^2/v_1\right)$
$\{R_1,R_3\}$	$x, \quad u_1 - xu_2$	$v_t = v_1^{-3/2}\left(v_3 - \frac{3}{2} v_2^2/v_1 + 2v_1\right)$
$\{R_1,R_2,R_3\}$	$x, \quad u_3$	$v_t = v^3 v_3$

Table 4.

Projection $a_s = \{S_1\}$		
$\{S_1\}$	$u, \quad u_1$	$v_t = v^{3/4} v_2^{-3/2}\left(v_3 + \frac{3}{2} v_1 v_2/v\right)$
$\{S_1+R_3\}$	$u - x^3/3,$ $u_1 - x^2$	$v_t = \left(v_2 + v_1^2/v + 2/v^2\right)^{-3/2} \times \left(v_3 + 6v_1v_2/v + 3v_1^3/v^2 + 4v_1/v^3\right)$
$\{S_1,R_1\}$	$u_1, \quad u_2$	$v_t = v_1^{-3/2}\left(v_3 - \frac{3}{2} v_2^2/v_1 - 2v_1^3\right)$
$\{S_1+R_3,R_1\}$	$u_1 - x^2,$ $u_2 - 2x$	$v_t = (vv_1 + 2)^{3/2}\left(v^3 v_3 + 3vv_1v_2\right) - \frac{3}{2}(vv_1 + 2)^{-5/2} v^2 \left(vv_2 + v_1^2\right)^2$
$\{S_1,R_1,R_2\}$	$u_2, \quad u_3$	$v_t = v^3 v_3$
$\{S_1+R_3,R_1,R_2\}$	$u_2 - 2x, \quad u_3$	$v_t = v^3 v_3 + 24v^2 v_1 v_2/(v^2+8) - 24v^3 v_1^3/(v^2+8)^2$
$\{S_1,R_1,R_2,R_3\}$	$u_3, \quad u_4$	$v_t = v^3 v_3 + 3v^2 v_1 v_2 - \frac{1}{4} v^3 v_1$

Table 5.

Projection $a_s = \{S_2\}$		
$\{S_2\}$	$u/x, \quad u_1$	$v_t = \left(v^2 v_2 + vv_1 - v\right)^{-3/2} \times \left(v^3 v_3 + 6v^2 v_1 v_2 + 3vv_1^3 - 3vv_1\right)$
$\{S_2+R_2\}$	$u/x + \log x,$ $u_1 + \log x$	$v_t = \left(v^2 v_2 + vv_1 - v + 1\right)^{-3/2} \times \left(v^3 v_3 + 6v^2 v_1 v_2 + 3vv_1^3 - 6vv_1 + 2v_1\right)$
$\{S_2,R_1\}$	$u_1 - u/x,$ xu_2	$v_t = (vv_1 - yv)^{-3/2} \times \left(v^3 v_3 + 3v^2 v_1 v_2 + 3vv_1^2 - 3vv_1\right) - \frac{3}{2}(vv_1 - yv)^{-5/2} v^2 \left(vv_2 + v_1^2 - v_1\right)^2$
$\{S_2,R_2\}$	$u_1 - u/x,$ xu_2	$v_t = (vv_1 - y)^{-3/2}\left(v^3 v_3 + 3v^2 v_1 v_2\right) - \frac{3}{2}(vv_1 - y)^{-5/2} v^2 \left(vv_2 + vv_1^2 - 1\right)^2$

Table 5, continuation.

$\{S_2+R_2,R_1\}$	$u_1+\log x$, xu_2	$v_t=(vv_1-v+1)^{-3/2} \times \left(v^3v_3+3v^2v_1v_2+3vv_1^2-3vv_1+2v_1\right) - \frac{3}{2}(vv_1-v+1)^{-5/2}v^2\left(vv_2+v_1^2-v_1\right)^2$
$\{S_2,R_1,R_2\}$	xu_2, x^2u_3	$v_t=v^3v_3+24v^2v_1v_2/(v^2+8)-24v^3v_1^3/(v^2+8)^2 - 3v^3v_1/x^2+(3v^4+24v^2)/x^3$
$\{S_2,R_1,R_3\}$	u_1-xu_2, x^2u_3	$v_t=v^3v_3-3\frac{v_1}{v}$
$\{S_2+R_2,R_1,R_3\}$	$u_1-xu_2-\log x$, x^2u_3	$v_t=v^3v_3-12v^2v_1v_2/(v^2-4)+12v^3v_1^3/(v^2-4)^2 - \frac{1}{2}v^3v_1+\left(\frac{1}{2}v^2+1\right)(v^2-4)^{1/2}v_1$
$\{S_2,R_1,R_2,R_3\}$	x^2u_3, x^3u_4	$v_t=v^3v_3+3v^2v_1v_2-\frac{1}{4}v^3v_1+\frac{3}{2}e^{-y}v^2$

Table 6.

Projection $\mathfrak{a}_s=\{S_1,S_2\}$		
$\{S_1,S_2\}$	u_1, uu_2	$v_t=(vv_1-yv)^{-3/2}\times\left(v^3v_3+3v^2v_1v_2+3yvv_1 - 6v^2v_1-3y^2vv_1+3yv^2\right) - \frac{3}{2}(vv_1-yv)^{-5/2}v^2\left(vv_2+v_1^2-yv_1-v\right)^2$
$\{S_1,S_2,R_1\}$	u_1, u_3/u_2^2	$v_t=v^3v_3-3\frac{v_1}{v}$
$\{S_1,S_2+R_2,R_1\}$	$u_1+\log u_2$, $u_3e^{2u_1}$	The same as $\{S_2+R_2,R_1,R_3\}$
$\{S_1,S_2,R_1,R_2\}$	u_3/u_2^2, u_4/u_2^3	$v_t=v^3v_3+3v^2v_1v_2-\frac{1}{4}v^3v_1+\frac{3}{2}e^{-y}v^2$
$\{S_1,S_2,R_1,R_2,R_3\}$	u_4^2/u_3^3, u_5/u_3^2	$v_t=v^3v_3+3v^2v_1v_2-\frac{3}{4}yv^2$

Table 7.

Projection $\mathfrak{a}_s=\{S_1,S_2,S_3\}$		
$\{S_1,S_2,S_3\}$	$uu_2-u_1^2/2$, u^2u_3	$v_t=v^3v_3+6yv_1/v-3$
$\{S_1,S_2,S_3,R_1,R_2,R_3\}$	$u_5/u_3^2-\frac{5}{4}u_4^2/u_3^3$, $u_6/u_3^{5/2}-\frac{9}{2}u_5u_4/u_3^{7/2}+\frac{15}{4}u_4^3/u_3^{9/2}$	$v_t=v^3v_3+3v^2v_1v_2 - \frac{3}{2}v^2$

4. A non-local involution

In the lists of the previous section there are some pairs of coincident equations, in spite of the fact that they come from factorizations corresponding to inequivalent subalgebras. Namely, the equations corresponding to the subalgebras $\{R_1,R_3\}$-$\{S_1,R_1\}$, $\{R_1,R_2,R_3\}$-$\{S_1,R_1,R_2\}$, $\{S_2,R_1,R_3\}$-$\{S_2,S_1,R_1\}$, $\{S_2+R_2,R_1,R_3\}$-$\{S_2+R_2,S_1,R_1\}$ and $\{R_1,R_3\}$-$\{S_1,R_1\}$ are pairwise equal. The reason is that equation (1.6) admits the following non-local involution:

$$\bar{x} = u_2, \qquad \bar{u} = \frac{1}{2}xu_2^2 - u_1u_2 + \frac{1}{2}D_x^{-1}(u_2^2), \tag{4.1}$$

$$\bar{u}_1 = xu_2 - u_1, \qquad \bar{u}_2 = x, \qquad \bar{u}_3 = \frac{1}{u_3},$$

where $\bar{u}_i = \frac{\partial^i \bar{u}}{\partial \bar{x}^i}$.

The differential substitution arising from the composition of the involution (4.1) and a factorization can be equivalent, modulo a point transformation, to another factorization. The corresponding factor equations are then equivalent via a point transformation. For example, the algebras $\{R_1, R_2, R_3\}$ yield the equation $v_t = v^3 v_3$, through the factoring differential substitutions $y = x$, $v = u_3$ and $y = u_2$, $v = u_3$ respectively. We can see after a short calculation that the last substitution is the composition of the involution, followed by the first substitution, followed by the point transformation $\bar{v} = 1/v$.

Another way of explaining the pairings is to study how the involution acts over the symmetry algebra. It can be seen, although lack of space precludes us from giving the details, that under (4.1) the symmetries (2.2)-(2.3) transform as:

$$R_1 \to R_1, \quad R_2 \to R_2, \quad R_3 \to 2S_1, \quad S_1 \to \frac{1}{2}R_3, \quad S_2 \to S_2, \tag{4.2}$$

being S_3 transformed into a new non-local symmetry. After the addition of a finite number of nonlocalities, the algebra closes. Now (2.4) shows that the paired subalgebras are no longer inequivalent as subalgebras of the enlarged algebra.

Acknowledgements This work has been partially supported by a Sabbatical grant from the Spanish Ministery of Education.

References

1. A. V. Mikhailov, A. B. Shabat and V. V. Sokolov, *The Symmetry Approach to Classification of Integrable Equations*, What is integrability?, Springer-Verlag (Springer Series in Nonlinear Dynamics), 1991, pp. 115–184.
2. R. H. Heredero, V. V. Sokolov and S. I. Svinolupov, *Toward the classification of third order integrable evolution equations*, J. Phys. A: Math. Gen. **27** (1994), 4557–4568.
3. V. V. Sokolov, *On the symmetries of evolution equations*, Russian Math. Surveys **43, 5** (1988), 165–204.
4. S. I. Svinolupov and V. V. Sokolov, *Factorization of evolution equations*, Russian Math Surveys **47, 3** (1992), 127–162.

5. S. I. Svinolupov, *On the analogues of the Burgers equations*, Phys. Lett. A **135**, 1 (1989), 32–36.
6. P. Olver, *Applications of Lie Groups to Differential Equations*, Second Edition, Graduate Texts in Mathematics, vol. 107, Springer-Verlag, New York, 1993.
7. P. Olver, *Differential Invariants*, Acta Appl. Math (to appear).

DEPARTAMENTO DE FÍSICA TEÓRICA II, Facultad de Ciencias Físicas, Univ. Complutense de Madrid, 28040 Madrid, Spain

Current address: School of Mathematics, Institute of Technology, 127 Vincent Hall, 206 Church Street S. E., Minneapolis, MN 55455, USA

E-mail address: rhh@fis.ucm.es

MATHEMATICAL INSTITUTE OF UFA CENTER OF RUSSIAN ACADEMY OF SCIENCES, CHERNYSHEVSKY STR. 112, 450000, UFA, RUSSIA.

E-mail address: sokolov@nkc.bashkiria.su, servs@nkc.bashkiria.su

INTEGRABLE TRILINEAR PDE's

J. HIETARINTA
Department of Physics, University of Turku, 20500 Turku, Finland

B. GRAMMATICOS
LPN, Université Paris VII, Tour 24-14, 5ème étage, 75251 Paris, France

A. RAMANI
CPT, Ecole Polytechnique, 91128 Palaiseau, France

ABSTRACT

In a recent publication we proposed an extension of Hirota's bilinear formalism to arbitrary multilinearities. The trilinear (and higher) operators were constructed from the requirement of gauge invariance for the nonlinear equation. Here we concentrate on the trilinear case, and use singularity analysis in order to single out equations that are likely to be integrable. New PDE's are thus obtained, along with others already well-known for their integrability and for which we obtain here the trilinear expression.

1. Introduction

What is astonishing with integrable PDE's is that there are so many of them. Infinite hierarchies of equations are known to date and the domain is still expanding. Moreover, these equations are often associated to physical models [1]. Then, using integrability, one can construct particular solutions and compute the pertinent physical quantities for these models.

One of the essential tools in the study of integrable PDE's, over the past 20 years, has been the bilinear formalism of Hirota [2]. The main advantage of this approach lies in the fact that it allows one to obtain multisoliton solutions in a straightforward way [3]. Still, the proof of integrability, based on the existence of an arbitrary number of solitons, may present considerable difficulties. It is often easier to test the integrability of a given equation using singularity analysis (Painlevé method)[4], especially since this criterion (absence of singular solutions that exhibit branching) can be implemented algorithmically.

In a recent work [5] we have presented an extension of Hirota's bilinear formalism to higher multilinearities. This was motivated by the existence of integrable equations that cannot be bilinearized, like the Satsuma determinantal trilinear PDE's [6]. (It must be made clear at this point that the nonexistence of a bilinear form is not a rigorous statement; it simply means that no bilinear form has been obtained to date). Our method is based on the assumption of gauge invariance. In the first exploratory study [5] we studied only the leading part of one-dimensional trilinear equations with one dependent variable and found those that satisfy the Painlevé criterion. The aim of the present paper is to complete this work, and obtain the integrable higher dimensional generalizations with the non-leading parts. As we will show in the following sections some of the equations obtained are new, while others are well-known for their integrability without necessarily possessing a simple bilinear form.

2. From bilinear to trilinear operators.

A prerequisite to the application of the Hirota bilinear formalism is a dependent variable transformation that converts the nonlinear equation into a quadratic 'pre-potential' form. In order to make things more clear let us present the classical example of the KdV equation. Starting from

$$u_{xxx} + 6uu_x + u_t = 0, \tag{2.1}$$

we introduce the transformation $u = 2\partial_x^2 \log F$ and obtain (after one integration):

$$F_{xxxx}F - 4F_{xxx}F_x + 3F_{xx}^2 + F_{xt}F - F_xF_t = 0. \tag{2.2}$$

This last equation can be written in a particularly condensed form using the Hirota D operator:

$$(D_x^4 + D_xD_t)F\cdot F = 0, \tag{2.3}$$

where the D-operator is defined by its antisymmetric derivative action on a pair of functions (the 'dot product'):

$$D_x^n f\cdot g = (\partial_{x_1} - \partial_{x_2})^n f(x_1)g(x_2)\big|_{x_2=x_1=x}. \tag{2.4}$$

The crucial observation here is the relation of the 'physical' variable $u = 2\partial_x^2 \log F$ to the Hirota's function F: the gauge transformation $F \to e^{px+\omega t}F$ leaves u invariant. It turns out that this is a general property of bilinear equations. In fact, one can *define* Hirota's bilinear equations through the requirement of gauge invariance. This statement was proven in [5].

Having obtained the Hirota bilinear operators on the sole requirement of gauge invariance we investigated in [5] the possible extension of this formalism and the introduction of multilinear operators. This turned out to be possible. Our first step was the extension of the bilinear operators to the trilinear case. We found that one convenient basis for the representation of the gauge invariant N'th order derivative operators is given by $(\partial_1 - \partial_2)^n(\partial_1 - \partial_3)^{N-n}$ for $n = 0,\ldots,N$, where the subscripts tell on which of the three functions each derivative acts. Thus the basic building blocks for the trilinear operators are again the Hirota bilinear D's: we must just specify the indices in this case. We thus have $D_{12} \equiv \partial_{x_1} - \partial_{x_2}$, $D_{23} \equiv \partial_{x_2} - \partial_{x_3}$, $D_{31} \equiv \partial_{x_3} - \partial_{x_1}$, but, of course the three are not linearly independent:

$D_{12}+D_{23}+D_{31}=0$. Their action on a 'triple dot product' is analogous to the bilinear case:

$$D_{12}f\cdot g\cdot h=(\partial_{x_1}-\partial_{x_2})f(x_1)g(x_2)h(x_3)\big|_{x_3=x_2=x_1=x}=(f'g-fg')h. \tag{2.5}$$

The choice of a particular pair of D's as the basic trilinear operators breaks the symmetry between the three coordinates x_i's. It is possible to restore this symmetry by introducing a different basis for the trilinear operators, T and T^*:

$$T=\partial_1+j\partial_2+j^2\partial_3\,,\quad T^*=\partial_1+j^2\partial_2+j\partial_3, \tag{2.6}$$

where j is the cubic root of unity, $j=e^{2i\pi/3}$. (Note that the star in T^* indicates complex conjugation for the coefficients in T but not for the independent variables). Note that $T^nT^{*m}F\cdot F\cdot F=0$, unless $n-m\equiv 0 \pmod 3$, which is the equivalent to the bilinear property $D^nF\cdot F=0$, unless $n\equiv 0 \pmod 2$. Moreover, a reality condition is satisfied: $T^nT^{*m}F\cdot F\cdot F=T^mT^{*n}F\cdot F\cdot F$.

The generalization to higher multilinear equations is straightforward. One can introduce the set of $n(n-1)/2$ operators D_{ij} acting on n-tuple dot-products $D_{ij}f_1\cdot f_2\cdot\ldots\cdot f_n$. Of course only $n-1$ of the D_{ij}'s are independent, a convenient basis being the D_{1j}, $j=2,\ldots n$. As in the trilinear case, one can also construct 'symmetric' operators:

$$M_n^m=\sum_{k=0}^{n-1}e^{2\pi ikm/n}\partial_{k+1} \tag{2.7}$$

for $0<m<n$. We have, for example, $D=M_2^1$, $T=M_3^1$, $T^*=M_3^2$ and so on.

3. General comments on multilinearization

Before dealing with specific cases let us present here some general considerations. Let us start with a nonlinear (in u) equation, of order $k+1$ having the form $u_t+\partial_xP(u,u_x,\ldots,u_{kx})=0$. Several well known integrable equations belong to this class. Let us assume that the leading part of P is weight-homogeneous in u and ∂_x, with u having the same weight as ∂_x^2. Then we can transform the equation into a multilinear expression through the transformation $u=\alpha\partial_x^2(\log F)$ and obtain, after one x-integration, generically an $(k+2)$-multilinear equation. The scaling factor α can then be chosen (perhaps in several ways) so as to make an F^2 term factor out: the resulting multilinear equation is at most k-linear. For example at order five ($k=4$) we have three integrable equations, and we should expect, in principle, these equations to have quadrilinear forms. Some unexpected cancellations, however, do occur. Thus the Sawada-Kotera equation [7] has a bilinear expression, the Lax-5 equation [8] a trilinear form, but the Kaup-Kuperschmidt equation [9] is quadrilinear.

In the process of multilinearization of a given nonlinear equation the following points should be noted:

1) Above we discussed only $u = \alpha\partial^2 \log(F)$ substitutions. If u always appears in the equation with at least one derivative then we can get a gauge invariant form also with $u = \alpha\partial \log(F)$, and if we always have at least two derivatives, $u = \alpha \log(F)$ is sufficient.

2) Any bilinear expression multiplied by F is cubic and gauge invariant and therefore has trilinear form, for example

$$F(D_x D_y)F\cdot F = \tfrac{2}{3}\, T_x T_y^* F\cdot F\cdot F, \tag{3.1}$$

$$F(D_x^3 D_y)F\cdot F = \tfrac{2}{3}\, T_x^2 T_x^* T_y^* F\cdot F\cdot F, \tag{3.2}$$

$$F(D_x^5 D_y)F\cdot F = \tfrac{2}{27}\,(10 T_x^3 T_x^{*2} T_y^* - T_x^5 T_y)F\cdot F\cdot F, \tag{3.3}$$

$$F(D_x^7 D_y)F\cdot F = \tfrac{2}{81}\,(35 T_x^4 T_x^{*3} T_y^* - T_x^7 T_y^* - 7 T_x^6 T_x^* T_y)F\cdot F\cdot F. \tag{3.4}$$

Therefore known one-component bilinear equations reappear in this trilinear study.

3) Taking a derivative of an equation having an n-linear form yields an equation that is $n+1$ multilinear, for example

$$\partial_x \left(F^{-2} D_x D_y F\cdot F\right) = \tfrac{2}{3} F^{-3}\, T_x^2 T_y F\cdot F\cdot F, \tag{3.5}$$

$$\partial_x \left(F^{-2} D_x^3 D_y F\cdot F\right) = \tfrac{2}{9} F^{-3}\, \left(T_x^4 T_y^* + 2 T_x^3 T_x^* T_y\right) F\cdot F\cdot F, \tag{3.6}$$

$$\partial_x \left(F^{-2} D_x^6 F\cdot F\right) \quad = \tfrac{2}{3} F^{-3}\, T_x^5 T_x^{*2} F\cdot F\cdot F. \tag{3.7}$$

The integrability of the derivative form is another matter: if for example the x-derivate form is also integrable then the original equation would be integrable with inhomogeneous x-independent term. Thus we will later find the x-derivative of Ito equation to be integrable, but not its t-derivative, therefore Ito's equation is integrable even with a inhomogeneous $g(t)$-term.

Since equations having a bilinear form will reappear in the trilinear analysis as mentioned before, let us collect here the integrable equations that can be bilinearized to the form $P(D)F\cdot F = 0$ [10,11]:

Kadomtsev-Petviashvili:

$$\left(D_x^4 - 4 D_x D_t + 3 D_y^2\right) F\cdot F = 0, \tag{3.8}$$

Ito:

$$\left(D_t D_x^3 + a D_x^2 + D_t D_y\right) F\cdot F = 0, \tag{3.9}$$

Hietarinta:

$$\left(D_x(D_x^3 - D_t^3) + a D_x^2 + b D_t D_x + c D_t^2\right) F\cdot F = 0, \tag{3.10}$$

Sawada-Kotera-Ramani:

$$\left(D_x^6 + 5 D_x^3 D_y - 5 D_y^2 + D_x D_t\right) F\cdot F = 0, \tag{3.11}$$

4. Singularity analysis of trilinear equations

One motivation behind the multilinear approach is to find new integrable equations. Eventually we would like to get something similar to the systematic classification of bilinear equations presented in [10,11]. In the following paragraphs we will limit ourselves to the singularity analysis of trilinear equations involving only one *dependent* variable, i.e. unicomponent equations $P(T_x, T_x^*, T_y, T_y^*, \ldots)F\cdot F\cdot F = 0$. (Let us recall here that in the bilinear case a complete classification of these simplest equations [10,11] was possible).

Since the dependent function u in a nonlinear equation is related to the multilinear F through $u = \alpha\partial^2 \log F$ it is clear that a zero in F induces a pole-like behavior in u. Let us here consider a concrete example: $T^{3k}F\cdot F\cdot F = 0$. Putting, for the dominant part, $F \sim \phi^n$, where ϕ is the singular manifold we find that the only possible leading behaviors have $n = 0, 1, \ldots, k-1$. The first corresponds to a nonsingular Taylor-like expansion, which is always possible. The second behavior $F \sim \phi$ corresponds to a simple zero of F that would give a (double) pole in u. Next the resonances r can be obtained if we substitute $F = \phi^n(1+\omega\phi^r)$ and collect terms linear in ω. For the equation to pass the Painlevé test the resonances must be integers and no incompatibilities must arise at any resonance. As an example let us take $k = 2$, i.e., the operator T^6. The leading behavior $F \sim \phi$ leads to the resonances $r = -1, 0, 1, 2$ and the roots of $r^2 - 13r + 60 = 0$ which are complex. Thus $T^6F\cdot F\cdot F = 0$ does not possess the Painlevé property.

4.1 Leading behaviors

The Painlevé analysis for the leading part of the equations was performed in [5], with just one independent variable. This is sufficient for the computation of dominant singularities and resonances. Note that for a given $N = n + m$ there may exist several pairs (n, m) such that $T^nT^{*m}F\cdot F\cdot F$ is not identically zero, namely those for which $n \equiv m \pmod 3$. The leading part of the general equation at order N is then given by a linear combination of all the non-vanishing $T^nT^{*m}F\cdot F\cdot F$'s. In each case we give below the precise combinations that lead to equations with the Painlevé property. The nonlinear forms of the leading parts of the equations are obtained by the standard substitution $F = e^g$ followed by $u = 2g''$. The results of [5] are summarized as follows:

N = 2 : operator : TT^* leading part : u.
In this case we can write the result also in bilinear form

N = 3 : operator : T^3 leading part : u'.

N = 4 : operator : T^2T^{*2} leading part : $u'' + 3u^2$.
This, of course, is just the leading part of the KdV equation in potential form, and can also be written in bilinear form, see (3.2).

N = 5 : operator : T^4T^* leading part : $u''' + 6u'u$.
This is the derivative of the expression obtained at $N = 4$.

N = 6 : operator : $\lambda T^6 + \mu T^3T^{*3}$. This is the first case where we have two possible (n, m) pairs. The λ, μ combinations that pass the Painlevé test are the following

a) $(\lambda = 7, \mu = 20)$ leading part : $u'''' + 10u''u + 5u'^2 + 10u^3$.
This is the leading part of the once integrated 5th order equation in the Lax hierarchy of KdV.

b) $(\lambda = -2, \mu = 20)$ leading part : $u'''' + 15u''u + 15u^3$.
This is the leading part of the once integrated Sawada-Kotera-Ramani equation (3.11), and can also be written in bilinear form, see (3.3).

c) $(\lambda = -\mu)$ leading part : $uu'' - u'^2 + u^3$.
This expression can be cast in determinantal equation

$$\begin{vmatrix} F'''' & F''' & F'' \\ F''' & F'' & F' \\ F'' & F' & F \end{vmatrix} = 0,$$

whose integrability was already noticed by Chazy [12].

N = 7 : operator : T^5T^{*2} leading part : $u^{(5)} + 15u'''u + 15u''u' + 45u'u^2$.
This is the Sawada-Kotera equation, i.e., the derivative of the expression obtained in $N = 6$b, c.f. (3.7).

N = 8 : operator : $\lambda T^7T^* + \mu T^4T^{*4}$ and nonlinearization with $u = 6g''$ instead of $u = 2g''$ used before.

a) $(\lambda = 4, \mu = 5)$ leading part : $u^{(6)} + 6u''''u + 10u'''u' + 5u''^2 + 10u''u^2 + 10u'^2u + \frac{5}{3}u^4$
This would correspond to a 7th order equation which, we believe, leads to a new integrable case.

b) $(\lambda = 4, \mu = 14)$ leading part : $u^{(6)} + 7u''''u + 7u'''u' + 7u''^2 + 14u''u^2 + 7u'^2u + \frac{7}{3}u^4$
This is the leading part of a once integrated higher Sawada-Kotera equation.

c) $(\lambda = -\mu)$ leading part : $u''''u - 3u'''u' + 2u''^2 + 4u''u^2 - 3u'^2u + \frac{2}{3}u^4$.
This new equation looks like an extension of Satsuma's equation given at N=6c above, but it cannot be written as a single determinant.

N = 9: No cases passing the Painlevé test exist at this order.

N = 10 : operator : $5T^8T^{*2} + 4T^5T^{*5}$ (and we take $u = 30g''$).
leading part : $u^{(8)} + 2u^{(6)}u + 4u^{(5)}u' + 6u''''u'' + 5u'''^2 + \frac{6}{5}u''''u^2 + 4u'''u'u + 2u''^2u + 2u''u'^2 + \frac{4}{15}u''u^3 + \frac{2}{5}u'^2u^2 + \frac{2}{375}u^5$
This is also a new equation.

Furthermore, no integrable candidates were found at orders $N = 11, 12$. The singularity analysis at these higher orders becomes progressively more difficult.(Already at $N = 12$ there exist three non-vanishing n, m combinations). It is, thus, not possible to extend our investigation to very high orders (as was done in our study of bilinear equations [11]), but we do believe that no further integrable candidates exist at higher orders.

4.2 Resonance conditions

In order to investigate the resonance conditions we must specify precisely the PDE we are working with, i.e., also the nonleading parts. Let us state from the outset that we are not going to consider equations of the form $N = 6c$ or $N = 8c$. These equations have a form that makes the *expected* leading singularity to vanish identically, which leads to certain complications in the singularity analysis: they will be discussed elsewhere. Also in every case examined, subleading parts of opposite parity (= odd vs. even number of T's) to that of the leading were found to violate the Painlevé property, but we did not make an exhaustive study of this. The technical details of the resonance conditions study are not particularly interesting, the analysis was performed on a computer using the REDUCE language [13] for symbolic manipulation. Although in some cases the resonances are particularly high (in the case of $N = 10$ we have a resonance at $r = 30$ for the leading behavior $F \sim \phi^{14}$) we were able in all cases to check the resonance conditions.

For low N the complete results are simple: for $N = 3$ one obtains only linear equations, and for $N = 4$ all the integrable equations could also be written in bilinear form.

For equations with $N \geq 5$ an important question was how to extend the leading-order trilinear operator to two dimensions and still keep the calculations manageable. Our choice has been to consider only 'monomial' leading parts, that is monomial in derivatives and not necessarily in T's. (From our experience on the bilinear case we do not expect nonmonomial leading parts to play a role for $N \geq 5$).

N = **5** :

The leading 1-dimensional operator is $T_x^4 T_x^*$ and three monomial generalizations to 1+1 dimensions are possible: i) $T_x^4 T_x^*$, ii) $T_x^4 T_y^* + \lambda T_x^3 T_y T_x^*$ and iii) $T_x^3 T_y T_y^* + \lambda T_x^2 T_y^2 T_x^*$.

The case i) supplemented by subleading terms leads only to the x-derivative of the KP equation (3.5-8):

$$(T_x^4 T_x^* - 4T_x^2 T_t + 3T_x T_y^2) F \cdot F \cdot F = 0.$$

In the case ii) (discarding the $\lambda = -1$ case that does not have a standard leading singularity) there exist the three Painlevé candidates $\lambda = -4$, 2 and 8. The first case does not satisfy the resonance condition. The case $\lambda = 2$ yields just the x-derivative of the Ito equation (3.9):

$$(T_x^4 T_t^* + 2T_x^3 T_x^* T_t + 3aT_x^3 + 3T_x^2 T_y) F \cdot F \cdot F = 0.$$

The final case $\lambda = 8$ is more interesting. The most general result passing the Painlevé test reads:

$$\left(T_y(T_x^{*4} + 8T_x^3 T_x^* + 9T_y^2) + 9T_x^2 T_t\right) F \cdot F \cdot F = 0. \tag{4.1}$$

Putting $F = e^{g/2}$ we obtain:

$$g_{xxxxy} + 4g_{xxy}g_{xx} + 2g_{xy}g_{xxx} + g_{yyy} + g_{xxt} = 0. \tag{4.2}$$

This equation is a generalization (the g_{yyy} term is new) of an equation obtained by Schiff [14] from a reduction of the self-dual Yang-Mills equations. Incidentally, for this equation we have checked that a nontrivial three soliton solution does exist and this is a further indication of its integrability. It is interesting to observe that this equation reduces to the derivative KP of Case i) if the first factor T_y is replaced by T_x in (4.1).

Finally the case iii) did not lead to any equations with the Painlevé property.

A general study of monomial leading parts was performed for all N's higher than 5. To make a long story short, only a monomial leading part of the form $T_x^{N-1}T_y$ has acceptable resonances for $N = 6, 7$. However, even in these cases, the resonance conditions are not satisfied. Thus, beyond $N = 5$ the leading part is not only monomial but also 1-dimensional.

N = 6 :

The leading operator is $\lambda T_x^6 + 20T_x^3T_x^*$ with λ =7 or -2. For $\lambda = 7$ the equation obtained is the fifth-order equation in the Lax hierarchy [8] plus the KdV:

$$(20T_x^3T_x^{*3} + 7T_x^6 + \alpha T_x^2T_x^{*2} + 27T_x^*T_y)F\cdot F\cdot F = 0 \tag{4.3}$$

or in nonlinear form after one integration ($u = 2\partial_x^2 \log F$):

$$u_{5x} + 10uu_{xxx} + 20u_xu_{xx} + 30u^2u_x + \alpha(u_{xxx} + 6uu_x) + u_y = 0 \tag{4.4}$$

Thus the Lax-5 equation does not possess a simple bilinear form like KdV but a trilinear one. For $\lambda = -2$ the resulting equation is the 3-dimensional Sawada-Kotera-Ramani equation,

$$(10T_x^3T_x^{*3} - T_x^6 + 45\alpha(T_x^2T_x^*T_y^* - \alpha T_yT_y^*) + 9T_x^*T_z)F\cdot F\cdot F = 0, \tag{4.5}$$

which also has a bilinear form (3.11).

In the case **N = 7** the only equation satisfying the Painlevé requirement is the x-derivative of the Sawada-Kotera-Ramani equation,

$$(3T_x^5T_x^{*2} + 5\alpha(T_x^4T_y^* - T_x^3T_x^*T_y) - 15\alpha^2T_xT_y^2 - 3T_x^2T_z)F\cdot F\cdot F = 0, \tag{4.6}$$

which means that the original equation is probably integrable even with an inhomogeneous $k(y,t)$ term.

For **N = 8** only the case 8a admits additive terms. Moreover it turns out that the first subleading term must also be 1-dimensional and it coincides with 6a. The full equation in this case reads:

$$\left(4T_x^7T_x^* + 5T_x^4T_x^{*4} + \alpha(20T_x^3T_x^{*3} + 7T_x^6) + 9\beta T_x^2T_x^*T_y^* + \frac{9\alpha\beta}{2}T_xT_y^*\right) F\cdot F\cdot F = 0, \tag{4.7}$$

and its nonlinear form with $u = 6\partial_x \log F$ is:

$$\begin{aligned} u_{7x} + 6u_{5x}u_x + 10u_{4x}u_{xx} + 5u_{xxx}^2 + 10u_{xxx}u_x^2 + \frac{5}{3}u_x^4 \\ +\alpha(3u_{5x} + 10u_{xxx}u_x + 5u_{xx}^2 + \frac{10}{3}u_x^3) \\ +\beta(u_{xxy} + u_xu_y) + \frac{\alpha\beta}{2}u_y = 0 \end{aligned} \tag{4.8}$$

To our knowledge this equation is a new integrable PDE.
Case N = **10**: Again one equation satisfies the integrability criterion. It reads:

$$\begin{aligned}\big(5T_x^8T_x^{*2} + 4T_x^5T_x^{*5} + \alpha(4T_x^7T_x^* + 5T_x^4T_x^{*4}) + \beta(20T_x^3T_x^{*2}T_y^* + 7T_x^5T_y)\\ +6\alpha\beta T_x^2T_x^*T_y^* + \frac{3\beta^2}{2}T_yT_y^*\big)F\cdot F\cdot F = 0 \qquad (4.9)\end{aligned}$$

This equation, too, has not been encountered before as an integrable equation.

5. Conclusion

In the preceding paragraphs we have presented an extension of Hirota's bilinear formalism that can encompass any degree of multilinearity. The main guide in our investigation has been the requirement that the equations be gauge-invariant. Since our objective is the study of integrability we have also presented a classification of one-component trilinear equations that pass the Painlevé test.

An interesting difference between the bilinear and the tri- (and multi-)linear cases is that now free parameters enter already at the leading part. In practice this means that the Painlevé analysis of the higher order unicomponent equations becomes increasingly difficult. Once the leading parts of these equations are fixed, we have studied the lower-order terms that can be added without destroying the Painlevé property. As a result new integrable equations were discovered. The fact that they have eluded discovery till now is understandable since these equations are of high order and there is no hope to discover them by chance: a solid guide and a systematic approach are needed.

Starting from a complete classification of unicomponent equations one can build up multicomponent ones following the approach we presented in [10,11] for the bilinear case. Further extensions can be presented and we can, of course, construct also higher multilinear equations (quadri-, penta-, etc.) equations.

Extension of our formalism to the discrete case is also possible. In this case the integrability requirement is just the property of singularity confinement [15]. Some preliminary results exist already in this direction.

Thus the multilinear extension to Hirota bilinear approach has a wide range of applicability. There are still many open problems, e.g., the computation of the multisoliton solutions of the trilinear equations. Hopefully also the τ-function formalism of the Kyoto school [16] can be extended in this direction.

Acknowledgements
The authors are grateful to R. Hirota, J. Satsuma, W. Oevel and R. Conte for illuminating discussions and exchange of correspondence, to W. Strampp for communicating his latest works on the subject and to J. Springael for comments on the manuscript.

References

[1] F. Calogero, in *Applications of analytical and geometric methods to nonlinear differential equations*, P.A. Clarkson ed., NATO ASI C413, Kluwer (1993), p.109.
[2] R. Hirota, Phys. Rev. Lett. **27** (1971) 1192.
[3] R. Hirota in *Soliton*, R. K. Bullough and P. J. Caudrey eds., Springer (1980), p. 157.
[4] M.J. Ablowitz, A. Ramani and H. Segur, Lett. Nuovo Cim. **23** (1978) 333.
[5] B. Grammaticos, A. Ramani and J. Hietarinta , Phys. Lett. **A190** (1994) 65.
[6] J. Matsukidaira, J. Satsuma and W. Strampp, Phys. Lett. **A147** (1990) 467. A extension of this trilinear approach, that preserves the 3×3 determinantal form, has been recently proposed by Strampp and coworkers: Y. Cheng, W. Strampp and B. Zheng: "Constraints of the KP hierarchy and multilinear forms", Comm. Math. Phys., to appear.
[7] K. Sawada and T. Kotera, Progr. Theor. Phys. **51** (1974) 1355; A. Ramani, in *Fourth International Conference on Collective phenomena*, J.L. Lebowitz ed., New York Academy of Sciences (1981) p. 54.
[8] H.C. Morris, J. Math. Phys. **18** (1977) 530.
[9] D.J. Kaup, Studies Appl. Math. **62** (1980) 189; B.A. Kupershmidt and G. Wilson, Invent. Math. **62** (1981) 403.
[10] J. Hietarinta, J. Math. Phys. **28** (1987) 1732, 2094, 2586; ibid **29** (1988) 628.
[11] B. Grammaticos, A. Ramani and J. Hietarinta , J. Math. Phys. **31** (1990) 2572.
[12] J. Chazy, Acta Mathematica, **34** (1911) 317.
[13] A. Hearn, REDUCE User's manual, Version 3.4 (Rand, Santa Monica, 1991).
[14] J. Schiff, in *Painlevé Transcendents, Their Asymptotics and Physical Applications*, D. Levi and P. Winternitz eds., Plenum (1992), p. 393.
[15] A. Ramani, B. Grammaticos and J. Satsuma, Phys. Lett. **A169** (1992) 323.
[16] M. Jimbo and T. Miwa, Publ. Res. Inst. Math. Sci **190** (1983) 943.

Nonlinear Difference Approximations for Evolutionary PDEs

DON A. JONES[1], LEN G. MARGOLIN[1], ANDREW C. POJE[1]

Abstract

We describe a procedure to improve both the accuracy and computational efficiency of finite difference schemes used to simulate nonlinear PDEs. The underlying idea is that of enslaving, which is the estimation of the small unresolved scales in terms of the larger resolved scales. We discuss details of the procedure and illustrate them in the context of PDEs in one spatial dimension. We present computational and analytic examples that demonstrate the predicted increases in accuracy and efficiency. We also show that the enslaved scheme can eliminate nonphysical oscillations in the numerical solution.

1. Introduction. In this paper we describe a procedure for improving both the accuracy and the computational efficiency of finite difference schemes that are used to solve dissipative PDEs. The increase in accuracy is the result of estimating the effects of the small scales that are unresolved on the mesh in terms of the larger resolved scales. The modified scheme is actually constructed to approximate the solution that the original scheme would generate on a finer mesh. Although the new scheme is more expensive to compute, we have found that this is more than compensated by the increase in accuracy.

Our procedure is similar in spirit to that used in constructing nonlinear Galerkin schemes [2], [3], [5], [6], [9]. That is, we derive a diagnostic relation between the unresolved small scales and the larger resolved scales. The main difference is that we are using finite difference methods rather than expansions in orthogonal polynomials. Our closure is similar to that employed in the Nonlinear Galerkin procedure, but is justified here by the assumption that the time derivative is not part of the principal balance of terms; thus our method is designed for flows that slowly evolve or that exhibit small scale fluctuations about a larger scale steady state.

The construction of enslaving relations in the context of finite difference schemes is not at all straightforward. In previous papers [8], [7] we addressed these difficulties, including the identification of the small scale modes and the derivation of the enslaving relations (see also [1] and the references therein). We applied our theory to the one-dimensional Burgers equation, and showed significant increases in accuracy and computational efficiency over the unimproved scheme in the flow regimes where our enslaving assumptions were valid.

We suppose we are given a discretized finite-difference equation that approximates the nodal values of the solution of a dissipative PDE to some order of accuracy. Our goal is to modify this discretization so as to produce solutions of higher accuracy. Of course, one can always improve the accuracy of a well-posed numerical scheme by increasing the resolution. Thus, an important aspect of our modification is that the new equations should

[1]IGPP, University of California, Los Alamos National Laboratory, Mail Stop C305, Los Alamos, New Mexico, 87544

be computationally more efficient – i.e., the modified scheme should require less CPU than the original scheme to produce a solution of a specified level of accuracy.

Our strategy is to derive a modified scheme on a coarse mesh that reproduces the accuracy of the original scheme on a twice-finer mesh. Note that modifications based on this strategy, even if completely successful, cannot improve the order of accuracy. That is, the error of the original scheme on the finer mesh will have the same dependence on the derivatives of the solution, but with smaller coefficients.

The increased efficiency is due to the relationship between the size of the allowable time step and the given grid spacing in a fully discretized finite difference scheme. Stability restrictions typically dictate that

$$\Delta t \sim \Delta x^p$$

where p can be as high as 2 for explicit time stepping algorithms in the presence of diffusion. In this case, refining the spatial resolution by a factor of 2 implies an eight-fold increase in the computational work for a one-dimensional problem. We show that the time step is governed by the resolution of the coarse mesh. Thus we improve the accuracy of simulation without reducing the computational time step.

Schematically, our implementation of the enslaving strategy follows the following steps.

1. Define a transformation of variables that maps the degrees of freedom of the fine mesh onto the coarse mesh. Half the variables on the coarse mesh represent the average solution in a cell – the large scales – and the other half represent smaller scales that will be unresolved on the coarse mesh.

2. Implement the original scheme on the twice-finer mesh. There are twice as many degrees of freedom on this fine mesh as there are on the coarse mesh.

3. Form the evolution equations for the new variables.

4. Apply a closure assumption to the new evolution equation for the small scales. This closure involves ignoring the time derivative, and yields a prognostic relation between the small scales and the large scales.

5. Solve the prognostic equation to express the small scales in terms of the larger scales. This is the enslaving relation.

6. Implement this relation in the new equation for the large scales. The result is our improved discretization.

2. Coordinate Transformations. As the first step, we apply the given difference scheme to a twice-finer mesh (Figure 1), whose cells are $\frac{\Delta x}{2}$ in length. On this fine mesh, there are approximately $2N$ cells (depending on the precise relation between the two meshes) and so $2N$ degrees of freedom. We label these $2N$ degrees of freedom alternately with either α or β defined by

$$\begin{aligned} \alpha_n &\approx u(n\Delta x) \\ \beta_n &\approx u\left((2n+1)\frac{\Delta x}{2}\right) . \end{aligned}$$

The idea is to define a change of variables that allows us to transform the calculation of $2N$ nodal values (α,β) of the solution on the fine grid to the calculation N nodal values a_n

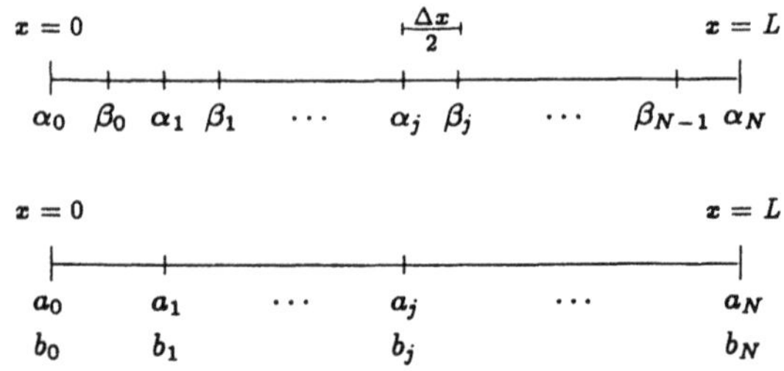

Figure 1: The fine mesh and the coarse mesh

and N approximations to the nodal value of the spatial derivative b_n on the coarse mesh (Figure 1). Namely, we want to define a_n and b_n such that

$$\begin{aligned} a_n &\approx u(n\Delta x) \\ b_n &\approx \Delta x \left.\frac{\partial u}{\partial x}\right|_{x=n\Delta x} . \end{aligned}$$

In order to implement the change of variables we need to express α and β as functions of a and b. Given the above interpretation of a and b, the first difference of b_n should approximate the second difference of the solution on the fine mesh, $\alpha_{n+1} + \alpha_n - 2\beta_n = b_{n+1} - b_n$. With this motivation, we choose

$$\begin{aligned} \alpha_n &= a_n \\ \beta_n &= \frac{(a_{n+1} + a_n)}{2} + \frac{(b_n - b_{n+1})}{2}, \end{aligned} \tag{2.1}$$

for $0 \le n \le N-1$. The relation (2.1) provides an explicit definition of a_n and an implicit definition b_n up to a constant (for instance, b_0 or b_N).

As shown in [7], for α and β_n approximating sufficiently smooth solutions, the above definition implies that

$$b_n = \frac{1}{4}\frac{\partial u(n\Delta x)}{\partial x}\Delta x + \mathcal{O}(\Delta x^3). \tag{2.2}$$

Note that $b \sim \mathcal{O}(\Delta x)$ compared with a. Thus we have verified that b is representative of the small scales (in both amplitude and spatial extent) while a represents the large scales of the solution.

Ultimately we will need to derive time evolution equations for a_n and b_n. This requires explicit relations for a_n and b_n in terms of α and β. For b_n, the choice of inverse relation is obvious based on the above discussion in particular (2.2).

$$b_n \approx \frac{\beta_n - \beta_{n-1}}{4} \tag{2.3}$$

The choice for an inverse relation for a_n is more subtle. One valid choice would be $a_n = \alpha_n$. With this choice the discrete Laplacian operator on the fine mesh is

$$\frac{4(\beta_{n+1} - 2\alpha_n + \beta_n)}{\Delta x^2} = \frac{2(a_{n+1} - 2a_n + a_{n-1})}{\Delta x^2} - \frac{b_{n+1} - b_{n-1}}{\Delta x^2} .$$

An alternate choice would be that which eliminates the appearance of b in the discrete Laplacian. Instead of choosing $a_n = \alpha_n$, we approximate a_n with

$$a_n \approx \frac{\alpha_n}{2} + \frac{\beta_n + \beta_{n-1}}{4} . \tag{2.4}$$

That is, we construct an evolution equation for a_n by taking linear combinations of the α_n, β_n equations suggested by (2.4). Implementing our change of variables, (2.1), we will see in the examples below that the discrete Laplacian for the a_n equation does not contain b. In the final analysis, b will only be constructed approximately as a function of a, we therefore have a choice (given by the choice of inverse for $a_n = F(\alpha, \beta)$) in which term to place this error; either in the dissipative term or in the time derivative. Since our approximation to b depends on small time derivatives, we prefer to choose an inverse relation that leaves the dissipative term in the a equation independent of b.

Since our original transformation has no exact local inverse we have introduced errors (A_n and B_n) given by

$$A_n = -\frac{1}{8}\{(a_{n+1} - 2a_n + a_{n-1}) - (b_{n+1} - b_{n-1})\} \tag{2.5}$$

and

$$B_n = b_n - \frac{1}{8}\{(a_{n+1} - a_{n-1}) - (b_{n+1} - 2b_n + b_{n-1})\} \tag{2.6}$$

obtained by composing (2.1) with (2.4) and (2.3) and taking the difference. In [7] it is shown that under appropriate assumptions,

$$A_n = \mathcal{O}(\Delta x^2), \qquad B_n = \mathcal{O}(\Delta x^3) .$$

Equations (2.5) and (2.6) are termed compatibility relations, and are useful in simplifying the derivation of the enslaving relations.

3. Examples. In this section we present three examples. In the first example we use the forced linear heat equation to show how the enslaving procedure naturally incorporates subgrid scale information about the forcing. In the second example we show that the enslaving procedure applied to an advection-diffusion equation eliminates spurious numerical oscillations. In the third example we analyze solutions of the forced Burgers equation and demonstrate the improvements in both accuracy and computational efficiency predicted by our theory. In this example we also discuss the effects of different time discretizations.

3.1 Heat Equation. We illustrate the above ideas with a linear equation containing the Laplacian operator.

$$\frac{\partial u}{\partial t} - \kappa \frac{\partial^2 u}{\partial x^2} = f(x) \tag{3.1}$$

$$u(0.t) = 0 = u(L, t) \qquad u(x, 0) = u_0(x),$$

where $\kappa > 0$ is given and $u_0(x)$ is a given smooth function satisfying the boundary conditions.

A standard second-order spatial discretization of (3.1) written on the α,β grid is given by

$$\begin{aligned}\frac{d\alpha_n}{dt} - \frac{4\kappa}{\Delta x^2}\left[\beta_n - 2\alpha_n + \beta_{n-1}\right] &= f_n \\ \frac{d\beta_n}{dt} - \frac{4\kappa}{\Delta x^2}\left[\alpha_{n+1} - 2\beta_n + \alpha_n\right] &= f_{n-1/2},\end{aligned} \tag{3.2}$$

where we do not explicitly show the time discretization. We will discuss the issue of time discretization in the last example.

Using the linear combinations given by (2.4) and (2.3) leads to

$$\begin{aligned}\frac{d}{dt}\left(\frac{\alpha_n}{2} + \frac{\beta_n + \beta_{n-1}}{4}\right) - \frac{\kappa}{\Delta x^2}\left(\alpha_{n+1} - 2\alpha_n + \alpha_{n-1}\right) &= \langle f\rangle_n \\ \frac{d}{dt}\left(\frac{\beta_n - \beta_{n-1}}{4}\right) - \frac{\kappa}{\Delta x^2}\left(\alpha_{n+1} - \alpha_{n-1} - 2\left(\beta_n - \beta_{n-1}\right)\right) &= \langle f'\rangle_n\,,\end{aligned} \tag{3.3}$$

where

$$\langle f\rangle_n = \frac{f_n}{2} + \frac{f_{n-1/2} + f_{n+1/2}}{4}, \qquad \langle f'\rangle_n = \frac{f_{n+1/2} - f_{n-1/2}}{4}\,.$$

We now use the change of variables (2.1) to rewrite the above in terms of a and b only.

$$\begin{aligned}\frac{da_n}{dt} - \frac{\kappa}{\Delta x^2}\left(a_{n+1} - 2a_n + a_{n-1}\right) &= \langle f\rangle_n + \frac{dA_n}{dt} \\ \frac{db_n}{dt} - \frac{\kappa}{\Delta x^2}\left(b_{n+1} - 2b_n + b_{n-1}\right) &= \langle f'\rangle_n + \frac{dB_n}{dt}\,.\end{aligned}$$

In order to arrive at an algebraic relation for b in terms of a we simply set

$$\frac{db_n}{dt} = 0.$$

Using the compatibility relations we neglect both dA_n/dt and dB_n/dt since, as shown in [7], both are at least $\mathcal{O}(\Delta x^2)$ in comparison to a_n and we have assumed that time derivatives do not enter the principal balance (*cf.* Equations (2.5), (2.6) above).

Rewriting the dissipative term in the b equation using (2.6), the enslaved scheme is

$$\begin{aligned}\frac{da_n}{dt} - \frac{\kappa}{\Delta x^2}\left(a_{n+1} - 2a_n + a_{n-1}\right) &= \langle f\rangle_n \\ \frac{8\kappa}{\Delta x^2}\left(\Phi_n(a) - \frac{1}{8}\left(a_{n+1} - a_{n-1}\right)\right) &= \langle f'\rangle_n\,.\end{aligned} \tag{3.4}$$

Note that due to the linear combination in Equation (3.3), the dissipative term in the a equation is the same as in the standard scheme written on the coarse mesh. That is, the discrete Laplacian in the a equation does not contain contributions from b.

If our assumptions concerning time derivatives are valid, the enslaved scheme, (3.4), which is a difference scheme on a Δx size mesh, should provide the same accuracy as the standard scheme, (3.3), which is on a mesh of size $\Delta x/2$. While general error estimates are given in [7] for the enslaved scheme, we here sketch a simpler proof for the steady-state case. In the standard scheme on the fine mesh, the truncation error at the nth node is $(\partial^4 u(n\Delta x)/\partial x^4/48)\Delta x^2 + \mathcal{O}(\Delta x^4)$. The truncation error in the averaged forcing term present in the enslaved scheme is

$$\langle f\rangle_n - f(n\Delta x) = -\frac{1}{16}\frac{\partial^2 f(n\Delta x)}{\partial^2 x}\Delta x^2 + \mathcal{O}(\Delta x^4).$$

Since the discrete Laplacian is second-order accurate, the truncation error for the dissipative term of the enslaved scheme on the coarse mesh is four times that of the original scheme or

$$\frac{\kappa}{\Delta x^2}(a_{n+1} - 2a_n + a_{n-1}) - \kappa\frac{\partial^2 u(n\Delta x)}{\partial x^2} = \frac{\kappa}{12}\frac{\partial^4 u(n\Delta x)}{\partial x^4}\Delta x^2 + \mathcal{O}(\Delta x^4).$$

However, at steady-state $\partial^2 f/\partial x^2 = -\kappa\partial^4 u/\partial x^4$. Thus, the error introduced by averaging the force over three grid points interacts with the error in the coarse-grained Laplacian to gives a total truncation error of

$$\left(\frac{1}{12} - \frac{1}{16}\right)\kappa\frac{\partial^4 u(n\Delta x)}{\partial x^4} = \frac{1}{48}\kappa\frac{\partial^4 u(n\Delta x)}{\partial x^4},$$

which is exactly the error of the standard scheme. Note that in this example the increased accuracy is due only to the use of subgrid scale information about the forcing.

3.2 Advection–Diffusion Equation. In addition to the increase in spatial accuracy for a given time step, we have found that the enslaved scheme acts to damp the spurious numerical oscillations inherent in most second-order accurate spatial discretizations. To illustrate this property, we consider the linear advection-diffusion equation given by

$$\frac{\partial u}{\partial t} + C\frac{\partial u}{\partial x} = \nu\frac{\partial^2 u}{\partial x^2}, \tag{3.5}$$

where C is a constant positive velocity and ν is a small but positive diffusion coefficient.

Again, a typical differencing of (3.5) on the α, β grid is given by

$$\begin{aligned}\frac{d\alpha_i}{dt} &= \frac{4\nu}{\Delta x^2}(\beta_i - 2\alpha_i + \beta_{i-1}) - \frac{C}{\Delta x}(\beta_i - \beta_{i-1})\\ \frac{d\beta_i}{dt} &= \frac{4\nu}{\Delta x^2}(\alpha_{i+1} - 2\beta_i + \alpha_i) - \frac{C}{\Delta x}(\alpha_{i+1} - \alpha_i)\ .\end{aligned} \tag{3.6}$$

Following the program defined above, the enslaved scheme written in terms of a and b is

$$\begin{aligned}\frac{da_i}{dt} &= \frac{\nu}{\Delta x^2}(a_{i+1} - 2a_i + a_{i-1}) - \frac{C}{2\Delta x}(a_{i+1} - a_{i-1}) + \frac{C}{4\Delta x}\left(\delta^2 b_i\right)\\ \delta^2 b_i &= \frac{\Delta x C}{4\nu}\delta^2 a_i\ ,\end{aligned} \tag{3.7}$$

where

$$\delta^2 h_i = (h_{i+1} - 2h_i + h_{i-1})\ .$$

Unlike the enslaved equation in the previous example, we have not used the compatiblilty relation (2.6) to simplify the discrete Laplacian of b_i.

Using the enslaving relation, the first of (3.7) can be written as

$$\frac{da_i}{dt} = \frac{\nu_e}{\Delta x^2}(a_{i+1} - 2a_i + a_{i-1}) - \frac{C}{2\Delta x}(a_{i+1} - a_{i-1}), \tag{3.8}$$

where the augmented diffusion coefficient is given by

$$\nu_e = \nu\left(1 + \frac{C^2\Delta x^2}{16\nu^2}\right).$$

One can show that solutions of (3.6) written on a mesh of size Δx are monotonic (nonoscillatory) if the condition $Re_{\Delta x} = \frac{C\Delta x}{\nu} \leq 2$ is satisfied. (*cf.* [4]) Given the above definition of the numerical diffusion, the grid Reynolds number for the enslaved scheme is given by

$$Re_e = Re_{\Delta x}\left(\frac{1}{1 + \left(\frac{Re_{\Delta x}}{4}\right)^2}\right).$$

As seen in Figure 2a, $Re_e \leq 2$ for all values of $Re_{\Delta x}$ which implies that the enslaved scheme preserves monotonicity for all spatial resolutions. An example of the regularization of the solution due to enslaving is given in Figure 2b.

3.3 Burgers Equation. To this point we have dealt only with linear PDEs. In order to incorporate nonlinear effects in the enslaving relation we consider Burgers equation under external forcing.

$$\frac{\partial u}{\partial t} - \lambda\frac{\partial^2 u}{\partial x^2} + u\frac{\partial u}{\partial x} = f(x,t), \tag{3.9}$$

$$u(0,t) = 0 = u(L,t) \qquad u(x,0) = u_0,$$

where $\lambda > 0$ is given and $u_0(x)$ is a given smooth function satisfying the boundary conditions.

Consider the second-order spatial discretization on a mesh with grid spacing Δx approximating (3.9) given by

$$\frac{du_n}{dt} - \frac{\lambda}{\Delta x^2}(u_{n+1} - 2u_n + u_{n-1}) + \frac{u_{n+1}^2 - u_{n-1}^2}{4\Delta x} = f_n. \tag{3.10}$$

The particular differencing of the nonlinear (advective) term is neïther unique nor optimal but is second-order accurate in space and is chosen for simplicity.

The corresponding enslaved scheme is

$$\frac{da_n}{dt} - \frac{\lambda}{\Delta x^2}\delta^2 a_n + \frac{3(a_{n+1}^2 - a_{n-1}^2) + 2a_n(a_{n+1} - a_{n-1})}{16\Delta x} + (N_a(a,\Phi(a)))_n = \langle f\rangle_n, \tag{3.11}$$

$$\frac{8\lambda}{\Delta x^2}\left(\Phi_n(a) - \frac{1}{8}(a_{n+1} - a_{n-1})\right) + \frac{1}{8\Delta x}\delta^2 a_n^2 = \langle f'\rangle_n \tag{3.12}$$

where

$$(N_a(a,\Phi(a)))_n = -\frac{1}{8\Delta x}\left(a_n\delta^2\Phi_n + (\Phi_{n+1}a_{n+1} - \Phi_n(a_{n+1} + a_{n-1}) + \Phi_{n-1}a_{n-1}\right).$$

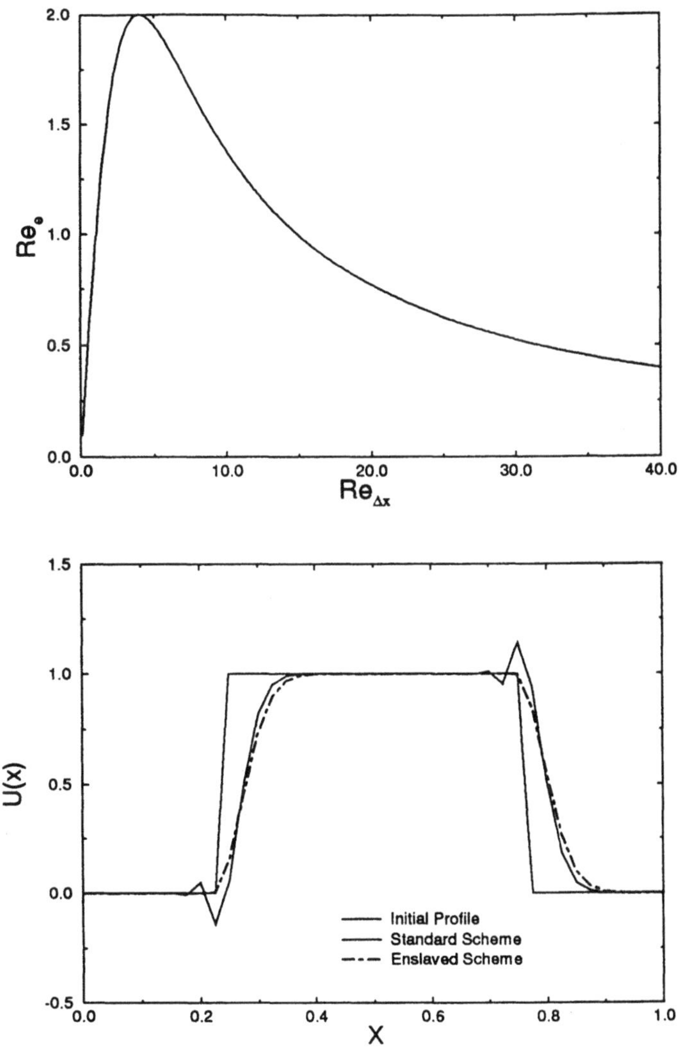

Figure 2: The relation between the grid Reynolds number in the standard and enslaved schemes (above). Forward Euler integration of the enslaved and standard schemes for the advection-diffusion equation with constant advective velocity. The grid Reynolds number in this is example is $Re_{\Delta x} = 3.15$ corresponding to a enslaved grid Reynolds number of 1.94

Again, (2.6) has been used to arrive at an explicit enslaving relation for Φ in terms of a. If we had not taken this step, we would arrive formally at a relation similar to that seen in the case of the linear advection diffusion equation, namely

$$\delta^2 b_n = \frac{\Delta x}{8\lambda}\delta^2 a_n \,.$$

This implies a new, damping term in the enslaved a equation with a coefficient that behaves nonlinearly with the solution amplitude and curvature. Although the proof from the linear case does not apply directly to the nonlinear Burgers equation, numerical experiments with the enslaved scheme exhibit a marked reduction in spurious oscillations about a shock front when compared with original difference scheme. Notice that had we chosen a different discretization for the advective term, higher-order differences of b would appear in the (3.12). Thus to solve for the function $\Phi(a)$ in (3.12) requires an iterative process. Such iteration may slightly increase the cost of the enslaving, but presents no new difficulties (see [7]).

The enslaving relation (3.12) is independent of the details of the time integration since it is derived by ignoring the time derivative. However the order of accuracy of the time integration can affect the formulation of our improved algorithm. We note that the maximum time step permitted by (3.10), when solved on the twice-finer mesh, is necessarily smaller than could be used on the coarse mesh. After the change of variables leading to (3.11), the system is still equivalent to (3.10) on the twice-finer mesh and so is also governed by the smaller time step of the fine mesh. It is the closure, setting the time derivative of b to zero, that allows us to increase the time step. Use of this larger time step is necessary if the new algorithm is to be more computationally efficient than the old.

It is important to remember that the algorithm (3.10) has both spatial errors and temporal errors. Our strategy has been to reduce the spatial errors of the coarse mesh to those of the fine mesh. However the larger time steps permitted by our algorithm will certainly lead to larger temporal errors than would be present in the fine-mesh solution integrated with the smaller time step. These errors may become noticeable as the magnitude of the time derivative increases, and indicate the need for a more accurate time integration scheme.

To illustrate this point, we present the following computational example. We consider the forced Burgers equation on the interval $[0,1]$ with Dirichlet boundary conditions. We choose an exact solution

$$u_e(x,t) = \sin(\pi x)\left(1 + \varepsilon\left(x\cos(\omega t) + (1-x)\sin(\omega t)\right)\right) \tag{3.13}$$

where ε and ω are positive constants that measure the importance of the time variation. As these parameters are increased, the time derivative of the solution grows and the validity of our enslaving relation decreases. By substituting (3.13) into the analytic Burgers equation, we can derive the force that generates this solution. For our example, we discretize this force, and use it to drive the enslaved system (3.11) and (3.12) as well as the original system (3.10). The accuracy of (3.11) (3.12) is compared to (3.10) using two different time-integration schemes. We measure the accuracy of each numerical solution by comparing it with the exact solution in the L^∞ norm – i.e., the maximum deviation of the computed nodal values from the true nodal values of the exact solution.

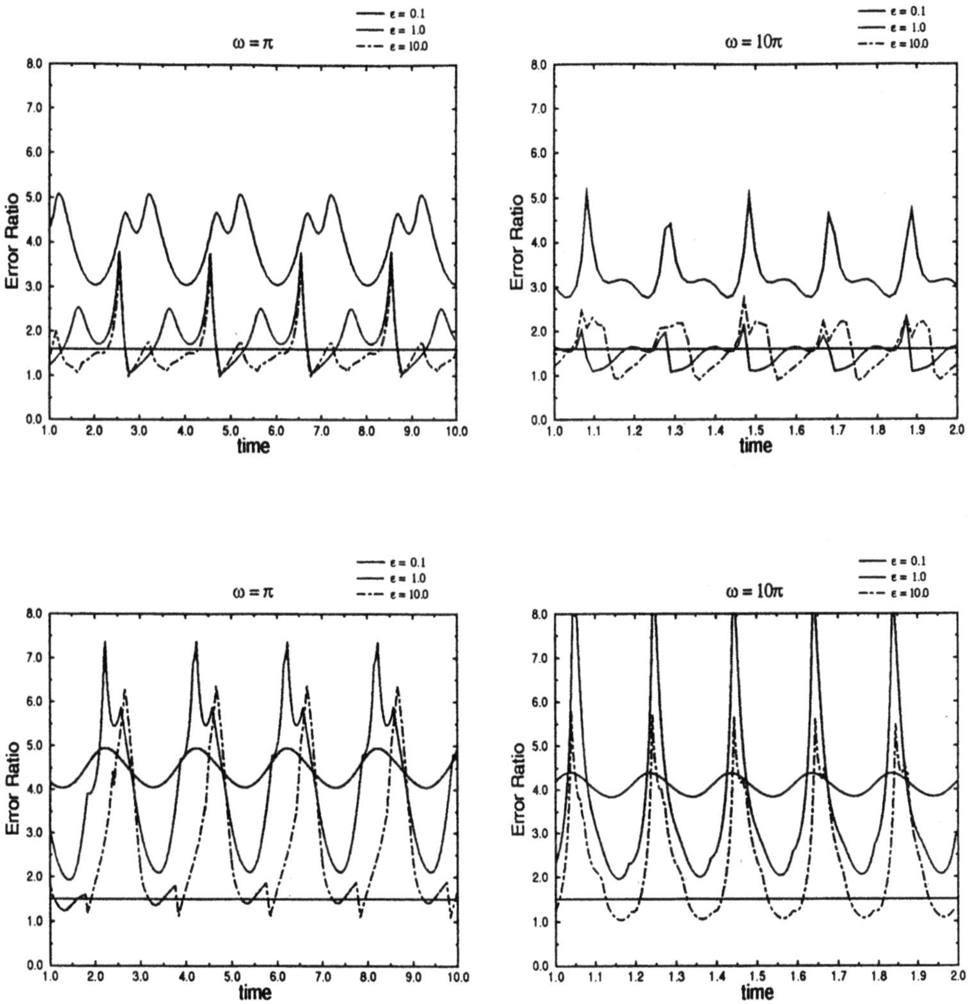

Figure 3: Error ratio of the original to the enslaved scheme for the case of forward Euler time stepping (above) and second order MacCormick integration (below) for the exact solution given by equation (4.15). The solid line on each plot at $Y \approx 1.6$ gives the 'breakeven' point below which the enslaved scheme becomes computationally less efficient than the original scheme.

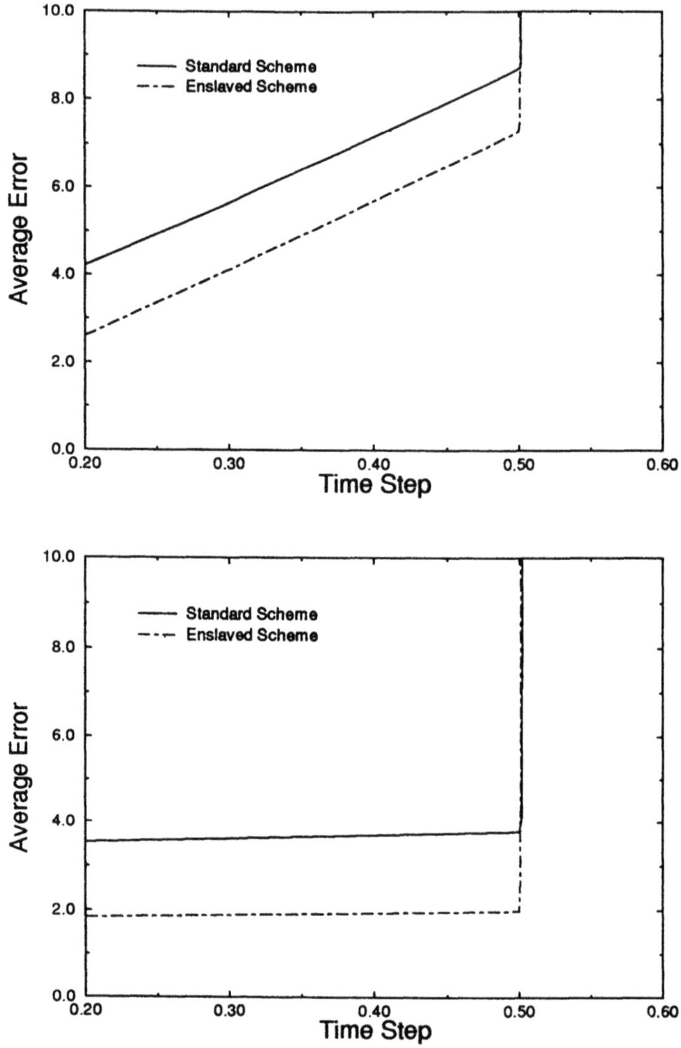

Figure 4: Accuracy versus time step for the enslaved and original schemes. Forward Euler scheme (above). MacCormick scheme (below).

In Figure 3a, we show the error ratio of the original scheme to the enslaved second-order scheme using forward Euler time integration. Each of the panels contains three calculations with a distinct value of ω and three values of ε. In Figure 3b, we repeat these six calculations, now using a second-order accurate MacCormick integration scheme. The solid line on each plot of Figures 3a and 3b, at the error ratio $R = 1.6$ represents the breakeven level, below which the enslaved scheme becomes computationally less efficient than the original scheme. The total amount of time spent above (below) this line indicates whether the enslaved scheme is more(less) efficient than the original scheme. In Figure 3a, there are significant regions where the error ratio drops below this line. These regions become much smaller when the higher-order MacCormick scheme is used. Note that in both figures, the error ratio is greater than unity almost everywhere. These points are discussed in more detail in [7]. All calculations in Figures 3a and 3b use the same mesh and the same time step.

In Figures 4a and 4b we illustrate the dependence of the error on the order of the integration scheme. Both figures are graphs of the average error (integrated over a problem cycle) as a function of the time step used. The numerical data in Figure 4a was generated with a forward Euler scheme and shows that the error of both the original and the enslaved schemes increases significantly with time step. The data in Figure 4b was generated with a MacCormick scheme, and shows no sensitivity to the time step. In both cases, the extreme values of $\omega = 10\pi$ and $\varepsilon = 10$ are used. Note also the original and the enslaved schemes allow the same maximum time step for both time integration schemes.

Acknowledgments. This work was supported in part by the U.S. Department of Energy's CHAMMP program. The authors also acknowledge the support of the Institute for Geophysics and Planetary Physics (IGPP) and the Center for Nonlinear Studies (CNLS) at Los Alamos National Laboratory.

References

[1] M. Chen, R. Temam, *Nonlinear Galerkin methods in the finite difference case and wavelet-like incremental unknowns*, Numer. Math., **64**, (1993), 271-294.

[2] C. Foias, O.P. Manley, R. Temam, *Modelization of the interaction of small and large eddies in two dimensional turbulent flows*, Math. Mod. and Num. Anal. M^2AN, **22** (1988), 93-114.

[3] C. Foias, G. Sell, E.S. Titi, *Exponential tracking and approximation of inertial manifolds for dissipative nonlinear equations*, J. Dynamics and Diff. Eq., **1** (1989), 199-243.

[4] A. Harten, *High resolution schemes for hyperbolic conservation laws*, J. Comp. Phys., **49**, (1983), 357-393.

[5] M.S. Jolly, I.G. Kevrekidis, E.S. Titi, *Approximate inertial manifolds for the Kuramoto-Sivashinsky equation: analysis and computations*, Physica, **D44** (1990), 38-60.

[6] D.A. Jones, L.G. Margolin, E.S. Titi, *On the effectiveness of the approximate inertial manifold- a computational study*, Theor. Comp. Fluid Mech. (to appear).

[7] D.A. Jones, L.G. Margolin, A.C. Poje, *Enslaved finite difference schemes for nonlinear PDEs*, submitted.

[8] L.G. Margolin, D.A. Jones, *An approximate inertial manifold for computing Burgers equation*, Physica **D60**, (1992), 175-184.

[9] R. Temam, *Attractors for the Navier-Stokes equations: localization and approximation*, J. Fac. of Sci., The Univ. of Tokyo, IA, **36** (1989), 629-647.

Exact Solutions for the Discrete Painlevé Equations

Kenji Kajiwara
Department of Electrical Engineering, Doshisha University, Tanabe, Kyoto 610-03, Japan
Yasuhiro Ohta
Department of Applied Mathematics, Faculty of Engineering, Hiroshima University, 1-4-1 Kagamiyama, Higashi-Hiroshima, Hiroshima 724, Japan
and
Junkichi Satsuma
Department of Mathematical Sciences, University of Tokyo, 3-8-1 Komaba, Meguro-ku, Tokyo 153, Japan

Abstract

Exact solutions of special function type for the discrete Painlevé II and III equations are investigated by using the bilinear formalism. It is shown that they are expressed by Casorati determinants of discrete Airy function and discrete Bessel functions, respectively. For the case of discrete Painlevé III equation, the discrete Bessel functions in this result are easily transformed to the q-Bessel functions. Based on this observation, we present a q-difference analogue of the Painlevé III equation.

1 Introduction

Recently, the discrete Painlevé equations have been attracting attention, together with so-called "singularity confinement"(SC) property which can be regarded as the discrete analogue of the Painlevé property [1],[2],[3],[4],[5],[6]. In continuous integrable systems, the Painlevé equations play an important role. Hence, it may be natural to expect that the discrete Painlevé equations should also be fundamental in discrete integrable systems.

It is now known that the discrete Painlevé equations admit several properties analogous to continuous ones, such as Lax pairs, Bäcklund transformations, coalescence cascade, and so on [7],[8]. As for the solutions, it is well known that the Painlevé equations admit exact solutions expressed by log derivative of determinants for special values of parameters [9], although they are transcendental in general. If the discrete Painlevé equations are "good" discretizations of original continous ones, then the structures of solutions should be preserved. Conversely, we may use this philosophy to check the "integrability" for discrete systems, since the validity of SC is not fully established although it is quite a powerful tool to detect "integrability".

In this contribution, we report some recent results on the exact solutions of the discrete Painlevé equations. We first investigate solutions of special function type for the discrete Painlevé II equation(dP_{II}), which reduces to those for P_{II} in the continuous limit. Moreover, we give the result on the solutions of discrete Painlevé III equation(dP_{III}), which can be expressed by Casorati determinant of "discrete Bessel functions". Remarkably, these discrete Bessel functions are easily transformed to the q-Bessel functions. Based on this fact, we also present a q-difference analogue of the Painlevé III equation.

2 Exact Solutions for dP_{II}: Discrete Airy Function

In this section we present exact solutions for dP_{II},

$$x_{n+1} + x_{n-1} = \frac{(\alpha n + \beta)x_n + \gamma}{1 - x_n^2} , \tag{1}$$

where α, β, γ are parameters. Before going to the discrete case, we give a brief summary of the solutions of special function type for P_{II} [9],

$$\frac{d^2u}{dx^2} = 2u^3 - 2xu + \alpha . \tag{2}$$

In the case of $\alpha = 2N + 1$, where N is an integer, eq.(2) admits a solution expressed by the τ function,

$$\tau_N = \begin{vmatrix} Ai & \frac{d}{dx}Ai & \cdots & \frac{d^{N-1}}{dx^{N-1}}Ai \\ \frac{d}{dx}Ai & \frac{d^2}{dx^2}Ai & \cdots & \frac{d^N}{dx^N}Ai \\ \vdots & \vdots & \ddots & \vdots \\ \frac{d^{N-1}}{dx^{N-1}}Ai & \frac{d^N}{dx^N}Ai & \cdots & \frac{d^{2N-2}}{dx^{2N-2}}Ai \end{vmatrix} , \tag{3}$$

where Ai is the Airy function satisfying

$$\frac{d^2}{dx^2}Ai = x\ Ai . \tag{4}$$

Solution of eq.(2) is given by

$$u = \frac{d}{dx}\left(\log \frac{\tau_{N+1}}{\tau_N}\right) . \tag{5}$$

Now let us turn to the discrete case. To investigate the solutions of special function type for the discrete Painlevé equations, it is convenient to use the discrete Riccati equation,

$$x_{n+1} = \frac{a_n x_n + b_n}{c_n x_n + d_n} , \tag{6}$$

where a_n, b_n, c_n and d_n are arbitrary functions in n. Here we remark that eq.(6) has a form of Möbius transformation. Now in this case, it is easily shown that if x_n satisfies

$$x_{n+1} = \frac{x_n - (an + b)}{1 + x_n} , \tag{7}$$

where a and b are constants, then x_n gives a solution for

$$x_{n+1} + x_{n-1} = \frac{(2an - a + 2b + 2)x_n - a}{1 - x_n^2} , \tag{8}$$

which is a special case of eq.(1). By putting $x_n = \frac{F_n}{G_n}$, we get

$$x_n = \frac{G_{n+1}}{G_n} - 1 , \tag{9}$$

$$G_{n+2} - 2G_{n+1} + G_n = -(an + b)G_n . \tag{10}$$

We note that the right-hand side of eq.(9) can be regarded as a discrete analogue of log derivative, and eq.(10) as a discrete analogue of the Airy equation, respectively. Hence, this result is a discrete counterpart of the simplest solution of $\mathrm{P_{II}}$.

Now we extend this result to $N \times N$ determinant solutions. We consider the τ function in $N \times N$ determinant form,

$$\tau_N(n) = \begin{vmatrix} A_n & A_{n+2} & \cdots & A_{n+2N-2} \\ A_{n+1} & A_{n+3} & \cdots & A_{n+2N-1} \\ \vdots & \vdots & \ddots & \vdots \\ A_{n+N-1} & A_{n+N+1} & \cdots & A_{n+3N-3} \end{vmatrix} , \tag{11}$$

where A_n satisfies

$$A_{n+2} - 2A_{n+1} = -(pn + q)A_n . \tag{12}$$

Then we can show that eq.(11) gives the solution of $\mathrm{dP_{II}}$ (1) with

$$\alpha = 2p, \quad \beta = (2N-1)p + 2q, \quad \gamma = -(2N+1)p , \tag{13}$$

through the dependent variable transformation,

$$x_n = \frac{\tau_{N+1}(n+1)\tau_N(n)}{\tau_{N+1}(n)\tau_N(n+1)} - 1 . \tag{14}$$

Continuous limit is given by choosing $p = -\epsilon^3$, $q = 1$, $n = x/\epsilon$ and taking the limit of $\epsilon \to 0$. This procedure recovers $\mathrm{P_{II}}$ and its solutions (2)-(5).

This is, in some sense, an unexpected result. We may expect from the continuous case (3) that the τ function can be given by so-called Hankel determinant which has a form,

$$\tau_N(n) = \begin{vmatrix} A_n & A_{n+1} & \cdots & A_{n+N-1} \\ A_{n+1} & A_{n+2} & \cdots & A_{n+N} \\ \vdots & \vdots & \ddots & \vdots \\ A_{n+N-1} & A_{n+N} & \cdots & A_{n+2N-2} \end{vmatrix} . \tag{15}$$

However, the above result shows that it is not the case. Actually, we see that the structure of τ function (11) is asymmetric, namely, suffix n increases by one in one direction and by two in another direction.

Here we show the brief outline of the derivation. Applying the bilinear formalism, we can decompose eq.(1) with eq.(13) into the following three bilinear difference equations,

$$\tau_N(n+3)\tau_N(n) - \tau_N(n+2)\tau_N(n+1) = \tau_{N+1}(n)\tau_{N-1}(n+3) , \tag{16}$$

$$\begin{aligned} & \tau_{N+1}(n+2)\tau_N(n+1) - 2\tau_{N+1}(n+1)\tau_N(n+2) & (17) \\ = \; & -(pn+q)\tau_{N+1}(n)\tau_N(n+3) , & (18) \end{aligned}$$

$$\begin{aligned} = \; & (pn+q)\tau_N(n+3)\tau_N(n) - (p(n+2N)+q)\tau_N(n+2)\tau_N(n+1) & (19) \\ = \; & \tau_{N+1}(n+1)\tau_{N-1}(n+2) . & (20) \end{aligned}$$

All of the above bilinear difference equations can be reduced to the Plücker relations, which are quadratic identities of determinants. This is the similar technique used to investigate the solutions of both continuous and discrete soliton equations. However, we have to deal with linear but second order variable coefficient ordinary difference equations, while we only have to work on functions satisfying quite simple linear partial differential or difference equations in the case of usual soliton equations. This situation makes the calculation somewhat technical. Readers may refer ref.[10] for further detail of derivation.

3 Exact Solutions for $\mathrm{dP_{III}}$: Discrete Bessel Functions

In this section, we discuss the solutions of special function type for $\mathrm{dP_{III}}$,

$$w_{n+1}w_{n-1} = \frac{\alpha w_n^2 + \beta\lambda^n w_n + \gamma\lambda^{2n}}{w_n^2 + \delta w_n + \alpha} , \tag{21}$$

where α, β, γ, δ, and λ are parameters. In the continuous case, $P_{\rm III}$

$$\frac{d^2u}{dx^2} = \frac{1}{u}\left(\frac{du}{dx}\right)^2 - \frac{1}{x}\frac{du}{dx} + \frac{1}{x}(\alpha u^2 + \beta) + \gamma u^3 + \frac{\delta}{u} , \tag{22}$$

where α, β, γ and δ are parameters, admits the solutions expressed by the τ function [9],

$$\tau_N^\nu = \begin{vmatrix} J_\nu & D\ J_\nu & \cdots & D^{N-1}J_\nu \\ D\ J_\nu & D^2\ J_\nu & \cdots & D^N\ J_\nu \\ \vdots & \vdots & \ddots & \vdots \\ D^{N-1}J_\nu & D^N\ J_\nu & \cdots & D^{2N-2}\ J_\nu \end{vmatrix} , \tag{23}$$

where $D = x\frac{d}{dx}$ and J_ν is the Bessel function of degree ν. The dependent variable u of eq.(22) is expressed as

$$u = \left(\log \frac{\tau_N^{\nu+1}}{\tau_{N+1}^\nu}\right)_x + \frac{\nu + N}{x} . \tag{24}$$

Equation (24) gives the particular solutions of eq.(22) with the parameters,

$$\alpha = 2(N - \nu) , \quad \beta = 2(\nu + N + 1) , \quad \gamma = 1 , \quad \delta = -1 . \tag{25}$$

We construct discrete analogue of the above solutions for eq.(21). Similar to the case of the previous section, we start from the discrete Riccati equation. In this case, we can show that if w_n satisfies

$$w_{n+1} = -\frac{aw_n + \lambda^n}{w_n + d} , \tag{26}$$

then it gives a particular solution of (21). Putting $w_n = \frac{F_n}{G_n}$, we see that

$$w_n = \tfrac{G_{n+1}}{G_n} - d , \tag{27}$$

$$G_{n+2} + (a - d)G_{n+1} + (\lambda^n - ad)G_n = 0 . \tag{28}$$

On the analogy of the continuous case, it is expected that eq.(28) is a discrete version of the Bessel equation, however, it is not trivial to identify eq.(28) with the Bessel equation. Instead of eqs.(27) and (28), we consider

$$w_n = \frac{J_\nu(n+1)}{J_\nu(n)} - q^\nu, \tag{29}$$

$$J_\nu(n+2) - (q^\nu + q^{-\nu})J_\nu(n+1) + \{1 + (1-q)^2 q^{2n}\}J_\nu(n) = 0 . \tag{30}$$

We can show via constructing the discrete Riccati equation that this w_n gives a solution of $dP_{\rm III}$ with the parameters

$$\alpha = -1, \quad \beta = (q^\nu - q^{-\nu-2})(1-q)^2, \quad \gamma = \frac{(1-q)^4}{q^2}, \quad \delta = q^\nu - q^{-\nu}, \quad \lambda = q^2 . \tag{31}$$

The way to regard eq.(30) as the discrete Bessel equation is as follows. We put $q = 1 + \epsilon$, $n = \frac{z}{\epsilon}$ and take the limit of $\epsilon \to 0$. Then eq.(30) yields

$$\frac{d^2 J_\nu}{dz^2} + (e^{2z} - \nu^2) J_\nu = 0 \ . \tag{32}$$

Equation (32) reduces to the Bessel equation in ordinary form if we put $x = e^z$. We note that the same results for this simplest solution is also derived in ref.[15]. We can extend this result to $N \times N$ determinant solutions. We consider the following τ function,

$$\tau_N^\nu(n) = \begin{vmatrix} J_\nu(n) & J_\nu(n+1) & \cdots & J_\nu(n+N-1) \\ J_\nu(n+2) & J_\nu(n+3) & \cdots & J_\nu(n+N+1) \\ \vdots & \vdots & \cdots & \vdots \\ J_\nu(n+2N-2) & J_\nu(n+2N-1) & \cdots & J_\nu(n+3N-3) \end{vmatrix} , \tag{33}$$

where $J_\nu(n)$ satisfies contiguity relations,

$$\begin{aligned} J_\nu(n) - q^{-\nu} J_\nu(n+1) &= -(1-q) q^n J_{\nu+1}(n) \ , & (34) \\ J_\nu(n) - q^{\nu} J_\nu(n+1) &= (1-q) q^n J_{\nu-1}(n) \ . & (35) \end{aligned}$$

We notice that the discrete Bessel equation (30) is derived from above contiguity relations. Then we can show that

$$w_n = \frac{\tau_{N+1}^\nu(n+1)\tau_N^{\nu+1}(n)}{\tau_{N+1}^\nu(n)\tau_N^{\nu+1}(n+1)} - q^{\nu+N} \ , \tag{36}$$

gives the solutions of the dP_{III},

$$w_{n+1} w_{n-1} = \frac{\alpha w_n^2 + \beta q^{2n} w_n + \gamma q^{4n}}{w_n^2 + \delta w_n + \alpha} , \tag{37}$$

with the parameters,

$$\begin{aligned} \alpha &= -q^{4N} \ , & \beta &= (q^{\nu+N} - q^{-\nu-N-2})\, q^{8N} (1-q)^2 \ , \\ \gamma &= q^{2(6N-1)} (1-q)^4 \ , & \delta &= (q^{\nu-N} - q^{-\nu+N})\, q^{2N} \ . \end{aligned} \tag{38}$$

We note that the τ function (33) has asymmetric structure similar to the case of dP_{II}.

This result is derived through the following five bilinear difference equations,

$$\begin{aligned} &\tau_{N+1}^\nu(n)\, \tau_N^{\nu+1}(n+1) - q^{-\nu-N} \tau_{N+1}^\nu(n+1)\, \tau_N^{\nu+1}(n) \\ &\quad = -(1-q) q^{n+2N} \tau_{N+1}^{\nu+1}(n)\, \tau_N^\nu(n+1) \ , \end{aligned} \tag{39}$$

$$\begin{aligned} &\tau_{N+1}^{\nu+1}(n)\, \tau_N^{\nu}(n+1) - q^{\nu-N+1} \tau_{N+1}^{\nu+1}(n+1)\, \tau_N^{\nu}(n) \\ &\quad = (1-q) q^{n+2N} \tau_{N+1}^{\nu}(n)\, \tau_N^{\nu+1}(n+1) \ , \end{aligned} \tag{40}$$

$$\tau_{N+1}^{\nu}(n)\,\tau_N^{\nu+1}(n+3) - q^{-\nu-N}\tau_{N+1}^{\nu}(n+1)\,\tau_N^{\nu+1}(n+2) \\ = -(1-q)q^n\tau_{N+1}^{\nu+1}(n)\,\tau_N^{\nu}(n+3)\,, \tag{41}$$

$$\tau_{N+1}^{\nu+1}(n)\,\tau_N^{\nu}(n+3) - q^{\nu-N+1}\tau_{N+1}^{\nu+1}(n+1)\,\tau_N^{\nu}(n+2) \\ = (1-q)q^n\tau_{N+1}^{\nu}(n)\,\tau_N^{\nu+1}(n+3)\,, \tag{42}$$

$$\tau_{N+1}^{\nu}(n+2)\,\tau_N^{\nu+1}(n+1) - q^{2N}(q^{\nu-N}+q^{-\nu+N})\tau_{N+1}^{\nu}(n+1)\,\tau_N^{\nu+1}(n+2) \\ +q^{4N}\{1+(1-q)^2q^{2n}\}\tau_{N+1}^{\nu}(n)\,\tau_N^{\nu+1}(n+3) = 0\,. \tag{43}$$

We note that the above bilinear difference equations are reduced to the Plücker relations. Further detail of the derivation will be reported elsewhere.

4 q-difference Analogue of $\mathrm{P_{III}}$: q-Bessel Functions

In the previous section, we have investigated the solutions of $\mathrm{dP_{III}}$ and shown that they are expressed by Casorati determinants of "discrete Bessel functions" defined by eq.(30). It is remarkable that eq.(30) is easily transformed to the q-Bessel equation,

$$J_\nu(q^2x) - (q^\nu + q^{-\nu})J_\nu(qx) + \{1+(1-q)^2x^2\}J_\nu(x) = 0\,, \tag{44}$$

by putting $q^n = x$. From this fact it may be natural to consider the following equation,

$$w(qx)w(q^{-1}x) = \frac{\alpha w(x)^2 + \beta x^2 w(x) + \gamma x^4}{w(x)^2 + \delta w(x) + \alpha}\,. \tag{45}$$

The solutions of eq.(45), with the parameters being the same as eq.(38), are given by

$$w(x) = \frac{\tau_{N+1}^{\nu}(qx)\tau_N^{\nu+1}(x)}{\tau_{N+1}^{\nu}(x)\tau_N^{\nu+1}(qx)} - q^{\nu+N}\,, \tag{46}$$

and the τ function $\tau_N^\nu(x)$ is expressed as

$$\tau_N^\nu(x) = \begin{vmatrix} J_\nu(x) & J_\nu(qx) & \cdots & J_\nu(q^{N-1}x) \\ J_\nu(q^2x) & J_\nu(q^3x) & \cdots & J_\nu(q^{N+1}x) \\ \vdots & \vdots & \cdots & \vdots \\ J_\nu(q^{2N-2}x) & J_\nu(q^{2N-1}x) & \cdots & J_\nu(q^{3N-3}x) \end{vmatrix}\,. \tag{47}$$

Here, $J_\nu(x)$ is the q-Bessel function of degree ν.

We next mention briefly on the Lax pair for eq.(45). The Lax pair for $\mathrm{dP_{III}}$ has been discussed in refs.[12] and [15]. In this case, the linear system for eq.(45) is given by the following q-difference system,

$$\Phi(x;qh) = L(x;h)\Phi(x;h)\,, \tag{48}$$

$$\Phi(qx;h) = M(x;h)\Phi(x;h), \tag{49}$$

whose compatibility condition yields

$$M(x;qh)L(x;h) = L(qx;h)M(x;h)\,. \tag{50}$$

Choosing the matrices L and M as

$$L(x;h) = \begin{pmatrix} \lambda_1 & \dfrac{c}{w(x)} + \lambda_1 & \dfrac{c}{w(x)} & 0 \\ 0 & \dfrac{q^2\lambda}{x^2} & q^2\dfrac{w(q^{-1}x)+\lambda}{x^2} & q^2\dfrac{w(q^{-1}x)}{x^2} \\ \dfrac{w(x)}{h^4x^2} & 0 & \lambda_3 & \dfrac{w(x)}{x^2} + \lambda_3 \\ \dfrac{1}{h^4}\left(\dfrac{c}{q^4w(q^{-1}x)} + \dfrac{\lambda}{x^2}\right) & \dfrac{c}{h^4q^4w(q^{-1}x)} & 0 & \dfrac{\lambda}{x^2} \end{pmatrix}, \tag{51}$$

$$M(x;h) = \begin{pmatrix} \dfrac{(\lambda_1x^2 - q^2\lambda)w(x)}{q^2\lambda w(x) + cx^2} & x^2\dfrac{c+\lambda_1 w(x)}{q^2\lambda w(x) + cx^2} & 0 & 0 \\ 0 & 0 & 1 & 0 \\ 0 & 0 & \dfrac{\lambda_3x^2 - \lambda}{w(x)+\lambda} & \dfrac{w(x)+\lambda_3x^2}{w(x)+\lambda} \\ \dfrac{1}{h^4} & 0 & 0 & 0 \end{pmatrix}, \tag{52}$$

respectively, where c, λ, λ_1 and λ_3 are arbitrary constants, then we have from eq.(50),

$$w(qx)w(q^{-1}x) = \frac{c\,(\lambda w(x) + \frac{c}{q^2}x^2)(w(x)+\lambda_3x^2)}{(\lambda_1 w(x) + c)(w(x)+\lambda)}\,, \tag{53}$$

which is essentially the same as eq.(45).

Finally, we comment on the limit to $\mathrm{P_{III}}$. We put $w = (1-q)xu$ and take the limit $q \to 1$. Then eq.(45) with the parameters (38) reduces to the $\mathrm{P_{III}}$ eq.(22) with the parameters (25). Moreover, the τ function (47) reduces to that of $\mathrm{P_{III}}$ (23) up to a trivial factor and the dependent variable transformation eq.(46) yields eq.(24).

5 Concluding Remarks

In this contribution, we have discussed the solutions of special function type for $\mathrm{dP_{II}}$ and $\mathrm{dP_{III}}$, and shown that they are expressed by Casorati determinants of "discrete Airy function" and "discrete Bessel functions", respectively. From these results, we conclude that dP_{II} and dP_{III} are "good" dicretizations of P_{II} and P_{III}, respectively. Moreover, the "discrete Bessel functions" in our results can be easily transformed to the q-Bessel function. Based on this observation, we have presented a q-difference analogue of P_{III}.

Finally, we give several remarks.

1. The structures of τ functions of dP_{II} and dP_{III} are asymmetric, while those for continuous cases are symmetric. We do not know yet whether such structures are common and essential property among the discrete Painlevé equations or not.

2. It is an interesting problem to study whether the other discrete Painleve equations proposed by Grammaticos et.al [7],[8],[12],[13],[14],[15] admit similar solutions or not.

3. Besides this class of solutions, there seems to exist another class of solutions of special function type [11]. We call such solutions "molecule type", and those in this article "lattice type". In the case of "lattice type", size of determinant appears in parameters of the equation, and usually continuous limits exist. On the contrary, in the case of "molecule type", size of determinant appears as the lattice site in the equation and hence its continuous limit "collapses". In this sense, the molecule type solutions are peculiar to discrete systems. Moreover, the molecule type solutions are closely related to the theory of orthogonal polynomials. It may be an interesting problem to construct molecule type solutions and investigate the relation with the orthogonal polynomials.

4. In the continuous case, there exists a class of rational solutions. It is an important problem to investigate the rational solutions for discrete Painlevé equations together with their determinant structures.

The authors are grateful to Prof. Grammaticos and Prof. Ramani for fruitful discussions and useful comments.

References

[1] E. Brézin and V.A. Kazakov, Phys. Lett. B**236**(1990), 144.

[2] V. Periwal and D. Shevitz, Phys. Rev. Lett., **64**(1990), 1326.

[3] A. R. Its, A. V. Kitaev and A. S. Fokas, Usp. Mat. Nauk, **45**, No. 6 (1990) 155.

[4] A. Ramani, B. Grammaticos, and J. Hietarinta, Phys. Rev. Lett., **67**(1991), 1829.

[5] F.W. Nijhoff and V.G. Papageorgiou, Phys. Lett. A**153**(1991), 337.

[6] B. Grammaticos A. Ramani, and V. G. Papageorgiou, Phys. Rev. Lett., **67**(1991), 1825.

[7] B. Grammaticos and A. Ramani, in *Applications of Analytic and Geometric Methods to Nonlinear Differential Equations* ed. by P. A. Clarkson (NATO ASI series Vol.C413, Kluwer, 1993), 299.

[8] A S. Fokas, B. Grammaticos and A. Ramani, J. Math. An. Appl. **180** (1993) 342.

[9] K. Okamoto, Ann. Mat. Pura Appl. **146**(1987), 337; Japan J. Math. **13**(1987) 47; Math. Ann. **275**(1986), 221; Funkcial. Ekvac. **30**(1987), 305.

[10] K. Kajiwara, Y. Ohta, J. Satsuma, B. Grammaticos and A. Ramani, J. Phys.A: Math. Gen. **27**(1994), 915.

[11] Y. Ohta, K. Kajiwara and J. Satsuma, "Bilinear Structure and Exact Solutions of the Discrete Painlevé I Equation", to appear in *Proceedings of the Workshop on Symmetries and Integrability of Difference Equations*, CRM Proceedings and Lecture Notes Series, AMS, 1994.

[12] V. G. Papageorgiou, F. W. Nijhoff, B. Grammaticos and A. Ramani, Phys. Lett. A **164** (1992), 57.

[13] A. Ramani and B. Grammaticos, J. Phys. A **25** (1992), L633.

[14] K. M. Tamizhmani, B. Grammaticos and A. Ramani, Lett. Math. Phys. **29** (1993), 49.

[15] B. Grammaticos, F. W. Nijhoff, V. Papageorgiou, A. Ramani and J. Satsuma, Phys. Lett. A **185** (1994), 446.

DYNAMICS OF COMPLEX SINGULARITIES FOR BURGERS' EQUATION

YOSHIFUMI KIMURA
Program in Applied Mathematics, University of Colorado at Boulder
Boulder, Colorado 80309-0526, U.S.A.
and
National Center for Atmospheric Research
P.O. Box 3000, Boulder, Colorado 80307-3000, U.S.A.

ABSTRACT

Complex space and time singularities are investigated for solutions of Burgers' equation. Such singularities are defined as zero points of the Cole-Hopf variable, which are extended in a 4-D space of $(z_r, z_i, \tau_r, \tau_i)$ where $z = z_r + iz_i$ and $\tau = \tau_r + i\tau_i$ are complex space and time variables, respectively. The distribution of singularity is calculated numerically for the solution under a smooth initial condition, $u(z,0) = -\sin z$. When $\tau_r \to \infty$, the locations of singularities are explicitly obtained by an asymptotic analysis. It is observed that the distribution of singularity approaches the one predicted from the asymptotic analysis.

1. Introduction

Since proposed as a model for turbulence, Burgers' equation[1],

$$\frac{\partial u}{\partial t} + u\frac{\partial u}{\partial x} = \nu\frac{\partial^2 u}{\partial x^2}, \tag{1}$$

where $u(x,t)$ is one dimensional velocity and $\nu \geq 0$ is kinematic viscosity, has stimulated researchers in physics, mathematics and engineering, to produce a number of papers.[2,3] Often the equation is used as a template for studying Navier-Stokes turbulence because both have the same type of nonlinearity and dissipation. The biggest difference between them, however, is that Burgers' equation can be linearized by the celebrated Cole-Hopf transformation[4,5],

$$u(x,t) = -2\nu\frac{d}{dx}\ln\phi = -2\nu\frac{\phi_{,x}}{\phi}, \tag{2}$$

to a diffusion equation

$$\frac{\partial \phi}{\partial t} = \nu\frac{\partial^2 \phi}{\partial x^2}. \tag{3}$$

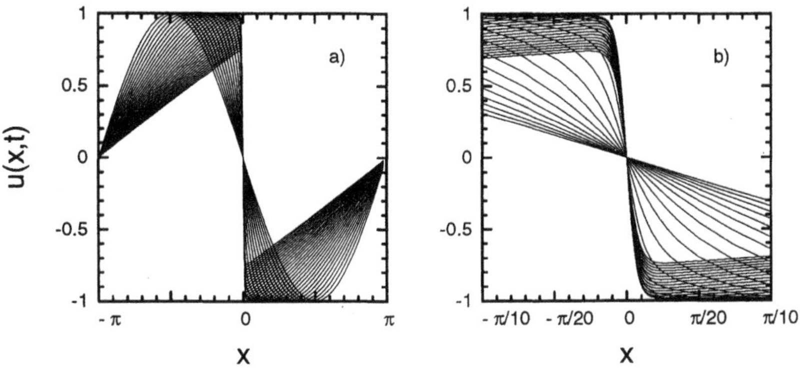

Fig. 1:
Numerical simulation of Burgers' equation with $\nu = 0.01$. Velocity profiles at $t = 0.0 \to 2.0$ with an interval of 0.1 are depicted.

a): The initial condition is $u(x,0) = -\sin x$. Two peaks collide at $x = 0$ to form a shock at around $t = 1.5$. Then the amplitude decreases due to viscosity.

b): The blow-up of a). The edge of the shock is not exactly vertical, and a diffusive boundary layer is formed between the two peaks. The thickness of the layer increases linearly in time.

When $\nu = 0$, nonlinear steepening leads to a singularity in a finite time even with a smooth initial condition. A singularity corresponds to divergence of derivatives of u with respect to x. If ν is nonzero, on the other hand, the problem is regularized and we cannot observe any singularity in real space or time.

Let us observe the above regularization in a result of numerical simulation of the equation with a simple initial condition

$$u(x,0) = -\sin x \ , \tag{4}$$

under a periodic boundary condition.† Fig. 1a shows $u(x)$ at various times t. For small t, by the effect of advection (the nonlinear term), the two peaks in the original sinusoidal shape approach to the line of $x = 0$. At around $t = 1.5$, they collide and make a triangular shock. The effect of diffusion becomes dominant after that, and the height of the shock decreases gradually [6]. Fig. 1b is a blow-up of Fig. 1a, and we can further observe the following things:

(1) There is a significant difference in the time scale of motion before and after the formation of the shock. The former corresponds to advection while the latter is due to diffusion.

† The pseudo-spectral method with 16384 points for space discretization is used. The calculation is de-aliased by the method of the 2/3-rule. $\nu = 0.01$ is used for viscosity.

(2) The edge of the shock is not exactly vertical, and we can define the separation of these peaks in the x-direction as a diffusive layer whose thickness is $\delta = O(\nu)$.[7]

(3) The thickness of the boundary layer grows, albeit slowly, linearly in time because of the effect of diffusion.

So far various regularization phenomena have been explained by using singularities in the complex plane of independent variables.[2,8,9] The concept of complex singularities works well for some (Hamiltonian) dynamical systems[10,11], then a real singularity may be understood as a limit of a pair of complex conjugate singularities when the imaginary parts approach zero. In spite of the success for ODEs, the extension of the concept to PDEs is still far-reaching, and one of the purposes of this paper is to present preliminary results on the analyses (asymptotic and numerical) of complex singularities for Burgers' equation and to look for the applicability of the concept for some PDEs. We try to understand the diffusion process after the (pseudo) shock is formed from the view point of motion of singularities.

The previous papers by Bessis & Fournier[12] and Kida[6] discussed the motion of complex *space* singularities for Burgers' equation while keeping the time real. In this paper both *space* and *time* variables are continued analytically into complex.† From the Cole-Hopf transformation, singularities for $u(z,\tau)$ are defined as zeros of $\phi(z,\tau)$ where z and τ are complex space and time variables, respectively. Then singularities naturally have extension in a sub-space in 4 dimensional real space $(z_r, z_i, \tau_r, \tau_i)$. One of the objectives of this paper is to visualize these extended singularities for Burgers' equation and to study the physical meaning of them.

2. Asymptotic solution

2.1 The region of $t >> 1$

The initial condition for heat equation corresponding to Eq. 4 is

$$\phi(x,0) = e^{\frac{1}{2\nu}\cos x} , \tag{5}$$

and the integral expression of the solution is given

$$\phi(x,t) = \frac{1}{\sqrt{4\pi\nu t}} \int_{\infty}^{\infty} \exp\left[-\frac{1}{2\nu}\left\{\frac{(x-x')^2}{2t} + \cos x'\right\}\right] dx' . \tag{6}$$

The asymptotic analysis of the solution when $t >> 1$ and $\nu << 1$ was given by Tatsumi & Kida.[13] They evaluated the dominant contribution from $x = -\pi$ and

† The possibility of this continuation is already pointed out by Frisch and Morf[2].

π at which $\cos x$ takes the minimum value -1. Then the integral is replaced by the summation of two terms,

$$\phi(x,t) \sim \frac{1}{\sqrt{4\pi\nu t}}\left[\exp\left\{-\frac{1}{2\nu}\left(\frac{(x-\pi)^2}{2t}-1\right)\right\}+\exp\left\{-\frac{1}{2\nu}\left(\frac{(x+\pi)^2}{2t}-1\right)\right\}\right]. \tag{7}$$

By substituting the above result into Eq. 2, we have the following asymptotic shock solution,

$$u(x,t) \quad \sim \quad \frac{x}{t}-\frac{\pi}{t}\tanh\left(\frac{\pi x}{2\nu t}\right) . \tag{8}$$

If we extend x and t into complex variables z and τ, respectively, the asymptotic solution has singularities at

$$\begin{aligned} \cosh\left(\frac{\pi z}{2\nu\tau}\right)=0 \quad &\leftrightarrow \quad z = i\nu(2n+1)\tau \\ &\leftrightarrow \quad \begin{cases} z_r = -\nu(2n+1)\tau_i \\ z_i = \nu(2n+1)\tau_r \end{cases} \qquad (n=0,\ \pm1,\ \pm2,\ ...). \end{aligned} \tag{9}$$

Thus singularities form a subspace in the 4-D space of $(z_r, z_i, \tau_r, \tau_i)$. It should be noted that if $\tau_i = 0$, i.e. time is restricted to real values, z should be isolated pure imaginary numbers. Previous papers[12,6] dealing with this case demonstrated that the motion of singular points is restricted on the imaginary axis. Singularities have real space extension when nonzero τ_i is introduced. Because of the linear relation between z_r and τ_i and between z_i and τ_r, we can expect a flat distribution of singularities (with a fixed τ_r) in this asymptotic time region.

2.2 The region of $t << 1$

In this asymptotic time region, Eq. 6 is not a suitable form because t is in the denominator. Instead we try to construct a power series solution in t for the heat equation (3) by assuming

$$\phi(x,t) = \sum_{n=0}^{\infty} a_n(x)t^n . \tag{10}$$

By substituting into Eq. 3, we have a recurrence relation for $a_n(x)$,

$$a_{n+1}(x) = \frac{\nu}{n+1}a_n''(x) , \quad (n=0,\ 1,\ 2,\ ...) . \tag{11}$$

By using the above recurrence relation with

$$a_0(x) = \phi(x,0) = e^{\frac{1}{2\nu}\cos x} \tag{12}$$

as the initial condition, a_ns can be calculated successively. The convergence property of Eq. 10 is an open problem.

3. Numerical method

Newton's method is used for detecting the zeros of ϕ. To determine the positions of zeros in the 4D space of $(z_r, z_i, \tau_r, \tau_i)$, τ_r and τ_i are specified and z_r and z_i are estimated using the following formula.

$$\begin{pmatrix} z_r \\ z_i \end{pmatrix} = J^{-1} \begin{pmatrix} F_r(z_{r0}, z_{i0}) \\ F_i(z_{r0}, z_{i0}) \end{pmatrix} + \begin{pmatrix} z_{r0} \\ z_{i0} \end{pmatrix} , \tag{13}$$

where F_r and F_i are the real and imaginary parts of the following integral,

$$\begin{aligned} \phi(z,\tau) &\sim \int_{-\infty}^{\infty} \exp\left[-\frac{1}{2\nu}\left\{\frac{(z-x')^2}{2\tau} + \cos x'\right\}\right] dx' , \\ &= F_r(z_r, z_i, \tau_r, \tau_i) \quad + \quad i\, F_i(z_r, z_i, \tau_r, \tau_i) , \end{aligned}$$

namely

$$\begin{aligned} F_r(z_r, z_i, t_r, t_i) &= \int_{-\infty}^{\infty} \exp\left[-\frac{1}{2\nu}(\alpha + \cos x')\right] \cos\frac{\beta}{2\nu}\, dx' , \\ F_i(z_r, z_i, t_r, t_i) &= -\int_{-\infty}^{\infty} \exp\left[-\frac{1}{2\nu}(\alpha + \cos x')\right] \sin\frac{\beta}{2\nu}\, dx' , \end{aligned} \tag{14}$$

where

$$\begin{aligned} \alpha &= \frac{\tau_r\left\{(z_r - x')^2 - z_i^2\right\} + 2\tau_i z_i (z_r - x')}{2(\tau_r^2 + \tau_i^2)} , \\ \beta &= \frac{2\tau_r z_i (z_r - x') - \tau_i\left\{(z_r - x')^2 - z_i^2\right\}}{2(\tau_r^2 + \tau_i^2)} . \end{aligned} \tag{15}$$

J is the Jacobian matrix,

$$J = \frac{\partial(F_r, F_i)}{\partial(z_r, z_i)} , \tag{16}$$

in which the Cauchy-Riemann relation

$$\begin{aligned} \frac{\partial F_r}{\partial z_r} &= \frac{\partial F_i}{\partial z_i} , \\ \frac{\partial F_r}{\partial z_i} &= -\frac{\partial F_i}{\partial z_r} , \end{aligned} \tag{17}$$

is satisfied. Eq. 13 which includes evaluation of indefinite integrals is iterated until $(z_r - z_{r0})^2 + (z_i - z_{i0})^2$ converges to values less than 10^{-6}. The Romberg method is used for an integration scheme. As the starting point, the value calculated by Kida for the case of real time $(z_r, z_i, \tau_r, \tau_i) = (0.0, 0.0415, 3.0, 0.0)$ is used.

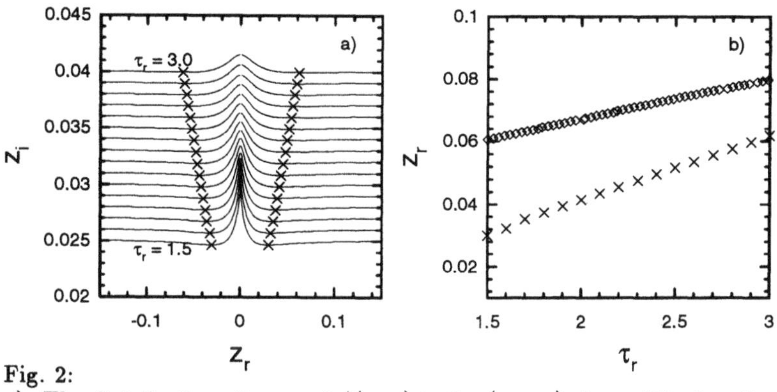

Fig. 2:
a): The distribution of zeros of $\phi(z, \tau)$ in the (z_r, z_i) plane. The locations of zeros are calculated by Newton's method, Eq. 13 with given τ_r $(1.5 \to 3.0)$ and τ_i $(0.0 \to 15.0)$. The interval of τ_i was small enough that the distribution becomes smooth lines. When $\tau_i = 0$, the singular points move on the axis of $z_r = 0$, and $\tau_r = 1.5$ is around the turning time. When $\tau_r \to \infty$ the layer approach the flat distribution predicted from Eq. 9. The cross marks $(\times)$ are points of the minimum imaginary part.
b): The motion of the points of the maximum curvature $(\diamond)$ and the points of the minimum imaginary part from a) $(\times)$ as functions of τ_r. They have the same tendency of linear growth, but do not match exactly. This means that the positions of the maximum curvature points ought to be determined not only by the contribution of the first layer of singularities but also upper ones.

4. Result and discussions

Figure 2a shows the distribution of the zeros of ϕ in the complex z plane. Different lines correspond to different τ_r. Because a small interval of τ_i is used, the distribution looks like continuous lines. Only the first layer of zeros nearest to the real axis is depicted at each τ_r. The cross marks are at the points where z_i takes the minimum value in each layer. Presumably these points play an important role in determining the overall analyticity of the solution. The following things are observed.

(1) The singularity distribution changes its structure with time τ_r. When τ_r becomes larger the distribution of singularities becomes flatter, and it seems that as τ_r goes to infinity the structure approaches to the one predicted from Eq. 9.† From the last equation in Eq. 9, we can calculate the asymptotic

† Eq. 8 has an arbitrariness with respect to a constant factor, which can be renormalized by introducing a new origin of time t_0. Thus Kida[6] used an expression $\frac{x}{t-t_0} - \frac{\pi}{t-t_0} \tanh\left(\frac{\pi x}{2\nu(t-t_0)}\right)$ for the asymptotic shock solution, and reported that t_0 has a value -1.

interval Δz_i of the first layers ($n = 0$) is $\nu\Delta\tau_r$. For the present calculation with $\nu = 0.01$ and $\Delta\tau_r = 0.1$, Δz_i should be 0.001 asymptotically, which is already attained by $\tau_r = 3.0$.

(2) When $\tau_i = 0$, singularities are pointwise and move on the imaginary axis as is reported in the previous papers.[12,6]†† The real time $\tau_r \sim 1.5$ is the turning time of the motion (the singular point moves downward before and upward after the time) at which the energy spectrum becomes shallowest and small scale eddies are most developed.

(3) The points closest to the z_r axis on the singular layers move outward almost linearly in time. Fig. 2b shows the positions of the nearest singular points and points of the maximum curvature in the velocity profile from the simulation. These two lines do not match exactly. This is perhaps because the most singular points in physical space should be determined not only by the first singular layer but also by the second or upper layers.

In this paper, only one value is used for viscosity. It may be interesting to study the motion of singularities according to the change of values of viscosity in particular as it goes to zero. Since the thickness of the diffusive boundary layer is $\delta = O(\nu)$, we can predict that the points of the minimum z_i (the cross points in Fig. 2a) move in a narrower band for smaller ν. If the band width also scales as $O(\nu)$, then such a scaling would be good evidence that the structure of complex space and time singularity plays a certain active role in determining the behavior of the solution in real space and time.

It may also be of interest to investigate the relation between the structure of complex space-time singularity and the so-called singular manifold which has been proposed as an extension of the Painlevé analysis to study integrability of given PDEs.[14] Burgers' equation belongs to a class of equations whose solutions have only meromorphic singularities. It is an open problem how the singular structure (i.e. Fig. 2a) reflects on the integrability of Burgers' equation.

As mentioned in the introduction, there is definite similarity between Burgers' and the Navier-Stokes equations. We hope that thorough investigations of the real and/or complex singularities for the former aid in understanding the counterparts of the latter.[15,16]

5. Acknowledgment

The author is grateful of the fruitful discussions and encouragements by Professors M. J. Ablowitz, J. D. Cole, S. Kida, R. B. Pelz, and H. Segur. Professor V. G. Makhankov and the Center for Nonlinear Studies at the Los Alamos National Laboratory are heartfully acknowledged for the partial support of my stay during the NEEDS '94 conference.

†† This may be only true for the time range after the generation of the shock.

All calculations are done on the IBM RS6000 machines at the National Center for Atmospheric Research (NCAR).

NCAR is sponsored by the National Science Foundation.

6. References

1. J. M. Burgers, *Adv. Appl. Mech.* **1** (1948) 171.
2. U. Frisch and R. H. Morf, *Phys. Rev. A* **23** (1981) 2673.
3. W. A. Woyczyński, in *Nonlinear Waves and Weak Turbulence*, ed. N. Fitzmaurice, D. Gurarie, F. McCaughan, and W. A. Woyczyński, (Birkhäuser, Boston, 1993).
4. J. D. Cole, *Quart. Appl. Math.* **9** (1951) 225.
5. E. Hopf, *Commun. Pure Appl. Mech.* **3** (1950) 201.
6. S. Kida, *J. Phys. Soc. Japan* **55** (1986) 1542.
7. J. D. Cole, *SIAM Review* **36** (1994) 425.
8. M. E. Brachet, D. I. Meiron, S. A. Orszag, B. G. Nickel, R. H. Morf, and U. Frisch, *J. Fluid Mech.* **130** (1983) 411.
9. C. Sulem, P. -L. Sulem, and H. Frisch, *J. Comp. Phys.* **50** (1983) 138.
10. Y. Kimura, *Physica* **D46** (1990) 439.
11. Y. Kimura, *Physica* **D51** (1991) 512.
12. D. Bessis, and J. -D. Fournier, *J. Physique Lett.* **45** (1984) L-833.
13. T. Tatsumi and S. Kida, *J. Fluid Mech.* **55** (1972) 659.
14. J. Weiss, M. Tabor, and G. Carnevale, *J. Math. Phys.* **24** (1983) 522.
15. R. M. Kerr, *Phys. Fluids* **A5** (1993) 1725.
16. Y. Kimura, and R. B. Pelz, in *Unstable and Turbulent Motion in Fluid*, ed. S. Kida, (World Scientific, Singapore, 1993).

NONCOMMUTATIVE INTEGRABLE SYSTEMS

B. A. Kupershmidt
The University of Tennessee Space Institute
Tullahoma, TN 37388
e-address: bkupersh@utsi.edu

1. In the hierarchy of things, the classical theory of integrable systems is the degenerate case $q = 1$ of the Quantum-groups-inspired theory of q-intebrable systems. The formulae in the latter are easy enough to guess, at least in the discrete case, by e.g. converting the linear, quadratic, and cubic Poisson brackets in [1] into the appropriate commutators. The proofs, however, are very hard and elusive. Technically, one of the difficulties (there are many others) can be traced to the following fact (familiar in the theory of Quantum matrix groups): If X and Y are matrices with noncommuting entries then, in general, $Tr(XYX^{-1}) \neq Tr(Y)$ even if the matrix entries of X commute with those of Y. Philosophically, the difficulties can be traced to the absence of a more general theory from which one could get the q-machine as a special case by imposing appropriate q-relations. In other words, one needs a theory of completely noncommutative integrable systems, together with necessary technical tools such as noncommutative versions of differential algebra, Lagrangian and Hamiltonian formalisms, etc. All this is done in [2]; some of the highlights are given below. The zero-dimensional case of Classical Mechanics is discussed in [3] and, in the super-version, in [4].

2. In the zero-dispersion limit, the classical KdV equation

$$u_t = (3u^2 + u_{xx})_x \tag{1}$$

becomes

$$u_t = const(u^2)_x,$$

the inviscid Burgers equation. The latter has the following noncommutative analogs (as abelian hierarchies):

$$u_t = const u^n u_x;\ u_t = const u_x u^n;\ u_t = const(u^{n+1})_x. \tag{3}$$

<u>There are no others</u>: The flow $u_t = O_1 u u_x + O_2 u_x u$ has a higher commuting flow iff $O_1 O_2 (O_1 - O_2) = 0$, where O_1 and O_2 are constants. Interestingly, only the last of the 3 alternatives in (3) survives to live in the KdV form: (1) is the only noncommutative KdV. (In its matrix version it was discovered in the 1970's by Calogero and Degasperis). The corresponding KdV Lax operators are $L = \partial^2 + u$ and $\mathcal{L} = \partial + u\partial^{-1}$[5]. The latter is the special case of the mKP Lax operator

$$\mathcal{L} = \partial + \sum_0^\infty a_i \partial^{-i}. \tag{4}$$

When $\{a_i = 0, i > 1\}$, one gets the Dispersive Water Waves (DWW) hierarchy, with the 1st nontrival flow

$$u_t = -u_{xx}/2 + h_x + u_x u + [u, h],$$

$$h_t = h_{xx}/2 + (uh)_x.\ [u = a_0, h = a_1] \tag{5}$$

For $h = 0$, a noncommutative Burgers hierarchy results:

$$u_t = (ad_u + \partial)(\hat{R}_u - \partial)^n(1),\ n \in \mathbf{N}, \tag{6}$$

with the noncommutative analog of the Hopf-Cole map

$$u = -U_x U^{-1}, \tag{7}$$

$$\partial_t(U) = (-1)^{n+1}\partial^n(U); \tag{8}$$

here $\hat{R}_u(v) = vu,\ \hat{L}_u(v) = uv,\ ad_u(v) = [u, v]$. (There exists also the mirror version of (6) - (8).)

3. The 1st mKdV equation is

$$v_t = v_{xxx} - 3(v_x v^2 + v^2 v_x),\ u = -v_x - v^2, \tag{9}$$

and it has also been known in its matrix form. Its potential forms are

$$V_t = V_{xxx} - 3V_{xx}V^{-1}V_x, \; v = -V_xV^{-1}, \tag{10}$$

$$W_t = W_{xxx} - 3W_xW^{-1}W_{xx}, \; v = -W^{-1}W_x. \tag{11}$$

The 2nd mKdV equation is

$$v_t = v_{xxx} - 6vv_xv + 3[v, v_{xx}], \;\; \mathcal{L} = \partial^2 + 2\partial v. \tag{12}$$

Here the story is a bit more complicated.

3. So, things <u>are</u> different in the noncommutative realm. How much? A lot survives: B^1; B^3 (except, possibly, in the Toda lattice case); B^2 becomes nonlocal. (B^i is the Hamiltonian structure $\#i$.) E.g., the KdV hierarcny (1) can be written in 2 Hamiltonian forms

$$u_t = B^1(\delta H_{n+1}/\delta u) = B^2(\delta H_n/\delta u), \tag{13}$$

$$B^1 = \partial, \; B^2 = \partial^3 + (\hat{L}_u + \hat{R}_u)\partial + \partial(\hat{L}_u + \hat{R}_u) + ad_u\partial^{-1}ad_u. \tag{14}$$

The catch is hidden in the definition of the noncommutative variational derivative $\delta/\delta u$. The latter is a new animal, and moreover, there are 2 of them. In the notation of Chapter 2 in [6], the noncommutative definitions are:

$$X(H) \approx \mathbf{X}^t \frac{\delta H}{\delta \mathbf{u}} \;\; (mod\ Im\partial\ \underline{\text{and}}\ \underline{\text{commutators}}), \tag{15}$$

$$X(H) \sim \frac{\delta H}{\delta \mathbf{u}}(\mathbf{X})(mod\ \text{“}divergencies\text{”}), \tag{16}$$

where X is an arbitrary evolution derivation. In (15), $\delta H/\delta \mathbf{u}$ is a vector-<u>function</u>, while in (16) it's a vector-<u>operator</u>. Only the 1st definition (15) appears in the theory; however, the 2nd one (16) provides the noncommutative generalization of the well- known commutative result: if H is x-independent then

$$\sum_j \frac{\delta H}{\delta u_j}(u_{j,x}) \sim 0. \tag{17}$$

4. Whenever a commutative result, construction, idea, etc. has a noncommutative analog, we can call the former stable. Stable things: Calculus of Variations; Hamiltonian formalism; Correspondence between linear Poisson brackets and Lie algebras, and between affine Poisson brackets and generalized 2-cocycles on Lie algebras; Integrable systems of Lax type in continuous and discrete spaces; (Lie-algebraic) Clelsch representations; Boussinesq, NLS, DNLS, FD (=ILW), KP and other fluid equations; Toda lattice, also relativistic; KP and mKP hierarchies (also relativistic), Miura maps,– continuous and discrete; Legendre transforms, Classical Mechanics; Classical r-matrices; super-versions; Hydrodynamical reductions; Gibbons forms; Discrete-time systems, or (Hamiltonian) Numerical schemes,– these are the most mysterious. And more. Some things are not stable, most notably zero-curvature equations, τ-functions (but $(\ell n\tau)_x$'s are stable), Lie albegras of vector fields, and (probably) compressible fluid dynamics.

A few stable examples: The Gardner deformation of KdV (1):

$$w_t = (3w^2 + w_{xx})_x - 3\epsilon^2(w^2 w_x + w_x w^2),$$

$$u = w + \epsilon w_x - \epsilon^2 w^2; \tag{18}$$

Hamiltonian matrix for the DWW hierarchy (5)

$$B^{DWW} = \begin{pmatrix} 0 & \partial + ad_u \\ \partial + ad_u & ad_h \end{pmatrix}; \tag{19}$$

The NLS equation, with $L = \partial + \mathbf{p}^t\partial^{-1}\mathbf{q}$:

$$\mathbf{p}_t = \mathbf{p}_{xx} + 2(\mathbf{p}^t\mathbf{q})\mathbf{p}, \ \mathbf{q}_t = -\mathbf{q}_{xx} - 2\mathbf{q}(\mathbf{p}^t\mathbf{q}), \tag{20}$$

with the canonical Hamiltonian form;

The DNLS equation, with $\mathcal{L} = \partial(1 + \mathbf{a}^t\partial^{-1}\mathbf{b})^{-1}$:

$$\mathbf{a}_t = [\mathbf{a}_x - 2(\mathbf{a}^t\mathbf{b})\mathbf{a}]_x, \ \mathbf{b}_t = -[\mathbf{b}_x + 2\mathbf{b}(\mathbf{a}^t\mathbf{b})]_x, \tag{21}$$

with the Hamiltonian matrix

$$B = \partial\begin{pmatrix} 0 & 1 \\ 1 & 0 \end{pmatrix};$$

KP and mKP Hamiltonian forms:

$$B^{KP} = -B^0, \ B^{mKP} = \begin{pmatrix} B^{DWW} & 0 \\ 0 & -B^1 \end{pmatrix}, \tag{22}$$

where B^r is the Hamiltonian matrix associated with the Lie algebra of differential operators $\{\sum_{i\geq 0} \partial^{i+r} X_i\}$.

Some things have no commutative analogs. E.g., Kontsevich-type formula

$$\sum_i [u_j, \frac{\delta H}{\delta u_j}] \sim 0, \quad \forall H. \tag{23}$$

On the other hand, one can always get back to the more familiar ground by treating all abstract noncommutative variables as matrices. The crucial formula for this reduction is

$$\left(\frac{\delta H}{\delta u}\right)^\beta_\alpha = \frac{\delta[Tr(H)]}{\delta u^\beta_\alpha}. \tag{24}$$

5. Consider the lattice KP hierarchy, with

$$L = \zeta + \sum_{i\geq 0} \zeta^{-i} q_i, \tag{25}$$

and with the 1st flow (notations are from [1])

$$q_{i,t} = (1 - \Delta^{-1})(q_{i+1}) + q_0^{(i)} q_i - q_i q_0, \quad i \in \mathbf{Z}_+. \tag{26}$$

The corresponding Hamiltonian form is

$$B^{KP}_{ij} = \hat{R}_{(\cdot)} \Delta^{-j} - \Delta^{-j} \hat{L}_{(\cdot)}, \quad (\cdot) = q_{i+j}. \tag{27}$$

Discretizing time, by the ansatz [7]

$$\tilde{L} = \Pi^{-1} L \, \Pi, \ \Pi = 1 + h(\zeta + \alpha), \tag{28}$$

$$\tilde{L} = L(t + \Delta t), \ h = const \Delta t, \tag{29}$$

we get an undetermined system

$$\tilde{q}_0 + \alpha^{(-1)} = q_0^{(-1)} + \alpha, \tag{30a}$$

$$[1 + h\alpha^{(i)}]\tilde{q}_i + h\tilde{q}_{i+1} = q_i(1 + h\alpha) + hq_{i+1}^{(-1)}, \ i \in \mathbf{Z}_+. \tag{30b}$$

All the other commutative troubles from [7] are present. To handle undeterminacy we regularize the system (30) by demanding that all the ideals I_N^q:

$$I_N^q = \{q_i = 0, i > N\}$$

are preserved by the time-step may (30) (as they are preserved by the continuous flow (26).) This results in the following solution

$$\frac{\alpha}{1 + h\alpha} = \sum_{s \geq 0} (-h)^s \frac{1}{1 + h\alpha} \cdots \frac{1}{1 + h\alpha^{(s)}} q_s. \tag{32}$$

The corresponding time-step map (30) is Hamiltonian. Let us consider different coordinates for the Lax operator L (25):

$$L = \zeta(1 + \sum_{i \geq 0} \zeta^{-i} q_i) = \zeta(1 + \sum_{j \geq 0} \zeta^{-j-1} r_j)^{-1}, \tag{33}$$

with the 1st flow

$$r_{j,t} = (\Delta - 1)(r_{j+1}) + r_j r_0^{(-j-1)} - r_0^{(1)} r_j, \ j \in \mathbf{Z}_+. \tag{34}$$

Time-discretizing the flow (34) in the r-coordinates, by the same ansatz (28):

$$[1 + h(\zeta + A)]\tilde{L} = L[1 + h(\zeta + A)], \tag{35}$$

we again get an undetermined system

$$\tilde{r}_0^{(1)} + A = r_0 + A^{(-1)}, \tag{36a}$$

$$[(1+hA)\tilde{r}_j + h\tilde{r}_{j+1}^{(1)} = r_j[1+hA^{(-j-2)}] + hr_{j+1},\ j \in \mathbf{Z}_+. \tag{36b}$$

To handle undeterminancy we regularize the system (36) by demanding that all the ideals J_N^r:

$$J_N^r = \{v_j = 0,\ \forall_j > N\} \tag{37}$$

are preserved by the time-step map (36) (as they are preserved by the continuous flow (34).) This results in the solution

$$-A = \sum_{s \geq 0} (-h)^s \frac{1}{1+hA^{(1)}} \cdots \frac{1}{1+hA^{(s+1)}} r_s. \tag{38}$$

Conjecture 39. $\alpha = A$ (even though the regularizations (31) and (37) are incompatible.) This is true modulo h^3, but is, probably, hopeless in general. When $\Delta = id$, i.e. when the lattice shrinks to a point, Conjecture 39 degenerates into

Conjecture 40. Suppose, with λ and h being commutative,

$$(1 + \sum_{i \geq 0} \lambda^{i+1} q_i)(1 + \sum_{j \geq 0} \lambda^{j+1} r_j) = 1, \tag{41}$$

$$q(x) = 1 + \sum x^{i+1} q_i,\ r(x) = 1 + \sum x^{j+1} r_j. \tag{42}$$

Let $x = 1 + O(h)$ and $y = 1 + O(h)$ be the unique solutions of the equations

$$xq(hx) = 1,\ y = r(hy). \tag{43}$$

Then $x = y$. This is obvious in the commutative realm. In the non commutative case, one has to develop a noncommutative version of the Lagrange- Bürmann series. The case of the mKP hierarchy leads to still more general conjectures.

References

[1.] Kupershmidt, B.A., "Discrete Lax Equations and Differential–Difference Calculus", Asterisque (Paris, 1985).

[2.] Kupershmidt, B.A., "KP or mKP: Noncommutative Aspects of Integrable, Lagrangian, and Hamiltonian Systems" (to appear).

[3.] Kontsevich, M., in "The Gel'fand Mathematical Seminars", pp. 173-189; L. Corwin et al. ed-s, Birkhäuser (Boston, 1993).

[4.] Gel'fand, I.M.and Smirnov, M.M., hep-th/9404103.

[5.] Kupershmidt, B.A., Comm. Math. Phys. 99 (1985) 51-73.

[6.] Kupershmidt, B.A., "The Variational Principles of Dynamics", World Scientific (Singapore, 1982).

[7.] Gibbons, J. and Kupershmidt, B.A., Phys. Lett. A 165 (1992) 105-110.

A DIRECT BILINEARIZATION SCHEME BASED ON THE USE OF PARTITION POLYNOMIALS

F. LAMBERT, I. LORIS, J. SPRINGAEL and R. WILLOX

Vrije Universiteit Brussel, Dienst Theoretische Natuurkunde
Pleinlaan 2, B-1050 Brussel

ABSTRACT

A systematic procedure for the bilinearization of classes of soliton equations (and related ordinary differential equations) is presented. The method, based on the use of generalized Bell polynomials, is illustrated with several examples.

1. Introduction

Bilinear forms[1] are an efficient tool to find solitary wave solutions to nonlinear partial differential equations (NLPDE's), to prove the existence of N-soliton solutions when such solutions are expected, and to relate the NLPDE's to Sato's framework of integrable hierarchies[2]. The success of Hirota's approach to soliton theory has given strength to the idea that every soliton system should be derivable from a fundamental equation, or system of equations, in bilinear form. These bilinear equations should be such that multisoliton solutions are represented by finite sums of exponentials.

A major problem in the application of the Hirota formalism is to find the appropriate set of variables (dependent and independent) and the "bilinearizing" dependent variable transformations. In this note we propose a direct bilinearization scheme based on the partitional properties of the Hirota D-operator and on a basic notion of scale invariance. The starting point is a partitional characterization of a class of NLPDE's which can be linearized by the Cole-Hopf transformation[3].

2. Cole-Hopf linearization

The linearizable equations of the Burgers-Hopf (BH) hierarchy[4] form a natural background for a discussion of the Hirota method. The point to be emphasized is that one may use a simple class of partition polynomials to verify whether a given NLPDE may be derived from an equation which is linearized through a logarithmic

or a Cole-Hopf transformation. The procedure is straightforward and may be illustrated by the following example.
Consider:

$$w_t + w_{3x} + \alpha\, w\, w_{2x} + \beta\, w_x^2 + \gamma\, w^2\, w_x = 0\ , \qquad w_{px} = \partial_x^p\, w\ . \tag{1}$$

We wish to examine which of these equations belong to (or are derivable from) the BH hierarchy, and to obtain the linearizing transformation.
It is important for this purpose to rewrite the Eq.(1) in terms of an appropriate *dimensionless* field variable. Invariance of the equation under a scale transformation (we assign a dimension 1 to x for definiteness and assume that α, β and γ are dimensionless):

$$x \to \lambda\, x\ , \quad t \to \lambda^r\, t\ , \quad w \to \lambda^d\, w \tag{2}$$

imposes the constraints: $r = 3$ and $d = -1$.
Introducing a dimensionless field $v(x,t)$ by setting $w = c\, v_x$, with $c =$ free dimensionless parameter, we try to rewrite:

$$v_{xt} + v_{4x} + \alpha\, c\, v_x\, v_{3x} + \beta\, c\, v_{2x}^2 + \gamma\, c^2\, v_x^2\, v_{2x} = 0 \tag{3}$$

as a linear combination of Bell polynomials[5] up to possible overall differentiations with respect to x or t. A guiding line is that these polynomials:

$$Y_{px}(v) \equiv \exp(-v)\partial_x^p \exp v\ , \quad Y_{pt}(v) \equiv \exp(-v)\partial_t^p \exp v\ , \tag{4}$$

obey a simple partitional recipe (each term corresponds to a partition of the order p with a coefficient equal to the combinatorial weight of the partition):

$$\begin{aligned} &Y_x(v) = v_x\ , \quad Y_t(v) = v_t \\ &Y_{2x}(v) = v_{2x} + v_x^2\ , \quad Y_{3x}(v) = v_{3x} + 3\, v_x\, v_{2x} + v_x^2 \\ &\vdots \end{aligned} \tag{5}$$

Thus, the Eq.(3) can be rewritten as:

$$\left[Y_t(v) + Y_{3x}(v) + (\alpha\, c - 3)v_x\, v_{2x} + \left(\frac{1}{3}\gamma\, c^2 - 1\right) v_x^3\right]_x + (\beta - \alpha)\, c\, v_{2x}^2 = 0\ . \tag{6}$$

If $\alpha = \beta$ and $\gamma = \frac{1}{3}\alpha^2$ it suffuces to choose $c = \frac{3}{\alpha}$ to see that Eq.(6) reduces to:

$$Y_t(v) + Y_{3x}(v) \equiv v_t + v_{3x} + 3\, v_x\, v_{2x} + v_x^3 \equiv \exp(-v)\left(\partial_t + \partial_x^3\right)\exp v = 0 \qquad (7)$$

Thus, it is found that the equation:

$$w_t + w_{3x} + \alpha\left(w\, w_{2x} + w_x^2\right) + \frac{1}{3}\alpha^2 w^2 w_x = 0 \qquad (8)$$

is derived from a *generating* equation (7) which is linearized by the logarithmic transformation. We say that the Eq.(8) is "linearized" by the transformation $w = \frac{3}{\alpha}\partial_x \ln\psi$.

3. Single field bilinearization

A similar procedure can be applied to decide whether a given NLPDE can be derived from a single field bilinear equation of polynomial type:

$$\mathcal{F}(D_x, D_t, \cdots)F \cdot F = 0\,, \quad \mathcal{F} : \text{polynomial} \qquad (9)$$

where the Hirota operators are defined by:

$$D_x^m\, D_t^n F \cdot G = (\partial_x - \partial_{x'})^m (\partial_t - \partial_{t'})^n F(x,t) G(x',t')\Big|_{x' = x,\, t' = t}\,. \qquad (10)$$

In this case one must look for a generating equation in the form of a linear combination of polynomials $P_{mx,nt}(q)$ set equal to zero, where:

$$P_{mx,nt}(q) = \exp(-q) D_x^m\, D_t^n \exp\left(\frac{q}{2}\right) \cdot \exp\left(\frac{q}{2}\right)\,, \quad m + n = \text{even}\,. \qquad (11)$$

These polynomials, defined in close analogy with the following two dimensional generalization of the Bell polynomials:

$$Y_{mx,nt}(v) = \exp(-v)\partial_x^m\, \partial_t^n \exp\left(\frac{v}{2}\right) \cdot \exp\left(\frac{v}{2}\right)\,, \qquad (12)$$

obey a similar partitional recipe restricted to *even part* partitions:

$$\begin{aligned} &P_{2x}(q) \equiv P_{2x,0}(q) = q_{2x}\,, \quad P_{x,t}(q) = q_{x,t} \\ &P_{4x}(q) \equiv P_{4x,0}(q) = q_{4x} + 3q_{2x}^2\,, \quad P_{3x,t}(q) = q_{3x,t} + 3q_{2x}\, q_{x,t} \\ &\vdots \end{aligned} \qquad (13)$$

To illustrate the procedure we first consider the KdV equation:

$$w_t + w_{3x} + w\, w_x = 0 \; . \tag{14}$$

Invariance under the scale transformation (2) requires $r = 3$ and $d = -2$. Introducing a dimensionless field q by setting $w = c\, q_{2x}$, with $c =$ dimensionless free parameter, we rewrite the equation:

$$q_{2x,t} + q_{5x} + c\, q_{2x}\, q_{3x} = 0 \tag{15}$$

as

$$\left(q_{xt} + q_{4x} + \frac{c}{2}\, q_{2x}^2\right)_x = 0 \; . \tag{16}$$

Choosing $c = 6$ we find that the Eq.(14) can be derived from the *generating* equation:

$$P_{x,t}(q) + P_{4x}(q) = 0 \; , \tag{17}$$

which through the formula (11) corresponds to the bilinear equation:

$$D_x(D_t + D_x^3)F \cdot F = 0 \; , \quad F = \exp\left(\frac{q}{2}\right) \tag{18}$$

In this case we say that the Eq.(14) is "bilinearized" by the transformation $w = 12\, \partial_x^2 \ln F$.

A more challenging set of equations is:

$$w_{xt} + w_{2x} + \alpha\, w_{3x,t} + \beta\, w_{2x}\, w_t + \gamma\, w_x\, w_{xt} = 0 \; . \tag{19}$$

Scale invariance of these Eq. cannot be preserved unless $\dim \alpha = 2$, $\dim \beta = \dim \gamma$ and $\dim w = 1 - \dim \beta$. Taking $\dim \beta = \dim \gamma = 2$ and $\dim w = -1$ we introduce a dimensionless $q(x,t)$ by setting $w = c\, q_x$, and try to rewrite:

$$\left(q_{xt} + q_{2x} + \alpha\, q_{3x,t}\right)_x + \beta c (q_{2x}\, q_{xt})_x + \frac{1}{2}(\gamma - \beta) c \left(q_{2x}^2\right)_t = 0 \tag{20}$$

in terms of linear combinations of P-polynomials:

$$\begin{aligned} &[P_{x,t}(q) + P_{2x}(q) + \alpha\, P_{3x,t}(q)]_x + (\beta c - 3\alpha)(q_{2x}\, q_{xt})_x \\ &\qquad + \frac{1}{6}(\gamma - \beta)\, [P_{4x}(q) - q_{4x}]_t = 0 \; . \end{aligned} \tag{21}$$

If $\gamma = \beta$ it suffices to choose $c = 3\frac{\alpha}{\beta}$ to obtain the *generating* equation:

$$P_{x,t}(q) + P_{2x}(q) + \alpha\, P_{3x,t}(q) = 0 \ , \tag{22}$$

indicating that the shallow water wave equation:

$$w_{xt} + w_{2x} + \alpha\, w_{3x,t} + \beta(w_{2x}\, w_t + w_x\, w_{xt}) = 0 \tag{23}$$

can be *derived* from the single field bilinear equation:

$$D_x(D_t + D_x + \alpha\, D_x^2\, D_t)F \cdot F = 0 \ , \quad F = \exp\left(\frac{q}{2}\right) \tag{24}$$

and that the "bilinearizing transformation" is $w = 6\frac{\alpha}{\beta}\, \partial_x \ln F$.

If $\gamma^2 \neq \beta^2$ we rewrite the Eq.(21) as:

$$\left[P_{x,t}(q) + P_{2x}(q) + \alpha\, P_{3x,t}(q) - \frac{1}{6}(\gamma - \beta)c \left[q_{3x,t} + 6\frac{(3\alpha - \beta c)}{(\gamma - \beta)c} q_{2x}\, q_{2t}\right]\right]_x + \frac{1}{6}(\gamma - \beta)c\, [P_{4x}(q)]_t = 0 \ . \tag{25}$$

Choosing $c = \frac{6\alpha}{\beta + \gamma}$ it becomes:

$$\left[P_{x,t}(q) + P_{2x}(q) + \frac{2\alpha\, \beta}{(\beta + \gamma)} P_{3x,t}(q)\right]_x - \alpha \left(\frac{\beta - \gamma}{\beta + \gamma}\right) [P_{4x}(q)]_t = 0 \ . \tag{26}$$

At this stage we should try to express $[P_{4x}(q)]_t$ as the x-derivative of another P-polynomial. Since P-polynomials are generalizable to more independent variables (in the same way as Y-polynomials), it is tempting to introduce an auxiliary independent variable τ, together with an appropriate constraint. If $q = q(x, t, \tau)$, possible auxiliary P-polynomials of lowest order are:

$$P_{x\tau}(q) = q_{x\tau} \ , \quad P_{t\tau}(q) = q_{t\tau} \ , \quad P_{2\tau}(q) = q_{2\tau} \ , \tag{27}$$

the first one of which contains an x-derivative. Hence a useful accompanying constraint is the condition $P_{4x}(q) \div P_{x,\tau}(q)$ which implies $[P_{4x}(q)]_t \div q_{x,\tau,t} = [P_{\tau,t}(q)]_x$.

This condition amounts (after rescaling of the auxiliary time τ) precisely to the generating Eq.(17) for KdV evolution with time τ:

$$P_{4x}(q) + P_{x,\tau}(q) = 0 \ . \tag{28}$$

We thus obtain the *generating* system:

$$\begin{cases} P_{x,\tau}(q) + P_{4x}(q) = 0 \\ P_{x,t}(q) + P_{2x}(q) + 2\,\dfrac{\alpha\,\beta}{(\beta+\gamma)}\,P_{3x,t}(q) + \alpha\left(\dfrac{\beta-\gamma}{\beta+\gamma}\right)P_{\tau,t}(q) = 0 \end{cases} \tag{29}$$

or the equivalent bilinear system ($F = \exp\frac{q}{2}$):

$$\begin{cases} D_x(D_\tau + D_x^3)F\cdot F = 0 \\ \left[D_x\,D_t + D_x^2 + 2\dfrac{\alpha\,\beta}{(\beta+\gamma)}D_x^3\,D_t + \alpha\left(\dfrac{\beta-\gamma}{\beta+\gamma}\right)D_t\,D_\tau\right]F\cdot F = 0 \end{cases} \tag{30}$$

which turns out to satisfy the Hirota two-soliton compatibility condition iff $\gamma = 2\beta$. This tells us that the AKNS equation[6]:

$$w_{xt} + w_{2x} + \alpha\,w_{3x,t} + \beta(w_{2x}\,w_t + 2\,w_x\,w_{xt}) = 0 \tag{31}$$

can be derived from a two-soliton generating bilinear system, and that the bilinearizing transformation is $w = 4\frac{\alpha}{\beta}\,\partial_x\ln F$.

4. Two field bilinearization

Application of the above procedure to the MKdV equation:

$$w_t + w_{3x} + w^2\,w_x = 0 \tag{32}$$

fails to produce a *generating* bilinear equation of type (9), or a corresponding system with an auxiliary time τ. However, one may look for a system of two equations of the form:

$$\mathcal{F}_i(D_x, D_t)F\cdot G = 0\ ,\quad \mathcal{F}_i = \text{polynomial}\ , \tag{33}$$

where G would play the role of an auxiliary field variable. It is usefull for this purpose to consider the following two field generalization of the $Y_{mx,nt}$:

$$\mathcal{Y}_{mx,nt}(u,v) \equiv Y_{mx,nt}(y)\Bigg|_{y_{r,s} = \begin{cases} u_{r,s} \text{ if } r+s = \text{ even} \\ v_{r,s} \text{ if } r+s = \text{ odd} \end{cases}} \tag{34}$$

with $y_{r,s} \equiv y_{rx,st}$.

These *binary* Bell polynomials obey a partitional recipe similar to that for $Y_{mx,nt}$, combined with a parity rule which governs the field labels u and v.

$$\begin{aligned}
&\mathcal{Y}_x(v) \equiv \mathcal{Y}_{x,0}(u,v) = v_x \ , \quad \mathcal{Y}_t(v) \equiv \mathcal{Y}_{0,t}(u,v) = v_t \\
&\mathcal{Y}_{2x}(u,v) \equiv \mathcal{Y}_{2x,0}(u,v) = u_{2x} + v_x^2 \\
&\mathcal{Y}_{x,t}(u,v) = u_{xt} + v_x\, v_t \\
&\mathcal{Y}_{3x}(u,v) \equiv \mathcal{Y}_{3x,0}(u,v) = v_{3x} + 3\, u_{2x}\, v_x + v_x^3 \ .
\end{aligned} \tag{35}$$

Their link with the Hirota terms is given by the formula[7]:

$$\mathcal{Y}_{mx,nt}(u,v)\Big|_{\begin{cases} u = \ln(FG) \\ v = \ln(F/G) \end{cases}} \equiv (FG)^{-1} D_x^m D_t^n F \cdot G \ . \tag{36}$$

Going back to the Eq.(32), with the dimensional constraint $\dim w = -1$, we set $w = c\, v_x$ with $c =$ free dimensionless parameter, and try to rewrite the equation:

$$v_{xt} + v_{4x} + c^2 v_x^2 v_{2x} = \left(v_t + v_{3x} + \frac{c^2}{3}\, v_x^3\right)_x = 0 \tag{37}$$

in terms of $\mathcal{Y}$-polynomials:

$$\left[\mathcal{Y}_t(v) + \mathcal{Y}_{3x}(u,v) - 3\mathcal{Y}_x(v)\left[u_{2x} + \frac{1}{3}\left(1 - \frac{c^2}{3}\right) v_x^2\right]\right]_x = 0 \ . \tag{38}$$

Choosing $c = \pm i\sqrt{6}$ we see that the *generating* MKdV equation:

$$v_t + v_{3x} - 2v_x^3 = 0 \tag{39}$$

may be expressed as:

$$\mathcal{Y}_t(v) + \mathcal{Y}_{3x}(u,v) - 3\mathcal{Y}_x(v)\mathcal{Y}_{2x}(u,v) = 0 \ . \tag{40}$$

The next step is to introduce a *linearizing constraint* on the auxiliary field $u(x,t)$ so as to be left with a system which involves only linear combinations of $\mathcal{Y}$-polynomials. An appropriate choice is the condition $\mathcal{Y}_{2x}(u,v) = \lambda$ (constant) which produces the system:

$$\begin{cases} \mathcal{Y}_{2x}(u,v) = \lambda \\ \mathcal{Y}_t(v) + \mathcal{Y}_{3x}(u,v) - 3\,\lambda\,\mathcal{Y}_x(v) = 0\,. \end{cases} \tag{41}$$

Setting $\lambda = 0$ for simplicity we find the generating systems:

$$\begin{cases} \mathcal{Y}_{2x}(u,v) = 0 \\ \mathcal{Y}_t(v) + \mathcal{Y}_{3x}(u,v) = 0 \end{cases} \quad \overset{u=\ln FG}{\underset{v=\ln(\frac{F}{G})}{\Longleftrightarrow}} \quad \begin{cases} D_x^2\, F \cdot G = 0 \\ (D_t + D_x^3) F \cdot G = 0\,. \end{cases} \tag{42}$$

On account of the separability property[7] of $\mathcal{Y}$-polynomials:

$$\mathcal{Y}_{mx,nt}(u = v + q, v) = \sum_{r=0}^{m} \sum_{s=0}^{n} \binom{m}{r} \binom{n}{s} P_{rx,st}(q) Y_{(m-r)x,(n-s)t}(v) \tag{43}$$

it is clear that by setting $u = v + q$, the generating system (42) transforms into the system:

$$\begin{cases} v_{2x} + v_x^2 + q_{2x} = 0 \\ v_t + v_{3x} + 3\,v_x\,v_{2x} + v_x^3 + 3\,v_x\,q_{2x} = 0 \end{cases} \tag{44}$$

which, on account of the definitions (4) and (12), is linearized by setting $v = \ln\psi$:

$$\begin{cases} \psi_{2x} + q_{2x}\psi = 0 \\ \psi_t + \psi_{3x} + 3q_{2x}\psi_x = 0 \end{cases} \Longleftrightarrow \begin{cases} L_1\psi \equiv \psi_{2x} + q_{2x}\psi \\ L_2\psi \equiv \psi_t + 2q_{2x}\psi_x - q_{3x}\psi = 0\,. \end{cases} \tag{45}$$

Hence, it is found that the MKdV equation (32) admits a linear generating system which produces solutions to the equation provided that q_{2x} satisfies the compatibility condition:

$$[L_1, L_2]\psi = 0\,. \tag{46}$$

Setting $q_{2x} = U$ one finds that this condition amounts precisely to the KdV equation:

$$U_t + U_{3x} + 6\,U\,U_x = 0\,. \tag{47}$$

It also follows from the generating system (44) that the transformation $U = -(v_{2x} + v_x^2)$ maps solutions of the primary MKdV Eq.(39) onto solutions of the

KdV equation (Miura transformation). Thus, it is the straightforward application of the bilinear scheme which tells us that the MKdV equation is the "modified" partner of another NLPDE which corresponds to the integrability condition of a linear system.
It is also clear that the Darboux property[8] of the linear system (45) (which allows the construction of multisoliton solutions to the Eq.(47) out of a trivial solution) can be regarded as a direct consequence of an explicit invariance of its bilinear counterpart (42) under the interchange $F \rightleftharpoons G$.

Application of the above procedure to an equation obtained by Hirota and Satsuma[9]:

$$w_{2t} - w_{4x} + \alpha\, w_{2x} w_t - \frac{\alpha^2}{2} w_x^2\, w_{2x} = 0\,, \tag{48}$$

and to the Broer-Kaup equation[10]:

$$w_{2t} - w_{2x} + w_{4x} + 2w_x\, w_{xt} + w_{2x}\, w_t + \frac{3}{2} w_x^2\, w_{2x} = 0\,, \tag{49}$$

produces similar bilinear generating systems which in both cases are linearized by the above transformation ($\psi = \frac{F}{G}$, $q = 2\ln G$). Both equations are thereby found to be the "modified" partner of a sech squared soliton system: the Boussinesq equation in the case of Eq.(48) and a nonlocal alternative[7] to the good Boussinesq equation in the case of Eq.(49).

Let us finally remark that the above bilinearization scheme can also be applied to an ordinary differential equation such as the Painlevé II equation:

$$w_{2x} - x\, w - 2w^3 - \alpha = 0 \tag{50}$$

for which it produces the generating system ($w = v_x$):

$$\begin{cases} \mathcal{Y}_{2x}(u,v) = 0 \\ \mathcal{Y}_{3x}(u,v) - x\mathcal{Y}_x(u,v) - \alpha = 0 \end{cases} \underset{v=\ln(\frac{F}{G})}{\overset{u=\ln FG}{\Longleftrightarrow}} \begin{cases} D_x^2 F \cdot G = 0 \\ (D_x^3 - xD_x - \alpha)F \cdot G = 0 \end{cases} \tag{51}$$

which is again transformed into a linear system by setting $u = v + q$, $v = \ln\psi$:

$$\begin{cases} L_1\psi \equiv \psi_{2x} + q_{2x}\psi \\ L_2\psi \equiv (2q_{2x} - x)\psi_x - (q_{3x} + \alpha)\psi = 0\,. \end{cases} \tag{52}$$

The compatibility of the system (52) is subject to the following equation for $U = q_{2x}$:

$$U_{3x} - x\,U_x - 2U - 6U\,U_x = 0 \qquad (53)$$

which is known[11] to be a similarity reduction of the KdV equation.
The point to be underlined is that it is shown by direct bilinearization that the Painlevé II equation is related to the Eq.(53) by the same Miura transformation as the one relating the MKdV equation to the KdV equation. The similarity of the bilinear systems (42) and (51) is a clear indication that Eq.(53) should be obtainable from KdV by a similarity reduction.

Acknowledgement

The authors wish to acknowledge the Belgian Government's support (UIAP III). R. Willox and I. Loris are Research Assistants at the National Fund for Scientific Research (Belgium). J. Springael acknowledges the financial support extended by the Solvay Institutes.

References

1. Y. Matsuno, *Bilinear Transformation Method*, Acad. Press, Orlando, (1984), 14
2. M. Jimbo and T. Miwa, *RIMS* Kyoto Univ. **19** (1983), 943
3. E. Hopf, *Comm. Pure Appl. Math.* **3** (1950), 201
4. D.V. Choodnovsky and G.V. Choodnovsky, *Nuovo Cimento B* **40** (1977), 339
5. E.T. Bell, *Ann. Math.* **35** (1934), 258
6. M. Ablowitz et al, *Stud. Appl. Math.* **53** (1974), 249
7. F. Lambert et al, *J. Phys. A* **27** (1994), 5325
8. V.B. Matveev and M.A. Salle, *Darboux transformations and solitons*, Springer Verlag, Berlin (1991), 31
9. R. Hirota and J. Satsuma, *Progr. Theor. Phys.* **57** (1977), 797
10. D. Kaup, *Progr. Theor. Phys.* **54** (1975), 369
11. P.G. Drazin, R.S. Johnson, *Solitons: an introduction*, Cambridge Univ. Press, Cambridge (1990), 172

The list of canonical discrete symmetries of Integrable Systems

A.N.Leznov

Institute for High Energy Physics, 142284 Protvino, Moscow Region, Russia

Abstract

The general properties of canonical discrete transformations of integrable systems are discussed. The list of discrete transformations with their solutions for most othen used in physical applications integrable systems is represented.

1 Introduction

At the recent time it became clear that the most important and character properties of integrable systems in most successive way may be understood and explained on the base of discrete symmetries of such systems. Under discrete transformation (local in the simplest case) we will understand the substitution (mapping) of following kind [1]

$$\tilde{u} \equiv \phi(u) = \phi(u, u', u'', ...) \qquad (1.1)$$

where u is vector function of unknown dynamical variables of the system under consideration, $u''^{\cdots}$ its arbitrary degree (but limited) derivatives with respect to independent arguments of the problem. In general case the relations of type (1.1) may be not local, this mean that for obtaining $\tilde{u}$ as function of u and its derivative it will be necessary to solve some differential (or integral, or some other kind) equation.

1.1 General properties of discrete substitution.

At first we want to enumerate the most important general properties of substitutions (1.1) as a result of observation on sufficiently large number of integrable systems [2].

1. All equations of the same hierarchy possess the same discrete transformation.

2. Substitution (1.1) are invertible: this means that equations (1.1) may resolved with respect to the "old" variables u and they can be expressed via the functions of the "new" variables $\tilde{u}$.

3. The substitution (1.1) is the canonical [3],[4]. This fact can be expressed in two equivalent forms. There exist single generating function from which by the rules of the theory of canonical transformations it is possible to write down the explicit formulae of substitution 1.1). In other words this means that with substitution (1.1) may be connected some Poisson structure [5] (and not single one) which is invariant with respect to transformations describing by substitution 1.1).

4. The conserved quantities of the theory are displaying on the whole divergence (with respect to space coordinates of the theory) under transformation (1.1).

5. The substitution (1.1) may be rewritten in the form of infinite chain of equations

$$\tilde{u}_{n+1} = \phi(u_n, u'_n, u''_n, ...) \tag{1.2}$$

where by u_n is denoted the result of n–time application transformation (1.1) to some initial function u_0 (possible solution of some integrable system). The general property of chains (1.2) consist in their integrability. This means that it is possible to obtain exact general solution of these chains under appropriate way of interrupting of the chain (by using some "good" boundary conditions) on its both ends. For all integrable systems with rational spectral parameter (in old terminology of inverse scattering method) the chain (1.2) coincide with equations of Toda lattice (Darbou transformation) or some its generalization. About situation in the case of eleptical spectral parameter see [6].

1.2 The problems what can be solved by help of discrete transformation.

Now we enumerate the most important results which can be obtained by help of discrete transformation (1.1).

1. It is possible to obtain the wide class of explicit solutions of integrable systems in determinant (Hirota) form. These solutions depend from some number of arbitrary functions [2].

2. By appropriate chose of these functions it is possible to extract those solutions which are invariant with respect to some inner authomorphism of dynamical system in consideration (partially multisoliton solutions) or obey some additional boundary conditions [7],[8].

3. The condition of invariance of Poisson structures with respect to transformation (1.1)

$$\phi'(u)J_n(u)\phi'^T(u) = J_n(\phi(u)$$

($\phi'(u)$– is the Frechet derivative) allows to obtain the explicit form nonlocal Hamiltonian operators of arbitrary order and after this to construct the whole hierarchy of integrable systems with given discrete substitution [3].

1.3 The equation for determining of discrete substitution

As it was discussed above that the knowledge of the discrete substitution allow to give positive solution for many problems of the theory of integrable systems. It remains only one "little" question how to choose the appropriate substitution among of infinite number possible ones?

We shall now formulate the equation solution of which is exactly the mapping (1.1) which satisfy all necessary conditions to exploit this substitution as discrete symmetry of some integrable system [9]. We will write this equation under the assumption of locality of all values in construction below.

Let $\phi'(u)$ is Frechet derivative corresponding to substitution (1.1)

$$\phi'(u) = \frac{\partial \phi}{\partial u} + \frac{\partial \phi}{\partial u'}D + \frac{\partial \phi}{\partial u''}D^2 .....$$

By $F(u)$ let us denote the vector colomun function components of which are some (local) functions of dynamical variables u and its derivatives $F_s(u) \equiv$

$F_s(u, u', u'', ...u^{(n}$.Then each solution of functional differential equation with the displacement arguments

$$F(\phi(u)) = \phi'(u)F(u) \tag{1.3}$$

may be connected with equation of evolution type

$$u_t = F(u)$$

which is invariant with respect to discrete transformation $\tilde{u} = \phi(u)$

Equation (1.3) is generalization of well known condition of integrability in the theory of differential equations. Indeed let us differentiate (1.1) with respect to some parameter and denote $\dot{\phi}(u) \equiv F(\phi(u)), \dot{u} \equiv F(u)$ then (with all necessary words) we come to (1.3).

So if the mapping (substitution) is integrable (in the above sense) then it is possible to consider it as discrete symmetry of some integrable system.

Below we represent the list of discrete transformation (with their solutions for some universal (but not single) method of interrupting of the corresponding chains (1.2)) for dynamical systems having wide applications in the different branches of the mathematical physic. By tradition we will call them as Backlund transformations. It is possible to convice by direct check that for all substitution below equation (1.3) indeed obey nontrivial solution.

2 The list of discrete substitutions.

Here we give a list of integrable systems together with Backlund transformations for them and the corresponding solutions.

2.1 Hirota equation.

$$\begin{aligned} &\dot{v} + \alpha(v''' - 6uvv') - i\beta(v'' - 2v^2u) + \gamma v' + i\delta v = 0, \\ &\dot{u} + \alpha(u''' - 6uvu') + i\beta(u' - 2u^2v) + \gamma u' - i\delta u = 0; \end{aligned} \tag{2.1}$$

$$\dot{} \equiv \frac{\partial}{\partial t}, \quad ' \equiv \frac{\partial}{\partial x};$$

$$\tilde{v} \equiv sv = \frac{1}{u}, \qquad \tilde{u} \equiv su = u(uv - (lnu)''), \qquad v_0 = 0,$$
$$\dot{u}_0 + \alpha u_0''' + i\beta u_0'' + \gamma u_0' - i\delta u_0 = 0.$$

In this and in the other cases the main role will be played by the principal minors of the following matrix:

$$\begin{pmatrix} \phi^s & \phi^{s+1} & \phi^{s+2} & \dots \\ \phi^{s+1} & \phi^{s+2} & \phi^{s+3} & \dots \\ \phi^{s+2} & \phi^{s+3} & \phi^{s+4} & \dots \\ \dots & \dots & \dots & \dots \end{pmatrix}$$

To denote the principal minors of these matrices we shall use the symbol D_r^n. Here n is the rank of the matrix and r is the symbol of its element of left upper corner. For the solution of the Backlund transformation we have

$$v_n = (-1)^n \frac{D_0^{n-1}}{D_0^n}, \qquad u_n = (-1)^{n+1} \frac{D_0^{n+1}}{D_0^{n+1}} \tag{2.2}$$

The methods of theory function of complex variables gives the same expression where the role D_0^n play the nonlocal integral

$$D_0^n = \int d\lambda_1 ... d\lambda_n c(\lambda_1) ... c(\lambda_n) W_n^2(\lambda_1, ..., \lambda_n), \tag{2.3}$$

where $W_n(\lambda)$ is the Vandermonde determinant; and $c(\lambda)$ is the integrand in the representation for u_0.

2.2 Nonlinear Schrödinger equations.

2.2.1 a.

$$\dot{q} + q'' - 2rq^2 = 0, \qquad \tilde{q} = \frac{1}{r}, \qquad \tilde{r} = r[rq - (\ln r)''];$$
$$-\dot{r} + r'' - 2qr^2 = 0\, q_o = 0 \qquad \dot{r}_0 = r_0''. \tag{2.4}$$

The solution of the Backlund transformation is the same as in the previous section.

2.2.2 b.

$$\dot{q} + q'' + 2(rq)q' = 0, \qquad \tilde{q} = \frac{1}{r}, \qquad \tilde{r} = r[(rq) + (\ln \frac{r}{r'})'];$$
$$-\dot{r} + r'' - 2(rq)r' = 0 \quad q_0 = 0 \quad \dot{r}_0 = r_0''. \tag{2.5}$$

The solution of the Backlund transformation is as follows

$$q_n = (-1)^n \frac{D_1^{n-1}}{D_0^n}, \qquad r_n = (-1)^{n+1} \frac{D_0^{n+1}}{D_1^{n+1}} \tag{2.6}$$

2.2.3 c.

$$\dot{q} + q'' - 2(rq^2)' = 0, \qquad \tilde{q} = r, \qquad \tilde{r} = q - (\frac{1}{r})';$$
$$-\dot{r} + r'' + 2(r^2 q)' = 0, \qquad q_0 = 0, \qquad \dot{r}_0 = r_0''. \tag{2.7}$$

The solution of the Backlund transformation is as follows

$$q_{2n} = \frac{D_1^{n-1} D_1^n}{(D_0^n)^2}, \qquad r_{2n} = \frac{D_0^{n-1} D_0^n}{(D_1^n)^2},$$
$$q_{2n+1} = \frac{D_0^{n-1} D_0^n}{(D_0^n)^2}, \qquad r_{2n+1} = \frac{D_1^{n-1} D_1^n}{(D_0^{n+1})^2}. \tag{2.8}$$

2.2.4 d.

$$q' + \ddot{q} + 2(r^2\dot{q} - q^3 r^2) = 0 \qquad -r' + \ddot{r} - 2(q^2\dot{r} + r^3 q^2) = 0$$

Backlund transformation in this case is some more complicated:

$$\tilde{q} = (r' + qr^2)^{-1}, \quad \tilde{r} = -(r' + qr^2)' + r^{-1}(r' + qr^2)^2$$

As in the last cases $q_0 = 0$, $-\dot{r}_0 + r_0'' = 0$ and the solution of Backlund transformation under this boundary conditions have the form:

$$q_n = \frac{D_{n-1}^1}{D_n^1}, \quad r_n = \frac{D_{n+1}^0}{D_n^0}, \quad D_{-1}^\alpha = 0, \quad D_0^\alpha = 0.$$

2.2.5 One- dimensional Heisenberg ferromagnetic in classical region (XXX - model).

$$\dot{S} = [S, S''], \qquad S = (S_-, S_0, S_+), \qquad S_0^2 + S_- S_+ = 1;$$

$$\tilde{S}_- = S_- + 2\left(\frac{1}{(\frac{s_+}{1+s_0})'}\right)', \quad \tilde{S}_+ = S_+ + 2\left(\frac{1}{(\frac{s_-}{1-s_0})'}\right)',$$

$$\tilde{S}_0 + 1 = -\tilde{S}_- \frac{S_+}{1+S_0}, \qquad S_-^0 = 0, \qquad S_0^0 = 1, \dot{S}_+ = 2S_+''$$

$$S_-^n = \frac{D_2^{n-1} D_2^n}{(D_1^n)^2}, \qquad S_0^n + 1 = 2\frac{D_0^n D_2^n}{(D_1^n)^2},$$

$$S_0^n - 1 = 2\frac{D_2^{n-1} D_0^{n+1}}{(D_1^n)^2}, \qquad S_+^n = -4\frac{D_0^{n+1} D_0^n}{(D_1^n)^2}. \tag{2.9}$$

2.2.6 XYZ-model in classical region. The Landau-Lifshits equation.

$$\dot{(\vec{S})} = \vec{S} \times S'' + \vec{S} \times (\vec{J S})$$

$$\vec{S} = (S_1, S_2, S_3), \quad (\vec{S})^2 = 1, \quad J = diag(J_1, J_2, J_3)$$

Under the steriographic projection

$$u = \frac{S_1 + iS_2}{1 + S_3} \qquad v = \frac{S_1 - iS_2}{1 + S_3}$$

and exchanging $-it \to t$ it became a system of the equations:

$$\dot{u} + u'' - 2v\frac{(u')^2 + R(u)}{1 + uv} + \frac{1}{2}\frac{\partial}{\partial u}R(u) = 0$$

$$\tag{2.10}$$

$$-\dot{v} + v'' - 2u\frac{(v')^2 + R(v)}{1 + uv} + \frac{1}{2}\frac{\partial}{\partial v}R(v) = 0$$

where $R(x) = \alpha x^4 + \gamma x^2 + \alpha$ $\frac{\partial R}{\partial x} = 4\alpha x^3 + 2\gamma x = 2\frac{R+\alpha(x^4-1)}{x}$ $\alpha = \frac{J_2 - J_1}{4}$ $\gamma = \frac{J_1 + J_2}{2} - J_3$ The system (2.10) is invariant under transformation $u \to U, v \to V$:

$$U = \frac{1}{v} \qquad \frac{1}{1 + \frac{V}{v}} - \frac{1}{1 + uv} = \frac{vv'' - (v')^2 + \alpha(v^4 - 1)}{(v')^2 + R(v)} \tag{2.11}$$

which is the Backlund transformation for this system. The corresponding solution reader can find in [6],[8].

2.2.7 Lund–Pohlmeyer- Regge model

$$\dot{y}' - 4y + 2(xy)y' = 0,\ \tilde{x} = (\dot{y} + xy^2)^{-1},$$
$$\dot{x}' - 4x - 2(xy)y' = 0,\ \tilde{y} = -(\dot{y} + xy^2) + y^{-1}(\dot{y} + xy^2)^2; \tag{2.12}$$

$$x_0 = 0\,, \qquad \dot{y}_0' = 4y_0.$$
$$x_n = (-1)^{n+1}\frac{D_1^{n-1}}{D_1^n}\,, \qquad y_n = (-1)^n\frac{D_0^{n+1}}{D_0^n}. \tag{2.13}$$

It is interesting to notice that discrete transformation of L–P–R is coinced with nonlinear Schrodinger equation (d). And indeed this two system are belonging to the same hierarchy.

2.2.8 The main chiral field problem in a space of n dimensions.

(the case of an algebra A_1). The main chiral field problem in n-dimensional space is described by the following system of equations:

$$(\theta_i - \theta_j)\frac{\partial^2 f}{\partial x_i \partial x_j} = [\frac{\partial f}{\partial x_i}, \frac{\partial f}{\partial x_j}], \tag{2.14}$$

where the function f takes values in A_1 algebra, θ_i are numerical parameters.

The last system is invariant with respect to transformation [10]

$$F_- = \frac{1}{f_+},$$
$$\frac{\partial F_0}{\partial x_i} = (f_0 - F_0 + \theta_i)\frac{\partial}{\partial x_i}\ln f_+ - \frac{\partial f_0}{\partial x_i},$$
$$\frac{\partial F_+}{\partial x_i} = (f_0 - F_0 + \theta_i)^2\frac{\partial f_+}{\partial x_i} - 2f_+(f_0 - F_0 + \theta_i)\frac{\partial f_0}{\partial x_i} - f_+^2\frac{\partial f_-}{\partial x_i}. \tag{2.15}$$

The last equations can be rewritten in the matrix form

$$\frac{\partial F}{\partial x_i} = \exp\left[-X^+(f_0 - F_0 + \theta_i)f_+\right]\exp[H\ln f_+]\, r\frac{\partial f}{\partial x_i}\, r^{-1}$$
$$\exp\left[-H\ln f_+\right]\exp[X^+(f_0 - F_0 + \theta_i)f_+], \tag{2.16}$$

where r is an automorphism of the algebra A_1 with the properties

$$rX^{\pm}r^{-1} = -X^{\mp},\ rHr^{-1} = -H;$$

$$f_-^0 = 0, \qquad f_0^0 = \tau, \qquad f_+^0 = \alpha^0,$$

where

$$\frac{\partial^2 \tau}{\partial x_i \partial x_j} = 0,\ (\theta_i - \theta_j)\frac{\partial^2 \alpha^0}{\partial x_i \partial x_j} = 2[\frac{\partial \tau}{\partial x_i}\frac{\partial \alpha^0}{\partial x_j} - \frac{\partial \tau}{\partial x_j}\frac{\partial \alpha^0}{\partial x_i}]. \tag{2.17}$$

To solve the Backlund transformation, let us consider the linear system of equations:

$$\theta_i \frac{\partial \alpha^l}{\partial x_i} - 2\frac{\partial \tau}{\partial x_i}\alpha^l = \frac{\partial \alpha^{l+1}}{\partial x_i}. \tag{2.18}$$

From the last equations it follows that each function α^l is a solution of the equation for α^0. We have an explicate expression for α^s:

$$\tau = \sum \phi_i(x_i),\ \alpha^s = \int d\lambda(\lambda)^s c(\lambda) \exp(\sum \frac{\phi_i(x_i)}{\lambda - \theta_i}). \tag{2.19}$$

In terms of the α^l the Backlund transformation have the solution

$$f_0^n = \frac{D_0^{n-1}}{D_0^n},\ f_0^n = \tau - \frac{\dot{D}_0^n}{D_0^n},\ f_+^n = \frac{D_0^{n+1}}{D_0^n}. \tag{2.20}$$

In the determinant $\dot{D}_0^n$ numbering of the indices of the last row is enlarged by unity. About application of the results of this subsection to the problem of construction multisoliton solutions of Sigma–chiral models see [7].

2.2.9 The main chiral field problem for an arbitrary semisimple Lie algebra.

For the case of a semisimple Lie algebra and for an element f being a solution of (12), the following statement takes place [10]: *There exists such an element S taking values in a gauge group that*

$$S^{-1}\frac{\partial S}{\partial x_i} = \frac{1}{f_-}[\frac{\partial f}{\partial x_i}, X_M^+] - \theta_i \frac{\partial}{\partial x_i}\frac{1}{f_-}X_M^+. \tag{2.21}$$

Here X_M^+ is the element of the algebra corresponding to its maximal root divided by its norm, i.e.,

$$[X_M^+, X^-] = H,\ [H, X^\pm] = \pm 2X^\pm,$$

$-f_-$ is the coefficient function in the decomposition of f of the element corresponding to the minimal root of the algebra. In this terms the Backlund transformation is as follows:

$$\frac{\partial F}{\partial x_i} = S\frac{\partial f}{\partial x_i}S^{-1} + \theta_i \frac{\partial S}{\partial x_i}S^{-1}, \tag{2.22}$$

2.2.10 The system of self-dual equations in four-dimensional space. (the case of the algebra A_1).

The self-dual equations for an element f with values in a semisimple Lie algebra have the following form:

$$\frac{\partial^2 f}{\partial y \partial \bar{y}} + \frac{\partial^2 f}{\partial z \partial \bar{z}} = [\frac{\partial f}{\partial y}, \frac{\partial f}{\partial z}]. \tag{2.23}$$

For this system the Backlund transformation has the form [11],[12]

$$F_- = \frac{1}{f_-}$$

$$\begin{aligned}
\frac{\partial}{\partial y}F_0 &= \frac{\partial}{\partial \bar{z}}\ln f_- - \frac{\partial}{\partial y}f_0 + (f_0 - F_0)\frac{\partial}{\partial y}\ln f_-, \\
\frac{\partial}{\partial z}F_0 &= -\frac{\partial}{\partial \bar{y}}\ln f_- - \frac{\partial}{\partial z}f_0 + (f_0 - F_0)\frac{\partial}{\partial z}\ln f_-, \\
\frac{\partial}{\partial y}F_+ &= -f_-\{(f_0 - F_0)\frac{\partial}{\partial y}(f_0 - F_0) + \frac{\partial}{\partial \bar{z}}(f_0 - F_0)\} - f_-^2\frac{\partial}{\partial y}f_+, \\
\frac{\partial}{\partial z}F_+ &= -f_-\{(f_0 - F_0)\frac{\partial}{\partial z}(f_0 - F_0) - \frac{\partial}{\partial \bar{y}}(f_0 - F_0)\} - f_-^2\frac{\partial}{\partial z}f_+ .
\end{aligned} \tag{2.24}$$

The substitution of (2.24) in the density of the topological charge gives the equality:

$$Q_F = q_f + \Box\Box \ln f_-$$

For the integration of the Backlund transformation we have the linear system of equation

$$\frac{\partial \alpha^l}{\partial \bar{y}} + 2\frac{\partial \tau}{\partial z}\alpha^l = \frac{\partial \alpha^{l+1}}{\partial z}, \frac{\partial \alpha^l}{\partial \bar{z}} - 2\frac{\partial \tau}{\partial y}\alpha = -\frac{\partial \alpha^{l+1}}{\partial y} \tag{2.25}$$

In this terms we have the solution of the self-dual system

$$f_-^n = \frac{D_0^{n-1}}{D_0^n}, \ f_0^n = \frac{\dot{D}_0^n}{D_0^n} + \tau, \ f_+^0 = \frac{D_0^{n+1}}{D_0^n}. \tag{2.26}$$

2.2.11 The system of self-dual equations in the case of an arbitrary semisimple algebra.

The following statement takes place [11],[12]:

There exists such an element S taking the values in the gauge group, that

$$S^{-1}\frac{\partial S}{\partial y} = \frac{1}{f_-}[\frac{\partial f}{\partial y}, X_M^+] - \frac{\partial}{\partial \bar{z}}(\frac{1}{f_-})X_M^+,$$
$$S^{-1}\frac{\partial S}{\partial z} = \frac{1}{f_-}[\frac{\partial f}{\partial z}, X_M^+] + \frac{\partial}{\partial \bar{y}}(\frac{1}{f_-})X_M^+. \tag{2.27}$$

Here X_M^+ is the element of the algebra corresponding to its maximal root, divided by its norm, i.e.,

$$[X_M^+, X^-] = H, \ [H, X^{\pm}] = \pm 2X^{\pm},$$

$-f_-$ is the coefficient function in the decomposition of f on the element corresponding to the minimal root of the algebra. The Backlund transformation have the form

$$\frac{\partial F}{\partial y} = S\frac{\partial f}{\partial y}S^{-1} + \frac{\partial S}{\partial \bar{z}}S^{-1}, \qquad \frac{\partial F}{\partial z} = S\frac{\partial f}{\partial z}S^{-1} - \frac{\partial S}{\partial \bar{y}}S^{-1}. \tag{2.28}$$

2.2.12 The main chiral field problem with the moving poles.

Many integrable systems arise from (2.23) by imposing symmetry requirements on its solutions. The cylindrically symmetric condition in four dimensional space restricts the form of the function f,

$$f = \frac{1}{\bar{y}}f(\xi, \bar{\xi}), \qquad \xi = \frac{z - \bar{z}}{2} + [(\frac{z + \bar{z}}{2})^2 + y\bar{y}]^{1/2}, \qquad \bar{\xi} = -\xi^*.$$

$$(\xi-\bar{\xi})\frac{\partial^2}{\partial\xi\partial\bar{\xi}}f=[\frac{\partial}{\partial\xi}f,\frac{\partial}{\partial\bar{\xi}}f]. \tag{2.29}$$

This is the equation for the main chiral field with moving poles.

The result of integration of equation (2.27) is given in the form

$$S=S(\xi,\bar{\xi})\exp X_M^+\frac{z}{f_-};$$

$$S^{-1}\frac{\partial}{\partial\xi}S=\frac{1}{f_-}[\frac{\partial}{\partial\xi}f,X_M^+]-\xi\frac{\partial}{\partial\xi}\frac{1}{f_-}X_M^+,$$
$$S^{-1}\frac{\partial}{\partial\bar{\xi}}S=\frac{1}{f_-}[\frac{\partial}{\partial\bar{\xi}}f,X_M^+]-\bar{\xi}\frac{\partial}{\partial\bar{\xi}}\frac{1}{f_-}X_M^+, \tag{2.30}$$

and the Backlund transformation has the following form:

$$\frac{\partial}{\partial\xi}F=S(\frac{\partial}{\partial\xi}f)S^{-1}-\xi\frac{\partial S}{\partial\xi}S^{-1},\quad \frac{\partial}{\partial\bar{\xi}}F=S(\frac{\partial}{\partial\bar{\xi}}f)S^{-1}-\bar{\xi}\frac{\partial S}{\partial\bar{\xi}}S^{-1}. \tag{2.31}$$

The relations (2.30), (2.31) realize the Backlund transformation for the main chiral field with moving poles.

2.2.13 The self dual equation under condition of cylindrical symmetry in three dimensional space.

The condition of cylindrical symmetry in three dimensional space leads to the following form of the solution:

$$f=\frac{1}{\bar{y}}f(\xi,\bar{\xi}),\qquad \xi=\frac{z+\bar{z}}{2}+i(y\bar{y})^{1/2},\qquad \bar{\xi}=-\xi^*;$$

$$(\xi-\bar{\xi})\frac{\partial^2 f}{\partial\xi\partial\bar{\xi}}=\frac{1}{2}(\frac{\partial f}{\partial\bar{\xi}}-\frac{\partial f}{\partial\xi})+[\frac{\partial f}{\partial\bar{\xi}}\frac{\partial f}{\partial\xi}]. \tag{2.32}$$

The Backlund transformation for equation (2.32) arising from (2.27), (2.28) has the form

$$S^{-1}\frac{\partial}{\partial\xi}S=\frac{1}{f_-}[\frac{\partial}{\partial\xi}f,X_M^+]+(\frac{1}{f_-}-\frac{\xi-\bar{\xi}}{2}\frac{\partial}{\partial\xi}\frac{1}{f_-})X_M^+,$$
$$S^{-1}\frac{\partial}{\partial\bar{\xi}}S=\frac{1}{f_-}[\frac{\partial}{\partial\bar{\xi}}f,X_M^+]+(\frac{1}{f_-}-\frac{\xi-\bar{\xi}}{2}\frac{\partial}{\partial\bar{\xi}}\frac{1}{f_-})X_M^+,$$

$$\frac{\partial}{\partial\xi}F = S(\frac{\partial}{\partial\xi}f)S^{-1} + \frac{\bar{\xi}-\xi}{2}\frac{\partial S}{\partial\xi}S^{-1}, \quad \frac{\partial}{\partial\bar{\xi}}F = S(\frac{\partial}{\partial\bar{\xi}}f)S^{-1} - \frac{\bar{\xi}-\xi}{2}\frac{\partial S}{\partial\bar{\xi}}S^{-1}.$$

In the case of algebra A_1 system (2.32) arises in the integration problem of the general relativity with two commuting Killing vectors [13].

2.2.14 The cylindrical symmetrically solution invariant under two orthogonal four-dimensional axises.

$$x_1\frac{\partial^2 F}{\partial x_1^2} + x_2\frac{\partial^2 F}{\partial x_2^2} = [\frac{\partial F}{\partial x_1}, \frac{\partial F}{\partial x_2}] \tag{2.33}$$

The case of the algebra A_1 the explicate form of the Backlund transformation is the following $F_- = \frac{1}{f_-}$

$$\frac{\partial F_0}{\partial x_1} = -1 + (f_0 - F_0)\frac{\partial ln f_-}{\partial x_1} + x_2\frac{\partial ln f_-}{\partial x_2} - \frac{\partial f_0}{\partial x_1}$$

$$\frac{\partial F_0}{\partial x_2} = 1 + (f_0 - F_0)\frac{\partial ln f_-}{\partial x_2} - x_1\frac{\partial ln f_-}{\partial x_1} - \frac{\partial f_0}{\partial x_2}$$

$$\tag{2.34}$$

$$\frac{\partial F_+}{\partial x_1} = -f_-[(f_0 - F_0)\frac{\partial(f_0 - F_0)}{\partial x_1} + x_2\frac{\partial(f_0 - F_0)}{\partial x_2}] - f_-^2\frac{\partial f_+}{\partial x_1}$$

$$\frac{\partial F_+}{\partial x_2} = -f_-[(f_0 - F_0)\frac{\partial(f_0 - F_0)}{\partial x_2} - x_1\frac{\partial(f_0 - F_0)}{\partial x_1}] - f_-^2\frac{\partial f_+}{\partial x_2}$$

The linear system of the equations has the form

$$x_1\frac{\partial\alpha^l}{\partial x_1} - \alpha^l + 2\alpha^l\frac{\partial\tau^l}{\partial x_2} = \frac{\partial\alpha^{l+1}}{\partial x_2}$$

$$x_2\frac{\partial\alpha^l}{\partial x_2} - \alpha^l - 2\alpha^l\frac{\partial\tau^l}{\partial x_1} = -\frac{\partial\alpha^{l+1}}{\partial x_1} \tag{2.35}$$

$$\tau^{l+1} = \tau^l + \frac{x_1 - x_2}{2}$$

The solution of the Backlund transformation coincide with the self-dual case (see (2.25) and (2.26)).

3 Examles of infinite integrable chains

Let us denote by u_i the solution of the integrable system, which arise after application i-times the Backlund transformation to some given one. The Backlund transformation connect the solutions with different values of the index i and so it arise the infinite chain of equations. In some cases this chain may be limited from "left" or from the "right" and as we have seen in the pervious section this possibility lead to some subclass of the solution of the integrable systems, which depend from the definite number of the arbitrary functions. The other possibility consists in the assuming of the periodical condition of the solution $\theta_{N+i} = \theta_i$, which choice some finite subclass of the solutions of the initial infinite chain. Let consider this possibility on the examples of the nonlinear Schrodinger equations (see (2.4)-(2.7)).

3.0.15 a

In this case we have only one variable for every value of the index i $r_i \equiv \exp x_i$. In this notations the Backlund transformations are rewritten as follows

$$q_{i+1} = \frac{1}{r_i} = \exp -x_i, \quad r_{i+1} = r_i[q_i r_i - (lnr_i)'']$$

or

$$x_i'' + \exp(x_{i+1} - x_i) - \exp(x_i - x_{i-1}) = 0$$

Under the condition $x_{i+N} = x_i$ this is the equations of the periodical Toda lattice for the algebra A_N

3.0.16 b.

$$(\ln x_i')' + \exp(x_{i+1} - x_i) + \exp(x_i - x_{i-1}) = 0$$

3.0.17 c.

In that case $q_{i+1} = \exp x_i$ and for the variable $N_i \equiv r_i$ we receive the chain of the equations

$$N_i' = N_i^2(N_{i-1} - N_{i+1})$$

This is on of possible variant of Lotky-Volterra chain.

3.0.18 d.

Inthis case (after trivial computations) it arised exactly Lotky–Volterra chain in its canonical form

$$N_i' = N_i(N_{i-1} - N_{i+1})$$

In the paper [14] the independent investigation of periodical integrable chains was used for the construction of explicite form of Backlund transformation for the definite class of the integrable systems. The solution of the Backlund transformation under the periodical condition may be connected with the double periodical solutions of such systems.

4 Concluding.

5. The main result of the present paper is in the formulas of the section 3 which realized the Backlund transformation for the integrable system under consideration. A more important consequences is in the assumption of a possibility to obtain the Backlund transformation on the base of the group theory. The Backlund transformation in question is the mapping whose invariants are the equations for integrable system. For this reason an independent construction of the Backlund transformation is equivalent to the problem of enumeration of all integrable systems or some class of them.

References

[1] A.N.Leznov,Parametrical (soliton like) Solutions of Integrable Systems as subclass of solutions depending on a set of arbitrary functions, the talk at "First International A.D.Sakharov Conference on Physic", Moskow (1991), Proceedings of conference World Sceintific, Singapur (1992).

[2] A. N. Leznov, Nonlinear Simmetries of Integrable Systems *J.of Sov.Lazer. Research***3-4**,278-288, (1992)

A.N.Leznov Backlund Transformation for Integrable Systems Preprint IHEP-92-112 DTP,(1992).

[3] A.N.Leznov, A.V.Razumov The Canonical Symmetry for Integrable Systems Preprint IHEP 92–146 DTP, (1992); Journal of Math. Phys.March

(1994). A.N.Leznov, A.V.Razumov The canonical Symmetry and Hamiltonian Formalism. II. Hamiltonian Operatrors Preprint IHEP 93–69 DPT Protvino (1993) (to be published in J.Math.Phys.).

[4] A.N.Leznov,A.B.Shabat and R.I.Yamilov Canonical transformations generated by shifts in nonlinear lattices Phys.Lett.A 174 (1993) 397-402.

[5] P.J.Olver Application of Lie Groups to differential equations (Springer, Berlin, 1986).

[6] B.N.Below, A.N.Leznov and W.J.Zakrzewski Generalization of Toda lattice to the Eleptic Curve Case Preprint IHEP-94-45 Protvino (1994).

[7] A.N.Leznov and W.J.Zakrzewski Soliton solutions of $SU(1,1)$ chiral sigma model Preprint DTP -93–33 Durham (1993); A.N.Leznov and W.J.Zakrzewski Multisoliton solutions for $SU(1,1)$ chiral sigma model with $SU(1,1)$ and $SU(2)$ groups Preprint IHEP 94–73 DTP Protvino (1994).

[8] B.N.Below, A.N.Leznov and W.J.Zakrzewski On the Solutions of the ingomogeneous Heisenberg Ferromagnet, Preprint DTP/93/15, Durham (1993).(to be published in J.Mod.Phys.).

[9] D.B.Fairlie, A.N.Leznov The Integrable Mapping as the Discrete Group of Internal Symmetry of Integrable Systems Preprint DTP/93/33, Durham (1993).

[10] A.N.Leznov Backlund Transformation for Main Chiral Field Problem for arbitrary Semisimple Algebra Preprint IHEP 91–136 DTP Protvino (1991).

[11] A.N.Leznov Backlund Transformation for Self-Dual Yang-Mills Equations in the case of arbitrary Semisimple Gauge Algebra Preprint IHEP 91–137 DTP Protvino (1991).

[12] Ch. Devchand, A.N.Leznov.,Backlund transformation for suppersimmetric self-dual theories for semisimple Gauge Groups and hierarchy of A_1 solutions Preprint IHEP DTP 92-170, (1992); Comm. Math. Phys.(March 1994).

[13] A. N. Leznov The Solution for axial–symmetry configurations in General Reletivity from the point of view of Discrete Symmetry Preprint IHEP 94-122, DTP Protvino (1994) .

[14] A.B.Shabat and R.I.Yamilov Discrete Transformations Generating by the shifts in the chains Leningrad Math.J. Vol.2(1991),No.2.

Some New Integrable Systems with Exact Solutions of the Pattern-Soliton Type

V.G. Makhankov* and V.I. Sanyuk**

**Center for Nonlinear Studies, Los Alamos National Laboratory, P.O.Box 1663, Los Alamos, N.M. 87545, USA, e-mail: mvg@goshawk.LANL.GOV and Joint Institute for Nuclear Research, LCTA, Head Post Office, P.O. Box 79, Dubna, Moscow 101000, Russia*

***Theoretical Physics Department, People's Friendship University of Russia Miklukho-Maklaya Str. 6, 117198 Moscow, Russia, e-mail: sanyuk@udn.msk.su*

Abstract

Some new integrable systems are described admitting localized pattern-like solutions (or soliton-like solutions) constructed of certain universal elements, the solibricks. The solibrics are, in fact, specific eigenfunctions of the linear time-dependent Schrödinger like operators and prove to be universal for a whole class (likely not single) of integrable systems such that any well localized solution can be constructed with the solibricks. A preliminary, obviously incomplete list of these systems is given.

Key words. time-dependent Schrödinger like equations, reflectionless potentials, new nonlinear integrable systems, etc.

AMS subject classifications. 35Q20, 47F05, 47H17, 58F07, 81C05

In this paper we develope a method proposed in [1], [2] and [3], and show that for a broad class of integrable Schrödinger like systems, localized solutions can be constructed in a unified way by means of some universal elements, the solibricks. Probably the most impressive feature of the construction proposed, is that passing from one integrable model to another, inside the definite class, conserves the form of solutions while changing their parameters, which are given by a set of spectral data. In other words, the type of "solution" generates, in a sense, a class of adjusted integrable models. Moreover, as far as the basis element of the construction is also a certain linear time-dependent equation (or system of them), defined by its dispersion relation, the construction can be extended to corroborate many well-known as well as unknown models. Some of the latter are listed later.

We show what the construction is by using as examples two Schrödinger like operators:

$$[i\partial_t - \partial_x^2 + u(x,t)]\psi(x,t,k) \equiv L\psi = 0 \; , \tag{1}$$

$$[i\partial_t - \partial_x^2 + i\hat{u}(x,t)\partial_x]\hat{\psi}(x,t,k) \equiv \hat{L}\hat{\psi} = 0 \ . \tag{2}$$

As was said earlier, it would have been, e.g. Dirac like operator as well [4]. The study consists of two different levels: linear and nonlinear. For the Schrödinger like operators, both have been discussed under certain restrictions adopted from the very beginning [1-3]. Since principal restrictions take effect only on the nonlinear level, and the difference appears *ibid*, below we briefly list the results of the first linear level of consideration, and then give more detailed elaboration of the second.

1 Linear level.

On this level we deal with obtaining localized solutions to eqs. (1) and (2) which are defined by certain **Spectral Data** (SD) consisting of a set of N distinct complex numbers $\kappa_i : \ i = 1, 2, ..., N$ and a complex valued $N \times N$ matrix b_{ij} (kinematic matrix). These will be used in order to solve a peculiar inverse problem.

By exploiting $k-$dependence of a wave-function $\psi(x, t; k)$ we can search for a solution of eq.(1) in the form of the plane wave Ansatz:

$$\psi(x,t,k) = [k^N + a_{N-1}(x,t)k^{N-1} + \cdots + a_0(x,t)]e^{ik(x+kt)} \tag{3}$$

for eq.(1) and

$$\hat{\psi}(x,t,k) = [\hat{a}_N(x,t)k^N + \hat{a}_{N-1}(x,t)k^{N-1} + \cdots + \hat{a}_0(x,t)]e^{ik(x+kt)} \tag{4}$$

for eq.(2).

We assume that wave functions (3) and (4) satisfy the following N conditions

$$\psi(x,t;k=\bar{\kappa}_i) = -\sum_{j=1}^{N} b_{ij}\psi(x,t;k=\kappa_j), \quad i,j = 1,\cdots,N \tag{5}$$

which along with the relations

$$u(x,t) = 2i\partial_x a_{N-1}(x,t) \tag{6}$$

for eq.(1) and

$$\hat{u}(x,t) = i\partial_x ln\hat{a}_N(x,t) \tag{7}$$

for eq.(2) uniquely determine the *w.functions* in both cases (see [1] and [2], respectively). Notice that these conditions, though not general, allow all the same to construct exponentially decaying (w.r.t. x) solutions. More general conditions, given in [1], are of the form

$$\sum_{j=1}^{M}\{b_{ij}^{(0)}\psi(x,t;\kappa_j) + b_{ij}^{(1)}\partial_k\psi(x,t;\kappa_j) + ... + b_{ij}^{(m_j)}\partial_k^{m_j}\psi(x,t;\kappa_j)\} = 0, \quad i = 1...N \tag{8}$$

for M poles and $N = M + \sum_{j=1}^{M} m_j$. This means that each pole gives a $(m_j + 1)$-fold contribution, which is typical for the case of degeneracy. If, for example, we consider the 2-fold degenerate first pole out of the total $N - 1$ poles, then we have

$$b_{ij}^{(m_j)} = 0, \quad for\ m_j \neq 1,\ j \neq 1$$

and

$$b_{ij}^{(1)} = \tilde{b}_i \delta_{1j}, \quad b_{ij}^{(0)} = b_{ij}$$

so that the equations are

$$\sum_{j=1}^{N-1} \{ b_{ij} \psi(x,t;k=\kappa_j) + \tilde{b}_i \delta_{1j} \partial_k \psi(x,t;k=\kappa_j) \} = 0 \quad i = 1...N\ .$$

A more comprehensive study of such a possibility is supposed to be published elsewhere. Here we just emphasize that solutions thereby obtained will contain rational x, t dependence rather than only exponential.

Eqs.(5) and (6) or (5) and (7) in the general case, define complex "*integrable*" potentials. In order to obtain real and nonsingular potentials one has to find some conditions that the poles κ_i and the matrix b_{ij} should obey. For the first equation such conditions were derived in [1], and for the second equation in [3].

To proceed further, it is convenient to consider another representation of the wave functions, the so-called pole-type representation (pole-type normalization) which involves the poles exactly.

In the first case we have

$$\Psi(x,t;k) = \frac{\psi(x,t;k)}{\prod_{i=1}^{N}(k-\kappa_i)} \equiv \left(1 + \sum_{j=1}^{N} \frac{r_j(x,t)}{k-\kappa_j}\right) e^{ik(x+kt)} \tag{9}$$

and in the second

$$\hat{\Psi}(x,t;k) = \frac{\hat{\psi}(x,t;k)}{\prod_{i=1}^{N}(k-\kappa_i)} \equiv \left(\hat{a}_N + \sum_{j=1}^{N} \frac{\hat{r}_j(x,t)}{k-\kappa_j}\right) e^{ik(x+kt)}\ . \tag{10}$$

Now for the first "hatless" case, equations (5) assume the form:

$$\Psi(x,t,\bar{\kappa}_i) = -\sum_{j=1}^{N} c_{ij} \Psi_j(x,t) \quad i = 1...N \tag{11}$$

where

$$\Psi_j = res_{k=\kappa_j} \Psi(x,t,k) = r_j e^{i\kappa_j(x+\kappa_j t)} \tag{12}$$

is a residue of $\Psi(x,t;k)$ at the pole κ_j and

$$c_{ij} = b_{ij} \frac{R'(\kappa_j)}{R(\bar{\kappa}_j)} \tag{13}$$

$$R(k) = \prod_{j=1}^{N}(k-\kappa_j), \quad R' = \frac{d}{dk} R(k)\ . \tag{14}$$

For the second "hatted" case, equations (11)-(14) preserve their form, but now we have $N+1$ unknown functions $\hat{a}_N, \hat{r}_j$ $(j = 1 ... N)$ and only N equations. This freedom can be eliminated by imposing an additional condition. Following [3] we define it at $k = 0$, as follows

$$\hat{\Psi}(x, t; k = 0) = \hat{a}_N - \sum_{j=1}^{N} \frac{\hat{r}_j}{\kappa_j} = 1$$

or

$$\hat{a}_N = 1 + \sum_{j=1}^{N} \frac{\hat{r}_j}{\kappa_j} \tag{15}$$

Thereby we have obtained $N + 1$ equations for $N + 1$ unknown functions, viz. (11) and (15).

Now we are ready to formulate conditions for the potentials u $(\hat{u})$ be real and nonsingular.

In the case of eq.(1), we have [1].

1) The matrix c_{ij} (13) is to be anti-Hermitian: $c = -c^+$.

2) Arrange the poles κ_j such that $Im\kappa_i > 0$ for $i = \overline{1,p}$ and $Im\kappa_j < 0$ for $j = \overline{p+1, N}$ then the Hermitian matrix $(-ic_{ij})$: $i, j = \overline{1,p}$ should be positively definite (not necessary strictly) and the $(-ic_{ij})$: $i, j = \overline{p+1, N}$ negative definite.

The first assertion is nearly staightforward if one consider the meromorphic function

$$\Omega = \overline{\Psi(x, t; \bar{k})}\Psi(x, t; k) \tag{16}$$

and calculate the sum of its residues at $k = \infty$ and $k = \kappa_i$, $k = \bar{\kappa}_j$. By use of (11) one has

$$-\sum_{i=1}^{N} [r_i(x, t) + \bar{r}_i(x, t)] + \sum_{ij} \bar{\Psi}_i \Psi_j (\bar{c}_{ji} + c_{ij}) = 0$$

thereby

$$\sum_{i=1}^{N} r_i(x, t) = -\sum_{i=1}^{N} \bar{r}_i(x, t)$$

and

$$u(x, t) = 2i\partial_x \sum_{i=1}^{N} r_i(x, t) = \bar{u}(x, t)$$

if

$$c_{ij} = -\bar{c}_{ji}$$

QED.

The proof of the second assertion needed for smoothness of u and Ψ is more sophisticated and can be found in [1].

In the case of eq.(2), let us consider the meromorphic function

$$\hat{\Omega} = \frac{1}{k}\overline{\hat{\Psi}(x, t; \bar{k})}\hat{\Psi}(x, t; k) \tag{17}$$

and calculate its residues

$$1-|\hat{a}_N|^2-\sum_{i,j=1}^{N}\overline{\hat{\Psi}_i}\hat{\Psi}_j\left(\frac{\hat{c}_{ij}}{\bar{\kappa}_i}+\frac{\hat{\bar{c}}_{ji}}{\kappa_j}\right)=0$$

whereby bearing in mind eq.(7) we have $\hat{u}$ be real and simultaneously nonsingular if

$$\hat{c}_{ij}=-\frac{\bar{\kappa}_i}{\kappa_j}\hat{\bar{c}}_{ji} \tag{18}$$

We should make here a reservation that if one will give up, for instance, the reality of the potential condition, new integrable systems occur. This point we'll turn to later on.

By considering asymptotic behaviour of the solutions at $x\to\pm\infty$ and $t=const$, or $t\to\pm\infty$ and $x=const$ in [1] it was shown that the potential u asymptotically decays into the direct sum of a few structural units:

$$u(x,t)_{t\to\infty}\to\sum solibricks+\sum breathers+\sum bions+\cdots \tag{19}$$

Those units are **solibricks** (Ψ_i or r_i) the lump-type solutions with $N\geq 1$, and $Re\kappa_i\neq Re\kappa_j$ for $i\neq j$ and $c_{ii}\neq 0$; **breathers** (strings) the periodic or quasiperiodic solutions with ($N\geq 2$) and $Re\kappa_i=Re\kappa_j$ and $c_{ii}\neq 0$; **bions** the periodic or quasiperiodic solutions of the special kind with $N=2$, $Re\kappa_i\neq Re\kappa_j$ for $i\neq j$ and $c_{ii}=0$. The first two are well-known yet from the pioneering work of Zakharov and Shabat [5]. The third type has been described in [2],[3],[6] and yet less known. Finally, one can easily show that the solibrics Ψ_i are the solutions of eq.(1) (see [1]) and $\hat{\Psi}_i$ of eq.(2), that is

$$L\Psi_i=0$$

$$\hat{L}\hat{\Psi}_i=0 \tag{20}$$

in both cases.

2 Nonlinear level. Trivial boundary conditions.

Now we proceed to obtain nonlinear systems for which nonlinear fields can be constructed by means of the solibrics $\Psi_i/\hat{\Psi}_i$ or $r_i/\hat{r}_i$. The play again will be with the residues of some meromorphic functions. As in the previous analysis, the residue at the point $k=\infty$ plays essential role therefore let us first consider expansions in that point of the fuctions $\Omega(x,t;k)$ eq.(16) and $\Psi(x,t;k)$ eq.(9):

$$\Omega=1+\sum_{n=1}^{\infty}I_n(x,t)k^{-n} \tag{21}$$

For real potentials $u = \bar{u}$, $I_1 = 0$.

$$\Psi = \left(1 + \sum_{n=1}^{\infty} \zeta_n(x,t)k^{-n}\right) e^{ik(x+kt)} \tag{22}$$

and

$$\zeta_1 = a_{N-1} = \sum_{i=1}^{N} r_i$$

One can readily find the relations coupling functions I_n and ζ_n. The first four are

$$I_1 = \zeta_1 + \bar{\zeta}_1 = \sum(r_i + \bar{r}_i) = 0, \qquad I_2 = \zeta_2 + \bar{\zeta}_2 - \zeta_1^2,$$

$$I_3 = \zeta_3 + \bar{\zeta}_3 = \zeta_1(\bar{\zeta}_2 - \zeta_2), \quad I_4 = \zeta_4 + \bar{\zeta}_4 + \zeta_1(\bar{\zeta}_3 - \zeta_3) + |\zeta_2|^2$$

Inserting (21) into eq.(1) one has the system of coupled equations:

$$i\dot{\zeta}_n = 2i\zeta'_{n+1} + \zeta''_n - u\zeta_n, \qquad n = 0, 1, \ldots \tag{23}$$

and $\zeta_0 = 1$.

Via playing with this system, when $\zeta_1 = -\bar{\zeta}_1$, it can be found that

$$\bar{\zeta}'_2 + \zeta'_2 - 2\zeta'_1\zeta_1 = \partial_x I_2(x,t) = \frac{1}{2}\partial_x u(x,t) \tag{24}$$

$$\partial_x[\bar{\zeta}_3 + \zeta_3 + \zeta_1(\bar{\zeta}_2 - \zeta_2)] = \partial_x I_3(x,t) = \frac{1}{2}\partial_t u(x,t) \tag{25}$$

and so on (see, e.g. [1]).

The procedure of getting these expressions is close to the one used in the early sixties for obtaining conservation laws in the frame of integrable systems. On the contrary, assuming the potential be purely imaginary, $u = iU$ we have $\zeta_1 = \bar{\zeta}_1$, $c_{ij} = \bar{c}_{ji}$ then

$$U = \partial_x I_1 = 2\zeta'_1 \tag{26}$$

$$\frac{1}{2}\dot{U} + \partial_x(U\int^x U d\xi) = \partial_x^2 I_2 \equiv \partial_x^2(\bar{\zeta}_2 + \zeta_2 + \zeta_1^2) \tag{27}$$

Finally, in the second case, eq.(2) we have

$$\frac{\hat{u}}{2} + C = -\hat{I}_1 \tag{28}$$

$$\hat{u}_t - C\hat{u}_x = -2\partial_x \hat{I}_2 \tag{29}$$

In order to close the system of equations, i.e. to find the coupling between u and Ψ_i we use again our trick with generating meromorphic functions by choosing them in the form that the poles wanted come to play. For instance, in the case of eq.(1) and $u = \bar{u}$, we can consider

$$\Omega_l = E_l(k)\overline{\Psi(x,t,\bar{k})}\Psi(x,t,k) \tag{30}$$

where the 'seed' $E_l(k)$ is one of the polinomials

$$\begin{aligned} E_1 &= k && (31)\\ E_2 &= k^2 + ak && (32)\\ E_3 &= k^3 + bk^2 + dk && (33) \end{aligned}$$

- -

where a,b and d are real constants.

For the first function E_1 (33) the residue at $k = \infty$ is defined by I_2, for E_2 (34) by I_2 and I_3, and for E_3 by I_2, I_3 and I_4. Further possibilities are still not considered. Via calculating the residue sum in the above three cases one obtaines:

$$u = -2F_1(x,t) \tag{34}$$

$$u_t + au_x = -2\partial_x F_2(x,t) \tag{35}$$

$$(\partial_t^2 - \frac{1}{3}\partial_x^4 + \frac{8}{3}d\partial_x^2)u + \frac{8}{3}bu_{xt} + 2(u^2)_{xx} = -\frac{8}{3}\partial_x^2 F_3(x,t) \tag{36}$$

where $F_l(x,t)$ is the following real quadratic form

$$F_l(x,t) = \sum_{ij}^{N} \overline{\Psi}_i E_{ij}^{(l)} \Psi_j \tag{37}$$

with

$$E_{ij}^{(l)} = c_{ij}[E_l(\bar{\kappa}_i) - E_l(\kappa_j)], \qquad l = 1,2,3 \tag{38}$$

is the Hermitian matrix.

For the purely imaginary potential $u = iU$ in the first case we have

$$\begin{aligned} E_1 &= 1 && (39)\\ E_2 &= k + a && (40) \end{aligned}$$

and

$$U = \partial_x F_1(x,t) \tag{41}$$

$$\frac{1}{2}U_t + \partial_x\left(U\int^x U d\xi + aU\right) = \partial_x^2 F_2(x,t) \tag{42}$$

In the second, "hatted" case, eq.(2), we can use

$$\begin{aligned} \hat{E}_1 &= 1 && (43)\\ \hat{E}_2 &= k + a && (44) \end{aligned}$$

to obtain

$$\hat{u} + C = 2\hat{F}_1(x,t) \tag{45}$$

$$\hat{u}_t + a\hat{u}_x = 2\partial_x \hat{F}_2(x,t) \tag{46}$$

with

$$\hat{E}_{ij}^{(l)} = \frac{\hat{c}_{ij}}{\bar{\kappa}_i}[\bar{\kappa}_i \hat{E}_l(\bar{\kappa}_i) - \kappa_j \hat{E}_l(\kappa_j)], \qquad l = 1, 2 \tag{47}$$

Such a search for "integrable" systems can be continued. We, however, put here the point and consider the systems already obtained. Since, as we have seen, the solibricks are the solutions of the linear equations (20), and relations (34)-(38),(41) and (42) give us the coupling between the potential $u(x,t)$ and the nonlinear (quadratic) form $F_l(x,t)$ w.r.t $\Psi_i(x,t)$ we obtain a series, in fact whole classes, of nonlinear integrable systems. Then the concrete form of the matrix $E_{ij}^{(l)}$ specifies a representative from the given class.

Examples

1. Let us consider

$$E_{ij}^{(1)} = \lambda \bar{\gamma}_i \gamma_j, \qquad i, j = \overline{1, N} \tag{48}$$

with $\lambda \in \mathbf{R}$ and $\vec{\gamma} = \{\gamma_1, ..., \gamma_N\}^{tr}$ being a constant complex-valued N-dimensional normalized vector such that

$$\sum_{i-1}^{N} |\gamma_i|^2 = 1 \tag{49}$$

Then the form $F_1(x,t)$ is

$$F_1 = \lambda \sum_{i,j=1}^{N} \bar{\Psi}_i \bar{\gamma}_i \, \Psi_j \gamma_j = \lambda \left| \sum_{j=1}^{N} \gamma_j \Psi_j(x,t) \right|^2$$

Introduce the nonlinear field

$$\varphi = \sqrt{\lambda} \sum_{j=1}^{N} \gamma_j \Psi_j(x,t) \tag{50}$$

and summarizing eqs.(20) over $i = \overline{1, N}$ with the weights γ_i, we have

$$(i\partial_t - \partial_x^2 - 2|\varphi|^2)\varphi = 0 \tag{51}$$

and

$$c_{ij} = \lambda \frac{\bar{\gamma}_i \gamma_j}{\bar{\kappa}_i - \kappa_j} \tag{52}$$

i.e. field (50) obeys the conventional NLSE and describes its N-soliton solution (if all $Im\kappa_i > 0$, $i = \overline{1, N}$ then $\lambda > 0$).

Let us underline that this solution is given as a direct sum over N solibricks.

Similarly N-soliton solutions can be constructed for the case (35) and (36) (for details, see [1]).

The same form of the matrix $E_{ij}^{(1)}$ (48) (more correctly c_{ij}) can be used for treating eq.(41) to give respectively

$$(i\partial_t - \partial_x^2 \pm i\partial_x|\varphi|^2)\varphi = 0 \tag{53}$$

with

$$c_{ij} = \frac{1}{2}E_{ij}$$

where again nonlinear field

$$\varphi = \sum_{i=1}^{N} \gamma_i \Psi_i \tag{54}$$

is a direct sum of the solibricks.

A simple modification of the matrix c_{ij} (at the same E_{ij}) allows one to handle eqs.(42) to obtain

$$(i\partial_t - \partial_x^2 + iU(x,t))\varphi = 0 \tag{55}$$

$$\frac{1}{2}U_t + \partial_x\left(U\int^x U d\xi + aU\right) = \partial_x^2|\varphi|^2 \tag{56}$$

with

$$c_{ij} = \frac{E_{ij}}{\bar{\kappa}_i + \kappa_j} \tag{57}$$

In this case, we can introduce a new field

$$\Phi_x = U \tag{58}$$

and obtain the system

$$(i\partial_t - \partial_x^2 + i\Phi_x)\varphi = 0 \tag{59}$$

$$\frac{1}{2}\Phi_t + \Phi\Phi_x = \partial_x|\varphi|^2 + C \tag{60}$$

with the solution given by eq.(54).

2. If instead of the vector $\vec{\gamma}$, we consider a $n \times N$ matrix γ_{pi} in order to define E_{ij}

$$E_{ij} = \sum_{p=1}^{n} \bar{\gamma}_{pi}\gamma_{pj}\varepsilon_p, \quad i,j = \overline{1,N} \tag{61}$$

then we come to the nonlinear vector field

$$\varphi_p = \sum_{j=1}^{N} \gamma_{pj}\Psi_j, \quad p = \overline{1,n} \tag{62}$$

and the equation for it

$$(i\partial_t - \partial_x^2 + 2 < \vec{\varphi}, \vec{\varphi} >)\varphi_p = 0, \tag{63}$$

with

$$\vec{\varphi} = \{\varphi_1, ..., \varphi_n\}^{tr},$$

$$< \vec{\varphi}, \vec{\varphi} > \equiv < \varphi, \varphi > = \sum_{p=1}^{n} \varepsilon_p |\varphi_p|^2 \tag{64}$$

and c_{ij} given by eq.(57). $\varepsilon_p = \pm 1$ so that the selection of ε_p's defines the type of the field space internal symmetry.

Field (62) can be easily shown to be a solution to the following system:

$$(i\partial_t - \partial_x^2 + u(x,t))\varphi_p = 0, \quad p = \overline{1,n} \tag{65}$$

$$u_t = -2\partial_x < \vec{\varphi}, \vec{\varphi} > \tag{66}$$

if

$$c_{ij} = \frac{E_{ij}}{\bar{\kappa}_i^2 - \kappa_j^2} \tag{67}$$

Then we have

$$(i\partial_t - \partial_x^2 + u(x,t))\varphi_p = 0, \quad p = \overline{1,n} \tag{68}$$

$$(\partial_x^2 - \frac{1}{3}\partial_x^4)u + 2(u^2)_{xx} = -\frac{8}{3}\partial_x^2 < \vec{\varphi}, \vec{\varphi} > \tag{69}$$

when

$$c_{ij} = \frac{E_{ij}}{\bar{\kappa}_i^3 - \kappa_j^3} \tag{70}$$

3. In the similar way, for the imaginary potential, one has

$$(i\partial_t - \partial_x^2 + i\partial_x < \vec{\varphi}, \vec{\varphi} >)\varphi_p = 0, \quad p = \overline{1,n} \tag{71}$$

when

$$c_{ij} = \frac{1}{2}E_{ij}$$

and

$$(i\partial_t - \partial_x^2 + i\Phi_x)\varphi_p = 0, \quad p = \overline{1,n} \tag{72}$$

$$\frac{1}{2}\Phi_t + \Phi\Phi_x = \partial_x < \vec{\varphi}, \vec{\varphi} > +C \tag{73}$$

when c_{ij} is given by eq.(57).

4. The above approach (item 1) is easily extended by considering two or more constant complex-valued N-dimensional normalized vectors: $\vec{\gamma}^{(1)} \equiv \vec{\gamma}, \vec{\gamma}^{(2)} \equiv \vec{\delta}, \ldots$ and

$$\sum_{i=1}^{N} |\gamma_i^{(p)}|^2 = 1 \tag{74}$$

Then we can construct a Hermitian matrix E_{ij} as follows

$$E_{ij} = \bar{\gamma}_i \delta_j + c.c., \qquad i, j = \overline{1, N} \tag{75}$$

or

$$E_{ij} = \sum_{p \neq q = 1}^{n} (\bar{\gamma}_i^{(p)} \gamma_j^{(q)} + c.c.), \qquad i, j = \overline{1, N} \tag{76}$$

The case (75) being a nonsymmetric generalization of the $U(2)$ case yields

$$[i\partial_t - \partial_x^2 + (\bar{\varphi}_1 \varphi_2 + \varphi_1 \bar{\varphi}_2)]\varphi_p = 0, \quad p = 1, 2 \tag{77}$$

$$\varphi_1 = \sum_{i=1}^{N} \gamma_i \Psi_i, \qquad \varphi_2 = \sum_{i=1}^{N} \delta_i \Psi_i \tag{78}$$

with

$$c_{ij} = \frac{\bar{\gamma}_i \delta_j + c.c.}{\bar{\kappa}_i - \kappa_j} \tag{79}$$

Expression (76) gives

$$[i\partial_t - \partial_x^2 + u(x,t)]\varphi_p = 0, \quad p = \overline{1, n} \tag{80}$$

$$u(x,t) = \sum_{p \neq q = 1}^{n} (\bar{\varphi}_p \varphi_q + c.c.) \tag{81}$$

$$\varphi_p = \sum_{i=1}^{N} \gamma_i^{(p)} \Psi_i \tag{82}$$

for formally the same c_{ij} (79).

Going further, one can obtain

$$[i\partial_t - \partial_x^2 + u(x,t)]\varphi_p = 0, \quad p = \overline{1, n} \tag{83}$$

$$u_t = -2\partial_x \sum_{p \neq q = 1}^{n} (\bar{\varphi}_p \varphi_q + c.c.) \tag{84}$$

if

$$c_{ij} = \frac{E_{ij}}{\bar{\kappa}_i^2 - \kappa_j^2} \tag{85}$$

and

$$(\partial_t^2 - \frac{1}{3}\partial_x^4)u + 2(u^2)_{xx} = -\frac{8}{3}\partial_x^2 \sum_{p \neq q = 1}^{n} (\bar{\varphi}_p \varphi_q + c.c.) \tag{86}$$

if

$$c_{ij} = \frac{E_{ij}}{\bar{\kappa}_i^3 - \kappa_j^3} \tag{87}$$

5. For the imaginary potential $u = iU$ we have eqs.(71)-(73) where instead of internal product $< \varphi, \varphi >$ we should substitute expression (81) and (76) instead of (61).

6. Let us notice that the matrix

$$E_{ij} = \sum_{p,q=1}^{n} (\bar{\gamma}_i^{(p)} \gamma_j^{(q)} + c.c.), \qquad i,j = \overline{1,N} \tag{88}$$

generates diagonal terms in the potential

$$u = |\varphi_1|^2 + \ldots + |\varphi_n|^2 + \bar{\varphi}_1\varphi_2 + \bar{\varphi}_1\varphi_3 + \ldots + c.c. \tag{89}$$

Doing the same way, we can treat the hatted case eq.(2) to obtain the folowing integrable systems of equations:

$$(i\partial_t - \partial_x^2 + i\hat{u}(x,t)\partial_x)\hat{\varphi} = 0 \tag{90}$$

with

1.

$$\begin{aligned} \hat{u}(x,t) &= 2|\hat{\varphi}|^2 \\ \hat{\varphi} &= \sum_{i=1}^{N} \hat{\gamma}_i \hat{\Psi}_i \\ \hat{c}_{ij} &= \bar{\kappa}_i \frac{\hat{E}_{ij}}{\bar{\kappa}_i - \kappa_j} \end{aligned} \tag{91}$$

2.

$$\begin{aligned} \hat{u}_t(x,t) &= 2\partial_x|\hat{\varphi}|^2 \\ \hat{\varphi} &= \sum_{i=1}^{N} \hat{\gamma}_i \hat{\Psi}_i \\ \hat{c}_{ij} &= \bar{\kappa}_i \frac{\hat{E}_{ij}}{\bar{\kappa}_i^2 - \kappa_j^2} \end{aligned} \tag{92}$$

3.

$$\begin{aligned} \hat{u}(x,t) &= 2 < \hat{\varphi}, \hat{\varphi} > \\ \hat{\varphi} &= \{\hat{\varphi}_1, \ldots, \hat{\varphi}_n\}^{tr} \end{aligned} \tag{93}$$

$$\hat{\varphi}_p = \sum_{i=1}^{N} \hat{\gamma}_{pi} \hat{\Psi}_i$$
$$\hat{c}_{ij} = \bar{\kappa}_i \frac{\hat{E}_{ij}}{\bar{\kappa}_i - \kappa_j} \equiv \sum_{p=1}^{n} \hat{\bar{\gamma}}_{pi} \hat{\gamma}_{pj} \varepsilon_p$$

with $< \hat{\varphi}, \hat{\varphi} >$ given by (64).

4.

$$\hat{u}_t(x,t) = 2\partial_x < \hat{\varphi}, \hat{\varphi} > \tag{94}$$
$$\hat{\varphi}_p = \sum_{i=1}^{N} \hat{\gamma}_{pi} \hat{\Psi}_i$$
$$\hat{c}_{ij} = \bar{\kappa}_i \frac{\hat{E}_{ij}}{\bar{\kappa}_i^2 - \kappa_j^2}$$

5.

$$\hat{u} = a_1 |\hat{\varphi}_1|^2 + a_2 |\hat{\varphi}_2|^2 + a_3(\hat{\bar{\varphi}}_1 \hat{\varphi}_2 + c.c.) \tag{95}$$
$$\hat{E}_{ij} = \sum_{p,q=1}^{n} (\hat{\bar{\gamma}}_i^{(p)} \hat{\gamma}_j^{(q)} a_{pq} + c.c.)$$
$$\hat{c}_{ij} = \bar{\kappa}_i \frac{\hat{E}_{ij}}{\bar{\kappa}_i - \kappa_j}$$

6.

$$\hat{u}_t = 2\partial_x \{ a_1 |\hat{\varphi}_1|^2 + a_2 |\hat{\varphi}_2|^2 + a_3(\hat{\bar{\varphi}}_1 \hat{\varphi}_2 + c.c.) \} \tag{96}$$
$$\hat{E}_{ij} = \sum_{p,q=1}^{n} (\hat{\bar{\gamma}}_i^{(p)} \hat{\gamma}_j^{(q)} a_{pq} + c.c.)$$
$$\hat{c}_{ij} = \bar{\kappa}_i \frac{\hat{E}_{ij}}{\bar{\kappa}_i^2 - \kappa_j^2}$$

and so on.

3 Nonlinear level. Nontrivial boundary conditions.

Nontrivial bondary conditions

$$|\Phi(x,t)|_{\to\infty} \to const.$$

can be included in the above scheme by adding in the functions E_l, (30)-(33), terms related to poles lying on the real axis, i.e.:

$$E_l(new) = E_l(old) + \sum_{r=1}^{q} \frac{\varepsilon_r b_r^2}{k - k_r}$$

with b_r and k_r being real numbers. More details the reader can find in the works [1],[3].

Here we put the final point by saying that the method originated in [1] and developed in [2],[3] allows one to create exact solutions and adjusted nonlinear equations, so that all such solutions are constructed via linear combinations of the universal elements, the **solibricks**, which are nothing but solutions of given linear systems of equations.

References

1. B.Dubrovin, I.Krichever, T.Malanynk, and V.Makhankov. Exact Solutions of Time–Dependent Schrödinger Equation with Self–Consistent Potentials. *Sov. J. Part. Nuclei,* **19 (3) (1988) 252-269.**

2. V.Makhankov. On Exact Solutions for Some Integrable Nonlinear Equations of the Schrodinger Type, in *NEEDS'92*, eds. V.Makhankov, I.Puzynin and O.Pashaev World Scientific, Singapore 1993, pp. 191-202.

3. V.Makhankov and B.Dubrovin. Time–Dependent Schrodinger-Like Operators and Related Integrable Models. Preprint LA-UR-93-4009, Los Alamos National Lab. Los Alamos 1993, submitted to SIAM Journal on Appl.Math.

4. J.Hronek and V.Makhankov. A Method to Generate Integrable Nonlinear Equations of the Dirac Type and Their N-Soliton Solutions. Preprint 851 Dip. di Fisica Univ. di Roma "La Sapienza", INFN - Sezoine di Roma, 1992.

5. V.Zakharov and A.Shabat. Exact theory of two-dimensional self-focusing and one-dimensional automodulation of waves in nonlinear media. *ZETF* **61 (1971) 118-134.**

6. V.Makhankov and S.Slavov. In *Solitons and Applications*, eds. V.Makhankov, V.Fedyanin and O.Pashaev World Scientific, Singapore 1990, pp.107-113.

VIRASORO CONSTRAINTS AND STRING EQUATIONS FOR THE JACOBIAN SUPER KP HIERARCHY*

Manuel Mañas,† Luis Martínez Alonso, and Elena Medina
Departamento de Física Teórica
Universidad Complutense
E28040 Madrid, Spain

Abstract

With the use of vacuum wave functions for the Jacobian super KP hierarchy we study the super Viraroso constraints in its two sectors, Ramond and Neveu-Schwarz, for such integrable hierarchy.

1 Introduction

The extension of the approach of [4] to the additional symmetries [10] of the Kadomtsev-Petviashvilii (KP) to some of its supersymmetric extensions (SKP), in particular to the the Jacobian SKP hierarchy introduced by Mulase and Rabin (JSKP) [9, 11] will be used two find the super Virasoro constraints for the JSKP. The method relies on the use of vacuum wave functions and the algebra of pseudosuperdifferential operators which leads to characterize a symmetry algebra of additional symmetries. As a consequence, we find centreless super Virasoro algebras of symmetry in their two versions, Neveu-Schwarz (NS) and Ramond (R) types, respectively. Furthermore, systems of constraints determined by super Virasoro generators and h-reductions under bosonic and fermionic flows are provided. Candidates for superstring equations are given.

The reader interested in a more detailed account is refered to [7] where a study of the Manin-Radul SKP is presented as well.

*Partially supported by CICYT proyecto PB92–0197.

†Partially supported by postdoctoral EC Human Capital and Mobility individual fellowship ERBCHBICT930440

2 The Jacobian super KP hierarchy and its additional symmetries

The JSKP hierarchy can be constructed with the aid of an algebra of pseudosuperdifferential operators:

$$\mathcal{E} = \left\{ X = \sum_{n \ll \infty} a_n D^n : a_n \in \mathcal{S} \right\}, \quad D = \frac{\partial}{\partial \theta} + \theta \frac{\partial}{\partial x},$$

where $\mathcal{S}$ is the algebra of superfields defined in terms of formal power series in (x, θ), an even and an odd variable respectively. Any pseudosuperdifferential operator X admits a decomposition $X = X_+ + X_-$ with X_+ being the differential part of X.

The JSKP hierarchy can be written as an infinite system of evolution equations

$$\frac{\partial K}{\partial t_n} = -(K A_n K^{-1})_- K, \quad n = 1, 2, \ldots \tag{1}$$

for an even super Volterra operator $K(x, \theta, t, D) = 1 + \sum_{n=1}^{\infty} w_n(x, \theta, t) D^{-n}$, where $t = \{t_n\}_{n \geq 1}$ includes even and odd variables t_{2n} and t_{2n+1}, respectively. Where $A_{2n} = D^{2n}$ and $A_{2n-1} = (D - \theta D^2) D^{2n-2}$.

The symmetries of system (1) we consider are as follows

$$\frac{\partial K}{\partial s} = -(K B K^{-1})_- K, \tag{2}$$

with homogeneous operators B depending on the supervariables t, and $p(s) = p(B)$. For (2) be a symmetry of the JSKP hierarchy (1) one needs the following supercommutation relation

$$\left[\frac{\partial}{\partial t_n}, \frac{\partial}{\partial s} \right] K = 0, \qquad n = 1, 2, \ldots.$$

Additional symmetries of the type (2) can be characterized from solutions B of $\frac{\partial B}{\partial t_n} = [A_n, B]$, $n = 1, 2, \ldots$.

The JSKP hierarchy has two vacuum wave functions corresponding to the NS and R sectors, satisfying $\frac{\partial f_0}{\partial t_n} = A_n f_0$, $n = 1, 2, \ldots$, which are connected through

$$f_0^{(\mathrm{R})}(k, \alpha, x, \theta, t) := f_0^{(\mathrm{NS})}(k, \alpha k^{-1/2}, x, \theta, t),$$

with

$$f_0^{(\mathrm{NS})}(k, \alpha, x, \theta, t) = \exp \left[kx + \alpha\theta + \sum_{n=1}^{\infty} (k^n t_{2n} + \alpha k^{n-1} t_{2n-1}) \right].$$

Here, k and α are respectively, bosonic and fermionic spectral parameters. A wave function f_0 allows us to stablish a correspondence $b \mapsto B$ between certain classes of superoperators

$b(k,\alpha,\frac{\partial}{\partial k},\frac{\partial}{\partial \alpha})$ and superoperators $B(t,x,\theta,D)$. This correspondence is determined by the condition

$$bf_0 = Bf_0,$$

and satisfies two fundamental properties. Firstly, given two homogeneous operators, say a and b, in (k,α) and their corresponding images, say A and B, one has

$$BAf_0 = Baf_0 = (-1)^{p(a)p(b)}aBf_0 = (-1)^{p(a)p(b)}abf_0.$$

On the other hand,

$$\frac{\partial B}{\partial t_n} = [A_n, B], \quad n = 1,2,\ldots.$$

It means, each operator B corresponding to one operator b determines an additional symmetry of (1). This is connected with a super $\mathcal{W}_\infty$ symmetry of the JSKP, for Manin-Radul see [3].

The domain and the form of the $b \mapsto B$ correspondence are not the same in the NS and the R sectors. Thus, for the NS sector we have

$$k^n \mapsto D^{2n}, \quad \alpha \mapsto B_0^{(\mathrm{NS})}, \quad \frac{\partial}{\partial k} \mapsto B_1^{(\mathrm{NS})}, \quad \frac{\partial}{\partial \alpha} \mapsto B_2^{(\mathrm{NS})}, \tag{3}$$

and for the R sector

$$k^n \mapsto D^{2n}, \quad \alpha k^{-1/2} \mapsto B_0^{(\mathrm{R})}, \quad \frac{\partial}{\partial k} \mapsto B_1^{(\mathrm{R})}, \quad k^{1/2}\frac{\partial}{\partial \alpha} \mapsto B_2^{(\mathrm{R})}. \tag{4}$$

Explicitly,

$$\begin{aligned}
B_0^{(\mathrm{NS})} &= \theta D^2 - D, &(5)\\
B_1^{(\mathrm{NS})} &= x + \sum_{n=1}^{\infty} n(t_{2n} + t_{2n+1}(D - \theta D^2))D^{2n-2}, &(6)\\
B_2^{(\mathrm{NS})} &= \theta + \sum_{n=1}^{\infty} t_{2n-1}D^{2n-2}, &(7)
\end{aligned}$$

and

$$\begin{aligned}
B_1^{(\mathrm{R})} = x - \frac{1}{2}\theta D^{-1} + \\
+ \sum_{n=1}^{\infty}(nt_{2n}D^{2n-2} + (n-3/2)t_{2n-1}(D - \theta D^2)D^{2n-4}).
\end{aligned} \tag{8}$$

Moreover,

$$B_0^{(\mathrm{R})} = B_0^{(\mathrm{NS})}, \quad B_2^{(\mathrm{R})} = B_2^{(\mathrm{NS})}. \tag{9}$$

3 Super Virasoro constraints and superstring equations

The pair (k, α) with $|k| = 1$ can be considered as an element in the supercircle $\widetilde{S^1}$ —$\widetilde{S^1}$ has as supersheaf $\tilde{\mathcal{F}}(\widetilde{S^1})$ generated by adjoining α to the usual angular variable k, [5]—. The algebra $\widetilde{\mathrm{diff} S^1}$ of superderivations in $\widetilde{S^1}$ determines the first-order additional symmetries of the JSKP. An element $\mathcal{X}$ in $\widetilde{\mathrm{diff} S^1}$ can be expressed in the basis $\{\mathcal{D} = \frac{\partial}{\partial \alpha} + \alpha \frac{\partial}{\partial k}, \mathcal{D}^2 = \frac{\partial}{\partial k}\}$ as $\mathcal{X} = (a(k) + \alpha b(k))\mathcal{D}^2 + (c(k) + \alpha d(k))\mathcal{D}$, the functions a and c are periodic in k and b, c are periodic or anti-periodic. The periodic and the anti-periodic cases are associated with the NS and the R vacuum wave functions, respectively. The supersymmetric extension, in its NS or R sectors, of the centreless Virasoro algebra is the subalgebra of superconformal derivations, i.e. those preserving $\mathcal{D}$, [5]. Being the generators

$$
\begin{aligned}
l_n &= k^{-n+1}\frac{\partial}{\partial k} + \frac{1}{2}(1-n)k^{-n}\alpha\frac{\partial}{\partial \alpha}, &\quad (10)\\
g_m &= k^{-m+\frac{1}{2}}\left(\frac{\partial}{\partial \alpha} + \alpha\frac{\partial}{\partial k}\right), &\quad (11)
\end{aligned}
$$

where $n \in \mathbb{Z}$ and $m \in \mathbb{Z} + 1/2$ or $m \in \mathbb{Z}$ depending whether we are considering the NS or the R sectors respectively. The supercommutation relations of these generators are

$$[l_n, l_m] = (n-m)l_{n+m}, \quad [l_n, g_m] = \left(\frac{n}{2} - m\right) g_{n+m}, \quad [g_n, g_m] = 2l_{n+m},$$

each element leads to an additional symmetry of the system (1). The corresponding operators L_n and G_m can be found by application of the $b \mapsto B$ correspondence to (10) and (11) obtaining

$$
\begin{aligned}
L_n^{(\mathrm{NS})} &= B_1^{(\mathrm{NS})} D^{-2n+2} + \frac{1}{2}(1-n) B_2^{(\mathrm{NS})} B_0^{(\mathrm{NS})} D^{-2n}, &\quad (12)\\
G_{n+1/2}^{(\mathrm{NS})} &= (B_2^{(\mathrm{NS})} + B_1^{(\mathrm{NS})} B_0^{(\mathrm{NS})}) D^{-2n}, &\quad (13)\\
L_n^{(\mathrm{R})} &= B_1^{(\mathrm{R})} D^{-2n+2} + \frac{1}{2}(1-n) B_2^{(\mathrm{R})} B_0^{(\mathrm{R})} D^{-2n}, &\quad (14)\\
G_n^{(\mathrm{R})} &= (B_2^{(\mathrm{R})} + B_1^{(\mathrm{R})} B_0^{(\mathrm{R})} D^2) D^{-2n}, &\quad (15)
\end{aligned}
$$

being B_0, B_1, B_2 the superoperators defined in Eqs. (5), (6), (7), (8), (9). In this way we conclude that the superoperators defined in Eqs. (12), (13), (14) and (15) supply a symmetry supersubalgebra of the JSKP hierarchy, which is isomorphic to NS or R supersymmetric extension of the centreless Virasoro algebra.

Self–similarity conditions under particular additional symmetries leads to associated constraints on the solutions of (1). Thus, given the flow corresponding to a symmetry $\frac{\partial K}{\partial s} = -(KBK^{-1})_- K$, the condition for K to be independent on s is $(KBK^{-1})_- = 0$, hence KBK^{-1} is a pure superdifferential operator.

Let us impose a pair of self-similarity conditions associated to a super Virasoro generator $G_{h/2}$ and a bosonic flow A_{2h} of (1), for a given integer $h \geq 1$. According to (13)-(15)

we have $G_{h/2-nh} = G_{h/2}(A_{2h})^n$, so that

$$(KG_{h/2-nh}K^{-1})_- = 0, \quad n = 0, 1, 2, \ldots, \tag{16}$$

and from the supercommutation relations of the super Virasoro algebra it follows also that

$$(KL_{-nh}K^{-1})_- = 0, \quad n = -1, 0, 1, \ldots. \tag{17}$$

It sould be noticed that the above constraints correspond to either the NS (h odd) or the R (h even) versions of the super Virasoro algebra. However, in both cases we may introduce a NS super Virasoro algebra by defining

$$\tilde{G}_n = \frac{1}{\sqrt{h}} G_{nh}, \; n \in \mathbb{Z} + \frac{1}{2}; \; \tilde{L}_n = \frac{1}{h} L_{nh}, \; n \in \mathbb{Z},$$

wich allows us to formulate the above constraints in the form

$$(K\tilde{G}_{1/2-n}K^{-1})_- = 0, \quad n = 0, 1, 2, \ldots, \tag{18}$$

and

$$(K\tilde{L}_{-n}K^{-1})_- = 0, \quad n = -1, 0, 1, \ldots. \tag{19}$$

In other words, the super Virasoro constraints can be understood, independently of h, in terms of a NS super Virasoro algebra.

An interesting aspect of the theory of additional symmetries is the analysis of string equations. In our case, the assumption of symmetry under the super Virasoro generator $G_{h/2}$ and the bosonic flow A_{2h} leads to an inmediate string equation of bosonic type given by

$$[KA_{2h}K^{-1}, Kh^{-1}L_hK^{-1}] = 1.$$

In what concerns string equations of fermionic type, a natural candidate underlies the JSKP hierarchy. It arises from the constrained solutions associated to self-similarity under $G_{h/2}$, A_{2h} and the fermionic flow A_h (A_{h+1}) for h odd (h even). Indeed, in terms of the $b \mapsto B$ correspondence, the generators of fermionic flows for the JSKP hierarchy satisfy

$$-k^n\alpha \mapsto A_{2n+1}, \; \text{(NS)},$$

$$-k^{n-1/2}\alpha \mapsto A_{2n+1}, \; \text{(R)},$$

and, as a consequence we deduce

$$[KG_{1/2+n}K^{-1}, KA_{2n+1}K^{-1}] = K[-k^n\alpha, g_{1/2+n}]K^{-1} = 1.$$

$$[KG_nK^{-1}, KA_{2n+1}K^{-1}] = K[-k^{n-1/2}\alpha, g_n]K^{-1} = 1.$$

That is to say, the pairs $(G_{h/2}, A_h)$ and $(G_{h/2}, A_{h+1})$ determine string equations for h odd and even, respectively. Solutions of the JSKP hierarchy satisfying the constraints

$G_{h/2}$, A_{2h} and A_h (A_{h+1}) for h odd (even) do not depend on $t_{h(1+2n)}, t_{2h(1+n)}, n \geq 0$ ($t_{h(1+2n)+1}, t_{2h(1+n)}, n \geq 0$).

That is to say, the pairs $(G_{h/2}, A_h)$ and $(G_{h/2}, A_{h+1})$ determine string equations for h odd and even, respectively. Solutions of the JSKP hierarchy satisfying the constraints $G_{h/2}$, A_{2h} and A_h (A_{h+1}) for h odd (even) do not depend on $t_{h(1+2n)}, t_{2h(1+n)}, n \geq 0$ ($t_{h(1+2n)+1}, t_{2h(1+n)}, n \geq 0$). An important example that might be described by the above situation for $h = 2$ is the one appearing in the super eigen-value model for 2D quantum supergravity of [1, 2]. Recall that in those papers a set of super-Virasoro constraints was proposed as characterising topological supergravity and the analysis of the corresponding superloop equations suggested the JSKP as the hierarchy underling the theory. The hierarchy is obtained from the JSKP hierarchy under the reduction given by the independence of the solutions upon $t_3, t_4, t_7, t_8, \ldots$ (a supersymmetric extension of the Korteweg-de Vries hierarchy). The algebra of super Virasoro is of R type but the Virasoro constraints maps to one of NS type. (In [6] recursion relations are found for the residues of fractional powers of the two-reduced JSKP —independece upon $t_4, t_8, \ldots$— but no super Virasoro structure is implemented.) When $h = 3$, that might be associated to $\mathcal{W}_3$ quantum supergravity, the reduction impossed to the JSKP hierarchy is the independence upon $t_3, t_6, t_9, t_{12}, \ldots$, being the super Virasoro algebra of NS type (this is a supersymmetric extension of the Boussinesq hierarchy). For a general h we are dealing with a supersymmetric extension of the h-th generalized Korteweg-de Vries hierarchy.

Finally, it should be noticed [7] that in the Manin-Radul SKP case the first super Virasoro constraint $G_{h/2}$ does not lead to any string equation of fermionic type under reductions with respect to the odd flows of the hierarchy.

References

[1] L.Alvarez-Gaumé, H.Itoyama, J.L.Mañes, and A.Zadra, Int.J.Mod.Phys. **A7** (1992) 5337.

[2] L.Alvarez-Gaumé, K.Becker, M.Becker, R.Emparan, and J.Mañes, Int.J.Mod.Phys. **A8** (1993) 2297.

[3] A. Das, E. Sezgi, and S. J. Sin, Phys. Lett. **B277** (1992) 435.

[4] L. A. Dickey, Mod. Phys. Lett. **A8** (1993) 1259.

[5] J. Harnad and B. A. Kupershmidt, Commun. Math. Phys. **132** (1990) 315.

[6] I. N. McArthur, Commun. Math. Phys. **159** (1994) 121.

[7] M. Mañas, L. Martínez Alonso, and E. Medina, *Additional Symmetries, Virasoro Constraints and String Equations for the Super KP hierarchies*, to appear in Phys. Lett. **B**.

[8] Yu I. Manin and A. O. Radul, Commun. Math. Phys. **98** (1985) 65.

[9] M. Mulase, J. Diff. Geom. **34** (1991) 651.

[10] A. Yu. Orlov and E. I. Schulman, Lett. Math. Phys. **12** (1986) 171; P. Grinevich and A. Yu. Orlov, *Virasoro action on Riemann surfaces, Grassmannians,* $\det\bar{\partial}_J$ *and Segal–Wilson* τ*-function* in *Research reports in Physics: Problems of modern quantum field theory* edited by A.Belavin, A.Klimyk, and A.Zalmolodchikov, Springer (1989), Berlin.

[11] J. M. Rabin, Commun. Math. Phys. **137** (1991) 533.

MULTI-DIMENSIONAL DAVEY-STEWARTSON SYSTEMS AND LINEAR FLOWS. EXPLICIT SOLUTIONS

Francisco Guil and Manuel Mañas*
Departamento de Física Teórica, Universidad Complutense
E28040-Madrid, Spain

Abstract

A multi-dimensional Davey-Stewartson system appears when finite-rank constraints are imposed over the right-derivatives of certain automorphisms solving the heat equation. Families of solutions encompasing Wronskian and Grammian determinants type solutions are found. The reduction to the Davey-Stewartson II is considered.

1 Introduction

In this communication we extend the approach presented in [8] to the Kadomtsev-Petviashvilii (KP) hierarchy to the multi-component case, and pay particular attention to a multi-dimensional Davey-Stewartson (DS). One of the main aspects of our construction is the finite-rank constraints impossed to the right-derivatives of some automorphisms of a linear space that solve a multi-dimensional heat equation.

As we shall show in Sec. 2, the rank one constraint is connected to the equations defining the Darboux or rotation coefficients of an n-orthogonal system of curvilinear coordinates. The heat equation implies a multi-dimensional DS system, and we give in Proposition 6 and in Theorem 1 the more general solution obtained with this approach. We extend to this case (Theorem 2) the Grammian type solutions found in [16, 15, 14, 13] for the KP hierarchy, and that, see for example [2], containing Wronskian and Grammian expressions (Theorem 3). The DSII reduction is considered in Sec. 3, where we give a solution of the reduction problem in this scheme. Theorems (4) and (5) provides Grammian and Wronskian-Grammian type expressions for solutions of DSII. This extend the Wronskian expressions found in [5].

For reasons of space all the results are presented without proof. The reader interested will find them in [10], where the DSI is considered as well, recovering the Grammian expressions found in [3, 7, 6] from basic principles and extending those solutions to include Wronskian expressions. In [11] we show how the Grammian expressions for the solutions of DSI fit into inverse scattering approach of [4].

*The second author acknowledges postdoctoral EC Human Capital and Mobility individual fellowship ERBCHBICT930440 and CYCT proyecto PB92-0197

2 A non-trivial reduction of a trivial system. Explicit solutions

In L(V), the set of linear operators of a linear space V, we consider the set of partial differential equations

$$\partial_i M_j + M_j M_i = 0, \tag{2.1}$$

for the n operators $M_i(x_1, \ldots, x_n) \in \mathrm{L}(V)$ depending on the n complex variables x_i. Because $(\partial_i\partial_j - \partial_j\partial_i)M_k = 0$, Eqs. (2.1) are a closed system that can be integrated quite easily.

Observe that

Proposition 1 *A solution $M_i(x_1, \ldots, x_n) \in L(V)$ is of the form $\partial_i\psi \cdot \psi^{-1}$, where*

$$\psi(x_1, \ldots, x_n) = \psi_1(x_1) + \cdots + \psi_n(x_n) \in GL(V). \tag{2.2}$$

To proceed further, we fix a linear decomposition of V, $V = V_1 \oplus \cdots \oplus V_n$, and we require that $\mathrm{im} M_i \subset V_i$. Moreover, if vectors $e_i \in V_i$ are choosen we ask whether the M_i can be of the form

$$M_i = A_i + e_i \otimes \alpha_i, \tag{2.3}$$

where $\alpha_i(x_1, \ldots, x_n) \in V^*$ is a linear functional over V and $A_i \in \mathrm{L}(V_i)$ is a constant linear operator. This is a natural requirement as is shown in [9, 10]. For the functions $p_{ij} := \langle \alpha_i, e_j \rangle$ we have

Proposition 2 *If the functions M_i, as given by Eq. (2.3) are a solution of (2.1) then*

$$\partial_k p_{ij} + p_{ik}p_{kj} = 0, \; i, j \neq k. \tag{2.4}$$

Thus, functions p_{ij} satisfy the equations defining the Darboux or rotation coefficients for an n-orthogonal system of curvilinear coordinates.

Notice that for $i = j$ the relations (2.4) implies $\partial_j p_{ii} - \partial_i p_{jj} = 0$, from where it follows the existence of a potential U such that

$$p_{ii} = \partial_i U.$$

Now, we shall consider one-parameter deformations of the p_{ij} obeying Eqs. (2.4) associated to the linear flow

$$\partial_t \psi = \Delta \psi, \tag{2.5}$$

where $\Delta = \sum_{i=1}^{n} \partial_i^2$. For the right-derivatives M_i one has $\partial_t M_i = \Delta M_i + \sum_j (\partial_j M_i) M_j$, hence p_{ij} with $i \neq j$ and the potential U satisfy

$$\partial_k p_{ij} + p_{ik} p_{kj} = 0, \qquad k \neq i, j; i \neq j, \tag{2.6}$$

$$\partial_i \partial_j U + p_{ij} p_{ji} = 0, \qquad i \neq j, \tag{2.7}$$

$$\partial_t p_{ij} - (\partial_i^2 - \partial_j^2) p_{ij} - 2 p_{ij} (\partial_i^2 - \partial_j^2) U - \sum_{k \neq i,j} (\partial_k p_{ik} \, p_{kj} - p_{ik} \partial_k p_{kj}) = 0, \ i \neq j. \tag{2.8}$$

The Eqs. (2.6), (2.7) and (2.6) can be understood as an integrable deformation of a system of curvilinear coordinates. This interpretation was pointed out for the Davey-Stewartson system in [12] in the case of two coordinates giving a geometrical meaning to the Baker function.

In order to find the general solution ψ to Eq. (2.5), having its right-derivatives as in (2.3), it is particulary useful to define new functionals $\beta_i = \alpha_i \psi$ in terms of which we obtain

Proposition 3 *Equations (2.5) and (2.3) are equivalent to*

$$\partial_i \psi = A_i \psi + e_i \otimes \beta_i, \ i = 1, \ldots$$
$$\partial_t \psi = \Delta \psi$$

where $\beta_i(x_i, t) \in V^*$ *satisfies*

$$\partial_t \beta_i(x_i, t) = \partial_i^2 \beta_i(x_i, t).$$

Now, define $\psi_0(x_1, \ldots, x_n, t) := \prod_{i=1}^{n} \exp[A_i x_i + A_i^2 t]$, and $b_i := \psi_0^{-1} e_i \in V_i$.

Proposition 4 *The solution to Eqs. (2.5) and (2.3) is given by*

$$\psi = \psi_0 \cdot \varphi,$$

with

$$\varphi(x_1, \ldots, x_n, t) = C + \sum_{i=1}^{n} \partial_i^{-1}(b_i(x_i, t) \otimes \beta_i(x_i, t)).$$

Here ∂_i^{-1} *is a primitive commuting with* ∂_i, *and* $C \in L(V)$.

We define $\xi_{ij} := \psi + e_j \otimes \beta_i - e_j \otimes \eta_j \psi$ and $\zeta_{ij} := \varphi + b_j \otimes \beta_i - b_j \otimes \delta_j \varphi$, where $\eta_j, \delta_j(x_1, \ldots, x_n, t) \in V^*$ are such that $\langle \eta_j, e_j \rangle = \langle \delta_j, b_j \rangle = 1$.

Now, we can state the main result of this paper.

Theorem 1 *If* φ *has a determinant, then*

$$U = \ln \det \varphi$$
$$p_{ij} = \langle \beta_i, \varphi^{-1} b_j \rangle = \frac{\det \xi_{ij}}{\det \psi} = \frac{\det \zeta_{ij}}{\det \varphi}$$

is a solution of Eqs. (2.6), (2.7) and (2.8).

Particular choices of V, A_i, e_i and β_i give different sets of solutions. For example, one can construct Grammian type expressions as follows.

Theorem 2 *If W_i is a linear space, we let $\mathcal{S}_i(x_i,t) \in W_i$, $i = 1,\dots,n$, be a solution of*

$$\partial_t \mathcal{S}_i = -\partial_i^2 \mathcal{S}_i,$$

define $\varsigma_i(x_i,t) \in (\oplus_{i=1}^n W_i)^*$ such that $\langle \varsigma_i, \mathcal{S}_i \rangle = 1$ and $\Upsilon_i(x_i,t) \in (\oplus_{i=1}^n W_i)^*$ is a solution of

$$\partial_t \Upsilon_i = \partial_i^2 \Upsilon_i.$$

For any constant $C \in L(\oplus_{i=1}^n W_i)$ we define

$$\Phi := C + \sum_{i=1}^n \partial_i^{-1}(\mathcal{S}_i \otimes \Upsilon_i),$$

and

$$\Phi_{ij} := \Phi + \mathcal{S}_j \otimes (\Upsilon_i - \varsigma_j \Phi).$$

Then,

$$U = \ln \det \Phi$$

$$p_{ij} = \frac{\det \Phi_{ij}}{\det \Phi},$$

is a solution of Eqs. (2.6), (2.7) and (2.8).

When W_i are finite-dimensional, for fixed bases $\mathcal{B}_i$ in W_i, $i = 1,\dots,n$, we can complete the preceding theorem as follows. Let $(\mathcal{S}_i)_m, m = 0,\dots,N_i - 1$, $N_i = \dim W_i$, be the components of $\mathcal{S}_i$ in the basis $\mathcal{B}_i$. Define $\Phi_{ij}^m, m = 0,\dots,N_j - 1$ as the matrix obtained from that corresponding to Φ by replacing the m-th row corresponding to the W_j-sector by Υ_i, then

Proposition 5 *For the p_{ij} of Theorem 2 one has*

$$p_{ij} = \frac{\sum_{m=0}^{N_j-1} (\mathcal{S}_j)_m \det \Phi_{ij}^m}{\det \Phi}.$$

To include in the formulas stated previously Wronskian type expressions we need the following observation.

Proposition 6 *The function ψ can be expressed as*

$$\psi = \sum_{i=1}^n \left[\sum_{m=0}^{M_i-1} e_i^m \otimes \partial_i^m \gamma_i + \psi_0 \left(C_i + \partial_i^{-1}(g_i \otimes \gamma_i)\right) \right],$$

where

$$e_i^m = A_i^{M_i-1-m} e_i, \; i = 1,\dots,n,$$

and

$$\beta_i = \partial_i^{M_i} \gamma_i, \quad g_i = A_i^{M_i} b_i = (-\partial_i)^{M_i} b_i.$$

With this proposition at hand we are able to formulate the following.

Theorem 3 *Let $\mathcal{S}_i(x_i, t)$ be as in Theorem 2 and $\overline{W}_i$, $i = 1, \ldots, n$, an M_i-dimensional space with basis $\{e_i^m\}_{m=0}^{M_i-1}$. Consider $\Upsilon_i(x_i, t) \in (\oplus_{i=1}^n(\overline{W}_i \oplus W_i))^*$ as in Theorem 2 and introduce the map*

$$\mathcal{W} : \oplus_{i=1}^n(\overline{W}_i \oplus W_i) \to \oplus_{i=1}^n \overline{W}_i$$

given by

$$\mathcal{W} := \sum_{\substack{m=0,\ldots,M_i-1\\ i=1,\ldots,n}} e_i^m \otimes \partial_i^m \Upsilon_i.$$

Now, we define

$$\Phi := \mathcal{W} + C + \sum_{i=1}^n \partial_i^{-1}(\mathcal{S}_i \otimes \Upsilon_i),$$

and

$$\Phi_{ij} := \Phi - e_j^{M_j-1} \otimes \left(\partial_j^{M_j-1}\Upsilon_j - \partial_i^{M_i}\Upsilon_i\right).$$

Then,

$$U = \ln \det \Phi$$
$$p_{ij} = \frac{\det \Phi_{ij}}{\det \Phi},$$

is a solution of Eqs. (2.6), (2.7) and (2.8).

3 The reduction to DSII

In this section we take the case of two coordinates, $n = 2$, and modify our symmetry to $\mathrm{i}\partial_t\psi = (\partial_1^2 - \partial_2^2)\psi$ so that Eqs. (2.6) dissapear and Eqs. (2.7) and (2.8) reads

$$\begin{aligned}\partial_1\partial_2 U + p_{12}p_{21} &= 0,\\ \mathrm{i}\partial_t p_{12} - \Delta p_{12} - 2p_{12}\Delta U &= 0,\\ \mathrm{i}\partial_t p_{21} + \Delta p_{21} + 2p_{21}\Delta U &= 0.\end{aligned}$$

Theorem 1 remains true if β_i is a solution of

$$\mathrm{i}\partial_t\beta_1 = \partial_1^2\beta_1, \quad -\mathrm{i}\partial_t\beta_2 = \partial_2^2\beta_2. \tag{3.1}$$

Now, Theorems 2 and 3 still hold if $\mathcal{S}_i$ and Υ_i, $i = 1, 2$, are solutions of

$$\begin{aligned}\mathrm{i}\partial_t\mathcal{S}_1 &= -\partial_1^2\mathcal{S}_1, \ \mathrm{i}\partial_t\Upsilon_1 = \partial_1^2\Upsilon_1\\ \mathrm{i}\partial_t\mathcal{S}_2 &= \partial_2^2\mathcal{S}_2, \ \mathrm{i}\partial_t\Upsilon_2 = -\partial_2^2\Upsilon_2.\end{aligned}$$

For the DSII case we take $x_1^* = x_2$, $\nabla U = (U_x, U_y) \in \mathbb{R}^2$ where $x_1 = z = (x + \mathrm{i}y)/\sqrt{2}$, $x, y \in \mathbb{R}$, and $\varepsilon p_{21}^* = p_{12} = p$, $\varepsilon = -1, 1$, being the resulting equations

$$\Delta U + 2\varepsilon |p|^2 = 0, \tag{3.2}$$
$$\mathrm{i}\partial_t p - (\partial_x^2 p - \partial_y^2 p) - 2p(\partial_x^2 U - \partial_y^2 U) = 0. \tag{3.3}$$

The reduction can be realized once we identify

$$V_1 \cong V_2 \cong V.$$

In $V \oplus V$ one has the ε-permutation operator

$$\begin{aligned} P : & V \oplus V \to V \oplus V \\ & v_1 \oplus v_2 \mapsto P(v_1 \oplus v_2) = v_2 \oplus \varepsilon v_1. \end{aligned}$$

Now, we can formulate

Proposition 7 *If*

$$\begin{aligned} e_2^* &= P e_1 \\ A_2^* &= \varepsilon P A_1 P \\ \beta_1^* &= \beta_2 T \end{aligned}$$

where $T \in SL(V \oplus V)$ *satisfies* $T^{-1} = \varepsilon T^*$, *and*

$$C_2 = P C_1^* T^*,$$

then

$$\begin{aligned} p_{21} &= \varepsilon p_{12}^* \\ \nabla U&(x, y, t) \in \mathbb{R}. \end{aligned}$$

Using Theorem 2 we can give examples of solutions of the DSII equation. First, those of Grammian type.

Theorem 4 *Let* $T \in SL(W_1 \oplus W_2)$ *be a linear operator over the complex space* $W_1 \oplus W_2$ *with* $W_i \cong W$, $i = 1, 2$, *such that* $T^{-1} = \varepsilon T^*$. *Denote by* $\mathcal{S}(z, t)$ *be a* W_1*-valued solution of*

$$-i\partial_t \mathcal{S} = \partial_z^2 \mathcal{S},$$

and let $\Upsilon(z, t) \in (W \oplus W)^*$ *be a solution of*

$$i\partial_t \Upsilon = \partial_z^2 \Upsilon. \tag{3.4}$$

Take $C \in L(\mathcal{W} \oplus \mathcal{W}, W_1)$ *and define*

$$\Phi := \phi + P\phi^* T^*$$

with

$$\phi := C + \partial_z^{-1}(\mathcal{S} \otimes \Upsilon),$$

and

$$\hat{\Phi} := \Phi + P\mathcal{S}^* \otimes (\Upsilon - \varsigma\Phi)$$

with $\varsigma(z, z^*, t) \in (\mathcal{W} \oplus \mathcal{W})^*$ *such that* $\langle \varsigma, P\mathcal{S}^* \rangle = 1$.

Then,

$$U = \ln \det \Phi$$

$$p = \frac{\det \hat{\Phi}}{\det \Phi}$$

is a solution of Eqs. (3.2) and (3.3).

When $\mathcal{W}$ is finite dimensional an alternative expression for p is

Proposition 8 *If, in a given basis of* $\mathcal{W}$, $\mathcal{S}$ *has as components* $\mathcal{S}_n$ *and* Φ_n *has as associated matrix that obtained from the one corresponding to* Φ *by replacing the* $(N + n)$*-th row* $(N = \dim \mathcal{W})$ *by* Υ, *then* p *in Theorem 4 can be expressed as*

$$p = \varepsilon \frac{\sum_n \mathcal{S}_n^* \det \Phi_n}{\det \Phi}.$$

Finally, we introduce Grammian-Wronksian expressions.

Theorem 5 *Take* $T \in SL(\overline{W}_1 \oplus W_1 \oplus \overline{W}_2 \oplus W_2)$, *such that* $T^{-1} = \varepsilon T^*$, *where* $\overline{W}_i \cong \overline{\mathcal{W}}$ *and* W_i *are as in Theorem 4. Choose* $\mathcal{S}$ *and* C *as in Theorem 4 and* $\Upsilon(z,t) \in (\overline{\mathcal{W}} \oplus \mathcal{W} \oplus \overline{\mathcal{W}} \oplus \mathcal{W})^*$ *a solution of Eq. (3.4). Given a basis* $\{e_n\}_{n=1}^M$ *of* $\overline{W}_1$ *we define*

$$\mathcal{W} := \sum_n e_n \otimes \partial_z^{n-1}\Upsilon,$$

$$\phi := \mathcal{W} + C + \partial_z^{-1}(\mathcal{S} \otimes \Upsilon),$$

and

$$\Phi := \phi + P\phi^* T^*$$

$$\hat{\Phi} := \Phi - e_M \otimes (\partial_z^{M-1}\Upsilon - (\partial_z^M \Upsilon)^*).$$

Then,

$$U = \ln \det \Phi$$

$$p = \frac{\det \hat{\Phi}}{\det \Phi}$$

is a solution of Eqs. (3.2) and (3.3).

The Wronskian expressions for solutions of DSII with $\varepsilon = 1$ calculated in [5] are contained in Theorem 5 when $\dim W = 0$ and $T = 1$.

The connection between the solutions presented in [1] will be analysed elsewhere.

References

[1] V. L. Arkadiev, A. K. Pogrebkov, and M. C. Polivanov, Inv. Prob. **5** (1989) L1, Physica **D36** (1989) 189.

[2] L. -L. Chau, J. C. Shaw, and H. C. Yen, Commun. Math. Phys. **149** (1992) 279.

[3] A. Degasperis, *The Davey-Stewartson I equation: A class of explicit solutions including as a special case the dromions* (1990).

[4] A. S. Fokas and P. M. Santini, Phys. Rev. Lett. **63** (1989) 1329, Physica **D44** (1990) 99; P. M. Santini, Physica **D41** (1990) 26.

[5] N. C. Freeman, C. R. Gilson, and J. J. Nimmo, J. Phys. A: Math. Gen. **23** (1990) 4793.

[6] C. R. Gilson, Phys. Lett. **A161** (1992) 423.

[7] C. R. Gilson and J. J. C. Nimmo, Proc. R. Soc. Lond. A **435** (1991) 339.

[8] F. Guil, M. Mañas, and G. Álvarez, Phys. Lett. **A190** (1994) 49.

[9] F. Guil and M. Mañas, *Constrained Linear Flows and the Davey-Stewarson Equations*, to appear in Physica **D**.

[10] F. Guil and M. Mañas, *Constrained Linear Flows and the Davey-Stewarson I and II Equations*, to appear.

[11] F. Guil and M. Mañas, *On the connection between Grammian determinant type expressions and inverse scattering of the Davey-Stewartson I Equation* to appear.

[12] B. Kopenelchenko, Phys. Lett. **A183** (1993) 153.

[13] S. Mikae, Y. Otha, and J. Satsuma, J. Phys. Soc. Jpn **59** (1990) 48.

[14] A. Nakamura, J. Phys. Soc. Jpn **58** (1989) 412.

[15] J. J. C. Nimmo, Inv. Prob. **8** (1992) 219.

[16] V. E. Zakharov and A. B. Shabat, Func. Anal. Appl. **8** (1974) 43.

MACROSCOPIC BEHAVIOR IN THE ABLOWITZ-LADIK EQUATIONS

PETER D. MILLER
*Program in Applied Mathematics, University of Arizona,
Tucson, AZ 85721, USA*

ABSTRACT

Modulation theory is used to study the Ablowitz-Ladik equations. Exact multiphase wavetrain solutions are found, and local conservation laws are averaged to obtain a macroscopic description of this lattice problem.

1. Transport in Nonlinear Lattices

Physical systems like organic polymers and waveguide arrays that are necessarily spatially discrete and yet translationally symmetric are *lattices* that may be modeled by an infinite number of coupled ODEs, indexed by integers n, that are invariant under the substitution $n \mapsto n+1$. One interesting question to ask about such systems is: to what degree is it possible for the *microscopic dynamics* (as described by the coupled ODEs) to contribute to collective phenomena on scales much larger than the lattice spacing? Such collective phenomena that can be described by closed sets of effective equations are called *macroscopic dynamics*. The hope is that robust macroscopic dynamics can be understood as a mechanism for coherent energy transport over very long scales. One way to seek macroscopic dynamics is to assume that all dynamical variables vary slowly along the lattice; this is the usual idea of the continuum limit. However, such a limit ignores the possibility that essential effects of the spatial discreteness can contribute to the macroscopic dynamics. We want to explore this possibility as well.

As a mathematical model of a lattice, we consider here the system of ODEs

$$i\partial_t A_n + f(|A_n|^2)(A_{n+1} + A_{n-1}) + F(|A_n|^2)A_n = 0\,. \tag{1}$$

The complex numbers $A_n(t)$ can be considered to be *complex mode amplitudes* of the form $q_n + ip_n$ for a system of oscillators with displacements q_n and momenta p_n. By choosing the real-valued functions $f(\rho)$ and $F(\rho)$ appropriately, it is possible to obtain a variety of effects. Nonconstant terms in the function F give rise to anharmonicity in the individual oscillators, and nonconstant terms in the function f introduce anharmonic coupling between these oscillators. For example, choosing $f(\rho) = 1$ and $F(\rho) = \pm 2\rho - 2$ results in the discrete nonlinear Schrödinger system (DNLS)

$$i\partial_t A_n + (A_{n+1} - 2A_n + A_{n-1}) \pm 2|A_n|^2 A_n = 0\,. \tag{2}$$

On the microscopic level, the DNLS is a description of a lattice of anharmonic oscillators, each linearly coupled to its two nearest neighbors. On the other hand, choosing $F(\rho) = -2$ and $f(\rho) = 1 \pm \rho$ results in the Ablowitz-Ladik system (AL)

$$i\partial_t A_n + (A_{n+1} - 2A_n + A_{n-1}) \pm |A_n|^2(A_{n+1} + A_{n-1}) = 0\,. \tag{3}$$

In contrast with the DNLS, AL describes on the microscopic level a lattice of harmonic oscillators, each coupled to its two nearest neighbors by nonlinear interactions. In spite of the differences on the microscopic level, these two systems of differential equations can both be scaled to yield NLS in the continuum limit. However, the fact that the two agree in this limit is no reason to believe that there should be any correspondence for general initial data.

The system of equations Eq. (1) is Hamiltonian. The conservation of energy can be written in local form, where the time derivative of the energy density is equal to a pure difference in n of an energy flux. There is also an additional local conservation law that follows from the gauge symmetry group $A_n \mapsto \exp(i\alpha)A_n$ for real α. There are many more local conservation laws for the special case of AL, which will allow us to obtain a very complete picture of macroscopic dynamics for this system.

2. Modulation Equations for Harmonic Plane Waves

As an initial example of the kind of transport mechanisms we have in mind, and to provide a demonstration of the way that spatial discreteness can contribute to macroscopic phenomena, we briefly describe the behavior of slowly varying harmonic wavetrains in the lattice system Eq. (1). The results of this section first appeared in the work of Hays, Levermore, and Miller.[5]

The lattice system Eq. (1) has exact solutions of the simple form

$$A_n(t) = \sqrt{\rho}\exp(i(kn - \omega t)), \tag{4}$$

for constants ρ, k, and ω satisfying the dispersion relation

$$\omega + 2f(\rho)\cos k + F(\rho) = 0. \tag{5}$$

We now consider approximate solutions, for which ρ and k are taken to be slowly varying, ω being determined locally from the dispersion relation Eq. (5). To obtain dynamical equations for the quantities ρ and k, we consider the solution Ansatz Eq. (4) with ρ, k, and ω depending on n and t, insert the Ansatz into the two local conservation laws for Eq. (1), and pass to the continuum limit, not in the field $A_n(t)$, but in the quantities ρ and k. This procedure yields two quasilinear PDEs for the quantities $\rho(X,T)$ and $k(X,T)$ on the scales $X = \varepsilon n$ and $T = \varepsilon t$ as $\varepsilon \downarrow 0$. These PDEs are called the *modulation equations* for the harmonic plane waves given by Eq. (4). If the characteristic speeds for these PDEs are real and distinct, the PDEs are *strictly hyperbolic*, and the corresponding Cauchy problem at $T = 0$ is well-posed, which is interpreted as indicating modulational stability of the underlying wavetrain. Otherwise, the PDEs are *elliptic*, and the Cauchy problem at $T = 0$ is ill-posed, which is interpreted as indicating modulational instability of the wavetrain. As a result of nonlinearity, the characteristic speeds depend on the values of ρ and k. Thus, it is possible for the modulation equations, posed for globally hyperbolic initial data, to evolve the fields $\rho(X,T)$ and $k(X,T)$ forward in T in such a way that the speeds cease to be real for some X and $T > 0$. We call this phenomenon *dynamic change of type*. It is also possible for $\partial_X \rho$ and $\partial_X k$ to become infinite for some X and $T > 0$, even though the speeds are real. Both of these catastrophes are interpreted as the local breakdown of the modulational Ansatz, Eq. (4).

These macroscopic limits, unlike the usual continuum limit, are able to distinguish qualitatively between different discretizations of the same equation. Figure 2 summarizes the results of the analysis described for the two discretizations of NLS discussed above, and gives the analogous results for harmonic plane wave modulation in NLS for comparison.

	DNLS	AL	NLS
Microscopic Equation	$i\partial_t A_n + (A_{n+1} - 2A_n + A_{n-1}) \pm 2\lvert A_n\rvert^2 A_n = 0$	$i\partial_t A_n + (A_{n+1} - 2A_n + A_{n-1}) \pm \lvert A_n\rvert^2 (A_{n+1} + A_{n-1}) = 0$	$i\partial_t A + \partial_x^2 A \pm 2\lvert A\rvert^2 A = 0$
Modulation Equations	$\partial_T \rho = \partial_X[2\rho \sin k]$ $\partial_T k = \partial_X[\mp 2\rho - 2\cos k]$	$\partial_T \rho = (1 \pm \rho)\partial_X[2\rho \sin k]$ $\partial_T k = \partial_X[-2(1 \pm \rho)\cos k]$	$\partial_T \rho = \partial_X[2\rho k]$ $\partial_T k = \partial_X[\mp 2\rho - k^2]$
Stability Criterion	$+$: $\cos k < 0$ $-$: $\cos k > 0$	$+$: No Stable Waves $-$: $\rho < 1$	$+$: No Stable Waves $-$: All Waves Stable
Dynamic Change of Type?	Sometimes	Never	Never

Fig. 1. Harmonic plane wave modulation for two discretizations of NLS.

One may ask whether the analysis described above can be generalized. Is is possible to replace the exact solutions, Eq. (4), with a more complicated family of exact solutions, and obtain modulation equations describing the slow evolution of the parameters that index these solutions? If one is content to work with the Ablowitz-Ladik equations, then the answer to this question is in the affirmative.

3. Wavetrains in AL

The Ablowitz-Ladik equations, Eq. (3), deserve special treatment in the study of macroscopic limits of Eq. (1) because in this case there exists an inverse scattering transform (IST) that can be used to linearize Eq. (3) to obtain exact solutions and also to identify an inexhaustible number of local conservation laws. Eq. (3) is the reduction of

$$
\begin{aligned}
-i\partial_t Q(n,t) &- \Big[Q(n+1,t) - 2Q(n,t) + Q(n-1,t)\Big] \\
&+ Q(n,t)R(n,t)\Big[Q(n+1,t) + Q(n-1,t)\Big] = 0\,, \\
-i\partial_t R(n,t) &+ \Big[R(n+1,t) - 2R(n,t) + R(n-1,t)\Big] \\
&- Q(n,t)R(n,t)\Big[R(n+1,t) + R(n-1,t)\Big] = 0\,,
\end{aligned}
\tag{6}
$$

where n has been promoted from a subscript to an argument, when $R = \mp\overline{Q}$ for $t \in \mathbb{R}$. These ODEs were found by Ablowitz and Ladik,[1] who were looking for lattice systems that could be solved with an IST. We refer to Eq. (6) as the complex Ablowitz-Ladik equations (CAL).

The goal of the rest of this paper is to describe the use of the machinery of the IST to find classes of exact wavetrain solutions that will play the role of generalizations of the plane wave Ansatz, Eq. (4), and to provide tools for describing the modulational behavior of these solutions. As we are only interested in producing functions that solve CAL and not in solving an initial value problem, we will use this machinery in a way that differs in appearance and structure, though certainly not in spirit, from its traditional usage in solving initial value problems. The details not mentioned here can be found in the paper of Miller, Ercolani, Krichever, and Levermore.[6]

We use the following Lax pair representation of CAL:

$$\mathbf{u}(n+1,t,z) = \mathbf{L}(n,t,z)\mathbf{u}(n,t,z)\,, \tag{7}$$

where

$$\mathbf{L}(n,t,z) = \begin{bmatrix} z & Q(n,t) \\ zR(n,t) & 1 \end{bmatrix}, \tag{8}$$

and

$$-i\partial_t \mathbf{u}(n,t,z) = \mathbf{B}(n,t,z)\mathbf{u}(n,t,z)\,, \tag{9}$$

where

$$\mathbf{B}(n,t,z) = \begin{bmatrix} z-1-Q(n,t)R(n-1,t) & Q(n,t)-z^{-1}Q(n-1,t) \\ zR(n-1,t)-R(n,t) & 1-z^{-1}+R(n,t)Q(n-1,t) \end{bmatrix}. \tag{10}$$

The consistency of these two linear problems is a condition on the potentials Q and R that is equivalent to CAL. This Lax pair is more useful for our purposes than that given for a vector $\mathbf{v}(n,t,\lambda)$ in the paper of Ablowitz and Ladik.[1] The latter Lax pair can be obtained from the one above by the transformation

$$\begin{aligned} v_1(n,t,\lambda) &= \lambda^{-n}u_1(n,t,z=\lambda^2)\,, \\ v_2(n,t,\lambda) &= \lambda^{-n-1}u_2(n,t,z=\lambda^2)\,. \end{aligned} \tag{11}$$

3.1. Constructing Finite Genus Solutions

The motivation for the Riemann surface method to be used below to produce multiphase wavetrain solutions to Eq. (6) stems from the study of the solutions satisfying periodic boundary conditions in n. If one uses Floquet theory to study Eq. (7) when Q and R are periodic in n with period N, it is possible to deduce that the Floquet multipliers are independent of both n and t. Because the spatial linear system for $\mathbf{u}$ is second order, the Floquet multipliers satisfy a quadratic equation with coefficients that turn out to have the simple structure of polynomials in z. The graph of this algebraic equation is the Floquet multiplier curve, Γ. It is a finite genus Riemann surface that is a two-sheeted branched cover of the z-plane. The vector $\mathbf{u}$ is most naturally considered as a function of n, t, and a point on Γ. For values of z different from the branch points, there are two linearly independent simultaneous solutions $\mathbf{u}$ of the linear problems making up the Lax pair. The two solutions fail to be linearly independent when z is a branch point of Γ. Formulas for $\mathbf{u}$ can be found using the Riemann theta function, which is the central object of analytic function theory on Riemann surfaces. From $\mathbf{u}$, it is easy to recover the potentials Q and R. As it happens in the focusing and defocusing cases, the corresponding formulas for the potentials can be written in such a way that it is clear that they depend periodically on a number of phase variables of the form $kn - \omega t$. Thus the potentials have the form of multiphase waves, but since spatial periodicity has been imposed the wavenumbers satisfy certain quantization conditions.

It is reasonable that to relax the constraint of spatial periodicity in an appropriate way should yield a family of multiphase wavetrain solutions of similar form, but without the quantization conditions on the wavenumbers. The analytic manifold of the spectral parameter should still be a finite genus Riemann surface, but we can no longer interpret this surface as a Floquet multiplier curve. Below, we will show how to construct simultaneous solutions $\mathbf{u}$ of the two linear problems of the Lax pair, Eq. (7) and Eq. (9), by first selecting an arbitrary set of branch points in the z-plane to determine an appropriate hyperelliptic Riemann surface. Because we are interested in general solutions, we must proceed without the Floquet scattering theory of the periodic problem. From $\mathbf{u}$ we will see exactly how to obtain a solution (Q, R) of Eq. (6).

The main idea here is that it is difficult to produce exact solutions to the nonlinear system Eq. (6), and it is easier to produce two linearly independent simultaneous solutions $\mathbf{u}(n,t,z)$ to the two linear problems, Eq. (7) and Eq. (9). The existence of two linearly independent solutions, $\mathbf{u}^{\pm}(n,t,z)$, even for an arbitrary value of the complex parameter z, guarantees the compatibility of the two linear problems, and thus guarantees that the potentials $Q(n,t)$ and $R(n,t)$ solve Eq. (6).

Let us proceed to find some solutions $\mathbf{u}$ of the Lax pair. In particular, we must make sure that $\mathbf{u}$ simultaneously satisfies Eq. (7) and Eq. (9) in the asymptotic limits $z \downarrow 0$ and $z \uparrow \infty$. When z is small, there are two possible dominant balances in the two problems:

$$\begin{pmatrix} u_1 \\ u_2 \end{pmatrix} = \exp(i(1 - z^{-1})t) \begin{pmatrix} \mathcal{O}(1) \\ \mathcal{O}(1) \end{pmatrix}, \tag{12}$$

$$\begin{pmatrix} u_1 \\ u_2 \end{pmatrix} = z^n \begin{pmatrix} \mathcal{O}(1) \\ \mathcal{O}(z) \end{pmatrix}. \tag{13}$$

In order to distinguish these two kinds of expansions, we refer to the former as 0^+ and the latter as 0^-. When z is large, there are also two possible dominant balances:

$$\begin{pmatrix} u_1 \\ u_2 \end{pmatrix} = z^n \exp(i(z - 1)t) \begin{pmatrix} \mathcal{O}(1) \\ \mathcal{O}(1) \end{pmatrix}, \tag{14}$$

$$\begin{pmatrix} u_1 \\ u_2 \end{pmatrix} = \begin{pmatrix} \mathcal{O}(1) \\ \mathcal{O}(z) \end{pmatrix}. \tag{15}$$

Once again, to distinguish these two kinds of expansions, we refer to the former as ∞^+ and the latter as ∞^-.

These facts, along with the analytic dependence of $\mathbf{u}$ on z suggest building the two linearly independent solutions $\mathbf{u}^{\pm}(n,t,z)$ as projections of a single vector function $\mathbf{u}(n,t,P)$, where P denotes a point on a two-sheeted branched cover, Γ, of the z-sphere. The manifold Γ consists of two copies of the Riemann sphere, ramified at an even number of fixed but arbitrary points $z_1,\ldots,z_{2g+2}$. Each copy of the sphere may be identified with the complex z-plane through stereographic projection. This is how we recover from $\mathbf{u}(n,t,P)$ two different vector functions $\mathbf{u}(n,t,z)$, both solving Eq. (7) and Eq. (9) for all complex z. As long as Γ does not branch at $z=0$ or $z=\infty$, there are two points of Γ over each of these z values. We may thus insist that the function $\mathbf{u}(P)$ have behavior determined by the expansions $0^{\pm}$ at the two points above $z=0$ and likewise that $\mathbf{u}(P)$ have behavior determined by the expansions $\infty^{\pm}$ at the two points above $z=\infty$. We will label these points on Γ as $0^{\pm}$ and $\infty^{\pm}$ according to the kind of expansion $\mathbf{u}$ has at these points. There is an ambiguity stemming from the way one chooses points $0^{\pm}$ above $z=0$ with respect to the choice of points $\infty^{\pm}$ over $z=\infty$ that results in two essentially different kinds of asymptotic behavior for the function $\mathbf{u}(P)$ for a given choice of branch points z_k. This is a feature of the construction of finite genus solutions to the Ablowitz-Ladik equations that distinguishes it from constructions in which the Lax pair becomes singular for only one value of the parameter z.

So let Γ be such a Riemann surface, branched at the n and t independent distinct z-values $z_1,\ldots,z_{2g+2}$, having the interpretation of constants of motion or action variables. Next, assume that $u_1(P)$ and $u_2(P)$ both have their poles in $\Gamma\setminus\{0^{\pm},\infty^{\pm}\}$ confined to the n and t independent points $P_1,\ldots,P_g$, having the interpretation of initial ($n=0$, $t=0$) values of angle variables. Finally, normalize the functions $u_1(P)$ and $u_2(P)$ so that

$$u_1 = z^n \exp(i(z-1)t)\Big\{1+\mathcal{O}(z^{-1})\Big\}, \tag{16}$$

near ∞^+, and

$$u_2 = \exp(i(1-z^{-1})t)\Big\{1+\mathcal{O}(z)\Big\}, \tag{17}$$

near 0^+. Then we have the following (for most sets of g points $P_1,\ldots,P_g$).

Lemma 1 *The function $\mathbf{u}(n,t,P)$ with the above properties exists and is constructed uniquely from the data $z_1,\ldots,z_{2g+2}$, $P_1,\ldots,P_g$, and the choice of expansions over $z=0$ and $z=\infty$. Furthermore, the function $\mathbf{u}(n,t,P)$ solves both the linear problems Eq. (7) and Eq. (9) globally in n, t, and P as long as the potentials $Q(n,t)$ and $R(n,t)$ are such that*

$$u_1 = \exp(i(1-z^{-1})t)\Big\{Q(n-1,t)+\mathcal{O}(z)\Big\},$$

near 0^+ and

$$u_2 = \exp(i(z-1)t)\Big\{R(n-1,t)+\mathcal{O}(z^{-1})\Big\},$$

near ∞^+.

The compatibility of the two linear problems Eq. (7) and Eq. (9) for these potentials then gives the following theorem.

Theorem 1 *The functions $Q(n,t)$ and $R(n,t)$ derived from the expansions for* **u** *solve the complex Ablowitz-Ladik equations Eq. (6).*

These solutions of CAL have the form of multiphase wavetrains:

$$Q(n,t) = A\exp(i\theta_0)\frac{\Theta((\theta_1,\ldots,\theta_g)^T)}{\Theta((\theta_1+\beta_1,\ldots,\theta_g+\beta_g)^T)}, \tag{18}$$

where $\theta_j = k_j n - \omega_j t - \delta_j$ are phase variables and Θ is the Riemann theta function defined by

$$\Theta(\mathbf{w}) = \sum_{\mathbf{m}\in\mathbb{Z}^g} \exp\left[\frac{1}{2}\mathbf{m}^T\mathbf{B}\mathbf{m} + \mathbf{m}^T\mathbf{w}\right], \tag{19}$$

and $\mathbf{B}$ is a symmetric matrix with negative definite real part obtained by integrating holomorphic differentials around closed loops on Γ. This indicates that, while the function $\mathbf{u}$ and the corresponding potentials Q and R are unique given the data, the representations of these in terms of theta functions are not well defined, depending on a choice of *homology cycles* on the surface Γ. This is a gauge symmetry that may be used to our advantage in simplifying final formulas.

If the genus g is less than 3, the procedure we have outlined above is equivalent to substitution of an Ansatz of the form given by Eq. (18) into Eq. (6) and solving for relations among the various parameters. If $g \geq 3$, it is no longer possible to find solutions without employing the complex geometry of the Riemann surface Γ; the dispersion relations become transcendental.

The solutions we have described all obey the constraint $Q(-1,0) = 1$. It is easy to free up this constraint by using the following symmetry group of CAL: if (Q,R) is a solution of CAL, then for each complex number ξ, $(\xi Q, \xi^{-1}R)$ is also a solution of CAL.

It is also possible to relax the constraint that the branch points be distinct, although this requires more complicated tools.[6] These degenerate solutions can correspond to homoclinic orbits and solitons as described by Ercolani and Miller.[3]

3.2. Focusing and Defocusing Solutions

Imposing *reality* means isolating data leading to solutions satisfying $R = \pm\overline{Q}$ for real t. These solve the restricted problem of interest, Eq. (3). The potentials as given by formulas like Eq. (18) may not be periodic functions of each of the θ_j individually unless reality is imposed on the data. We want to impose reality in order to have a family of multiphase wavetrains.

The difficulty in restricting the class of solutions found above to the focusing and defocusing subclasses is that the simple condition $R = \pm\overline{Q}$ must be translated into conditions on the data that we used to generate the potentials Q and R, namely the branch points z_k, the points P_j, the scaling parameter ξ, and the choice of the asymptotic behavior of $\mathbf{u}$ over $z = 0$ and $z = \infty$. A convenient tool for carrying out this translation is the system of squared eigenfunctions associated to the function $\mathbf{u}$. These squared eigenfunctions can be imagined as a kind of intermediate level between the potentials (Q, R), where the reality conditions are trivial, and the data for specifying the function $\mathbf{u}$, where the corresponding conditions are transcendental. In fact, at the level of the squared eigenfunctions, the reality conditions are algebraic.

Let us introduce the squared eigenfunctions. Recall the notation $\mathbf{u}^{\pm}(z)$ for the two vector solutions to the Lax pair obtained by stereographic projection from the function $\mathbf{u}(P)$ on Γ. Define

$$\begin{aligned}
\varphi(z) &= u_1^+(0,0,z)u_1^-(0,0,z)\,, \\
\chi(z) &= u_2^+(0,0,z)u_2^-(0,0,z)\,, \\
f(z) &= \frac{1}{2}\Big[u_1^+(0,0,z)u_2^-(0,0,z) + u_2^+(0,0,z)u_1^-(0,0,z)\Big]\,.
\end{aligned} \tag{20}$$

It is possible to deduce from properties of $\mathbf{u}(P)$ described in the last subsection that the squared eigenfunctions are polynomials that can be normalized so that f is monic. Thus, they take the form

$$\begin{aligned}
\varphi(z) &= \varphi_g z^g + \varphi_{g-1}z^{g-1} + \cdots + \varphi_0\,, \\
\chi(z) &= z\Big[\chi_g z^g + \chi_{g-1}z^{g-1} + \cdots + \chi_0\Big]\,, \\
f(z) &= z^{g+1} + f_g z^g + \cdots + f_1 z + f_0\,.
\end{aligned} \tag{21}$$

This construction can be viewed as a transformation from the problem data to the $3g+3$ complex coefficients of these polynomials. However, it is also possible to invert the mapping by operations no more complicated than root finding.

The advantage of this change of variables is that the reality conditions on the polynomial coefficients are easy to deduce from the linear problems Eq. (7) and Eq. (9). Namely for focusing $(-)$ and defocusing $(+)$ potentials,

$$f(z) = \frac{z^{g+1}}{\overline{f_0}}\overline{f(\overline{z}^{-1})}\,, \tag{22}$$

$$\chi(z) = \pm\frac{z^{g+1}}{\overline{f_0}}\overline{\varphi(\overline{z}^{-1})}\,. \tag{23}$$

These relations imply that in both the focusing and defocusing cases, the branch points must, as a set, be symmetric in reflection through the unit circle in the z-plane. In fact, more detailed investigations[6] suggest the following conjectures*:

Conjecture 1 *In the focusing case, the branch points cannot lie on the unit circle unless they have multiplicity greater than 1.*

Conjecture 2 *In the defocusing case, the branch points all lie on the unit circle as long as $|Q(n,t)| < 1$ for all n.*

---

*The first conjecture has been proved[6] when Q and R are periodic in n. The second conjecture is suggested by the fact that the defocusing system preserves the sign of $|Q(n,t)| - 1$ for each n, and that only the potentials for which $|Q(n,t)| < 1$ for all n contribute to the continuum limit, in which all branch points are fixed by the antiholomorphic involution $z \mapsto \overline{z}^{-1}$.

4. Modulation Equations for Multiphase Waves in AL

When we were concerned with modulation of harmonic plane waves, we inserted the plane wave Ansatz into local conservation laws and passed to the continuum limit in the parameters of the motion. We have a large set of oscillatory solutions available, as constructed in the last section. These exact solutions generalize the plane wave Ansatz Eq. (4). Let us now present a compact representation of the infinite number of local conservation laws implied by Eq. (6). Define

$$F(n,t,P) = d\log\frac{u_1(n+1,t,P)}{u_1(n,t,P)}\,, \tag{24}$$

$$G(n,t,P) = \partial_t d\log u_1(n,t,P)\,. \tag{25}$$

Then, it follows from the compatibility of differentiation in t with spatial differencing in n that

$$\partial_t F(n,t,P) = G(n+1,t,P) - G(n,t,P)\,, \tag{26}$$

which has the form of a local conservation law with parameter $P \in \Gamma$. As trivial as it may appear, this expression contains an infinity of nontrivial local conservation laws obtained by expanding with respect to P.

The idea in obtaining modulation equations in integrable systems (as described by Flaschka, Forest, and McLaughlin[4] for KdV and Bloch and Kodama[2] for the Toda lattice) is to imagine a slowly deforming Riemann surface $\Gamma(X=\varepsilon n, T=\varepsilon t)$ for $\varepsilon \downarrow 0$. Let z_k depend on the slow scales X and T. Substitute the theta function representation of u_1 into the conservation law generator and use angled brackets to denote an average over n and t sufficient to remove dependence on n and t. Then one obtains the modulation equations for the finite genus solutions of Eq. (6) in *homology gauge form*†:

$$\begin{aligned}\partial_T\omega_{(3)} + \partial_T\left\langle d\log\frac{\Theta(A(P)-\mathbf{Z}+\mathbf{U}(n+1)+\mathbf{V}t)}{\Theta(A(P)-\mathbf{Z}+\mathbf{U}n+\mathbf{V}t)}\right\rangle &= \\ \partial_X\omega_{(2)} + \partial_X\left\langle d\frac{\partial_t\Theta(A(P)-\mathbf{Z}+\mathbf{U}(n+1)+\mathbf{V}t)}{\Theta(A(P)-\mathbf{Z}+\mathbf{U}(n+1)+\mathbf{V}t)}\right\rangle &.\end{aligned} \tag{27}$$

The Abel mapping, $A(P)$, is a vector of integrals of normalized holomorphic differentials along a contour ending at the point P. The differentials $\omega_{(3)}$ and $\omega_{(2)}$ control the asymptotic behavior of the function $\mathbf{u}$ near the points $0^\pm$ and $\infty^\pm$. The vectors $\mathbf{U}$ and $\mathbf{V}$ contain wavenumbers and frequencies obtained from these two differentials. Only the vector $\mathbf{Z}$ depends on the initial phase information encoded in the points P_j.

In the focusing and defocusing cases, it is usually‡ possible to choose the homology cycles on Γ in such a way that the modulation equations take the simple form

$$\partial_T\omega_{(3)} = \partial_X\omega_{(2)}\,. \tag{28}$$

†We use this terminology because the appearance of these modulation equations may be altered by changing the basis of homology cycles on Γ.

‡Here, "usually" means in the case that none of the frequencies or wavenumbers of the underlying wavetrain are commensurate.

This kind of cosmetic improvement is an application of the homology gauge symmetry of the finite genus formulas mentioned above. All initial phase information in the vector $\mathbf{Z}$ is lost in this representation. It turns out that the branch points $z_k(X,T)$ are Riemann invariants of these equations, and that the characteristic speed of z_k is real if z_k is on the unit circle. Then, our conjectures of the previous section lead us to the following conclusions:

- In the focusing case, these modulation equations are never hyperbolic and there are no stable multiphase wavetrains.
- In the defocusing case, there exist both stable and unstable multiphase wavetrains. The stable wavetrains are those whose amplitudes are everywhere less than 1.

5. Resonances

If there are resonances in the underlying focusing or defocusing wavetrain, corresponding to commensurate frequencies or wavenumbers, then the modulation equations must be written in a form more general than Eq. (28). The terms involving the phase vector $\mathbf{Z}$ cannot be made to vanish by any choice of homology cycles, and in fact $\mathbf{Z}$ must be taken to depend on X and T, thus introducing the phenomenon of *phase modulation*. The modulation equations for the branch points $z_k(X,T)$ must be supplemented by modulation equations for certain functions of the pole positions P_j. In these resonant cases, the modulation equations need not have the stability properties given at the end of the last section. Indeed, Miller, Ercolani, and Levermore[7] discuss a simple modulationally stable resonance in the focusing Ablowitz-Ladik equations.

References

1. M.J. Ablowitz and J.F. Ladik. Nonlinear differential-difference equations and Fourier analysis. *J. Math. Phys.*, 17:1011–1018, 1976.
2. A.M. Bloch and Y. Kodama. Dispersive regularization of the Whitham equation for the Toda lattice. *SIAM J. Appl. Math.*, 52(4):909–928, 1992.
3. N.M. Ercolani and P.D. Miller. Singular solutions to the Ablowitz-Ladik equations and Bäcklund transformations. in preparation.
4. H. Flaschka, M.G. Forest, and D.W. McLaughlin. Multiphase averaging and the inverse spectral solution of the Korteweg-de Vries equation. *Comm. Pure Appl. Math.*, 33:739–784, 1980.
5. M.H. Hays, C.D. Levermore, and P.D. Miller. Macroscopic lattice dynamics. To appear in *Physica D*, 1994.
6. P.D. Miller, N.M. Ercolani, I.M. Krichever, and C.D. Levermore. Finite genus solutions to the Ablowitz-Ladik equations. Submitted to *Comm. Pure Appl. Math.*, 1994.
7. P.D. Miller, N.M. Ercolani, and C.D. Levermore. Modulation theory in the presence of phase locking. Submitted to *Physica D*, 1994.

DARBOUX TRANSFORMATIONS FROM REDUCTIONS OF THE KP HIERARCHY

J J C NIMMO
Department of Mathematics, University of Glasgow,
Glasgow G12 8QW, Scotland

ABSTRACT

The use of effective Darboux transformations for general classes Lax pairs is discussed. The general construction of "binary" Darboux transformations preserving certain properties of the operator, such as self-adjointness, is given. The classes of Darboux transformations found include the multicomponent BKP and CKP reductions of the KP hierarchy.

1. Introduction

Darboux transformations define a mapping between the solutions of a linear differential equations and a similar equation containing different coefficients. Since integrable nonlinear evolution equations frequently arise as the compatibility condition for a pair for such equations, Darboux transformations may be used to construct families of exact solutions of the nonlinear equations. Typically these are multi-soliton solutions. A good introduction to this topic, including the following example, is given in the monograph by Matveev and Salle[1].

As an example which motivates the work to be presented, consider the Lax pair*

$$L = i\partial_y + \partial^2 + u, \quad M = \partial_t + 4\partial^3 + 3\partial u + 3u\partial + 3i\partial^{-1}(u_y).$$

for the variant of the Kadomtsev-Petviashvili equation known as KPI

$$(u_u + 6uu_x + u_{xxx})_x) - 3u_{yy} = 0,$$

in which u is a real variable. This means that $[L, M] = 0$ if and only if u satisfies KPI.

For all non-zero θ such that $L(\theta) = M(\theta) = 0$, a Darboux transformation is defined by the operator $G = \theta\partial\theta^{-1}$ in the sense that

$$L(\psi) = M(\psi) = 0 \implies \widetilde{L}(G(\psi)) = \widetilde{M}(G(\psi)) = 0,$$

*Here and below $\partial = \partial/\partial x$ and $\partial_y = \partial/\partial y$ and so on.

where $\widetilde{L}$ and $\widetilde{M}$ are the operators obtained from L and M by replacing u by $\tilde{u} = u + 2(\log\theta)_{xx}$. This result is readily proved by observing that

$$\widetilde{L}G = GL \quad \text{and} \quad \widetilde{M}G = GM,$$

i.e.

$$\widetilde{L} = GLG^{-1} \quad \text{and} \quad \widetilde{M} = GMG^{-1}.$$

In this way the Darboux transformation manifests itself as a (differential) gauge transformation. It also follows that

$$[L, M] = 0 \implies [\widetilde{L}, \widetilde{M}] = 0,$$

i.e. $\tilde{u}$ satisfies KPI whenever u does. Hence the Darboux transformation induces an auto-Bäcklund transformation.

There is a problem with this transformation however. For almost all u, since θ is the solution of a complex equation, $\tilde{u}$ is not real and so we do not obtain solutions of KPI. At the root of the problem is the fact that, while L and M are self-adjoint, $\widetilde{L}$ and $\widetilde{M}$ are not. In order to overcome this problem one may use a "binary" transformation (to be defined in the next section) which does preserve the self-adjointness of L and M.

This paper is concerned with the use of binary transformations to preserve the structure of two classes of operators with matrix coefficients and arbitrary order.

2. The structure of the binary transformation

For an (matrix) operator L, let $S = \{\theta,\ \text{non-singular} : L(\theta) = 0\}$ (and define $\widetilde{S}$, $S^\dagger$ for operators $\widetilde{L}$, $L^\dagger$ etc.). A (formally invertible) gauge transformation G_θ, for $\theta \in S$, defines a mapping

$$G_\theta\colon S \to \widetilde{S}, \quad \text{where} \quad \widetilde{L} = G_\theta L G_\theta^{-1}.$$

Consider also the (formal) adjoint operator $G_\theta^\dagger$. (Taking the formal adjoint is, as usual, the linear operation defined by

$$(a\partial^i)^\dagger = (-1)^i \partial^i a^\dagger,$$

for a matrix a, where $a^\dagger$ denotes the Hermitian conjugate of a.) Since $\widetilde{L}^\dagger = G_\theta^{\dagger^{-1}} L^\dagger G_\theta^\dagger$, we have

$$G_\theta^\dagger\colon \widetilde{S}^\dagger \to S^\dagger.$$

By determining the kernel of $G_\theta^\dagger$ we obtain some nontrivial solution in $\widetilde{S}^\dagger$. Typically, we may identify this subset in terms of θ and denote a member as $\imath(\theta)$. For example, in the classical case when $G_\theta = \theta\partial\theta^{-1}$ and $G_\theta^\dagger = -\theta^{\dagger^{-1}}\partial\theta^\dagger$, we find that

$$G_\theta^\dagger(\rho) = 0 \iff \rho = \theta^{\dagger^{-1}} c,$$

where c is independent of x.

We represent this situation in the diagram below.

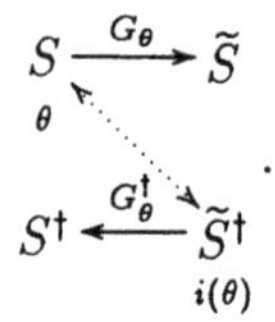

To describe the general form of the binary transformation we consider operators L, $\widetilde{L}$ and $\widehat{L}$ and the corresponding sets of non-singular solutions matrices S, $\widetilde{S}$ and $\widehat{S}$. Let $\theta \in S$ and $\widehat{\theta} \in \widehat{S}$ be such that $G_\theta: S \to \widetilde{S}$ and $G_{\hat{\theta}}: \widehat{S} \to \widetilde{S}$. Then we get the mapping

$$G_{\hat{\theta}}^{-1} G_\theta: S \to \widehat{S}.$$

The difficulty with this definition of a transformation is that to define it we need one of the solutions we are trying to determine, namely $\widehat{\theta}$! To overcome this, we use the fact that there corresponds to $\widehat{\theta} \in \widehat{S}$ a solution $i(\widehat{\theta}) \in \widetilde{S}^\dagger$ and then use the mapping $G_\theta^{\dagger^{-1}}: S^\dagger \to \widetilde{S}^\dagger$ to obtain $\widehat{\theta} = i^{-1}(G_\theta^{\dagger^{-1}}(\phi))$ for any $\phi \in S^\dagger$. This is shown in the diagram below.

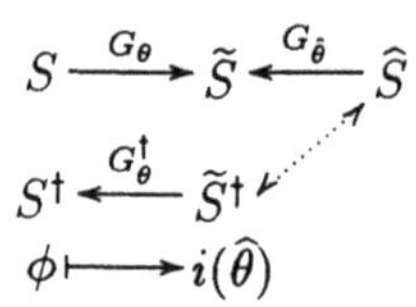

In this way we obtain the definition of a general binary transformation.

Definition Consider an operator L and gauge operator G_θ, where $\theta \in S$, such that $G_\theta^\dagger(i(\theta)) = 0$. For each $\phi \in S^\dagger$, define

$$G_{\theta,\phi} = G_{\hat{\theta}}^{-1} G_\theta,$$

where $\widehat{\theta} = i^{-1}(G_\theta^{\dagger^{-1}}(\phi))$. Then

$$G_{\theta,\phi}: S \to \widehat{S},$$

where $\widehat{L} = G_{\theta,\phi} L G_{\theta,\phi}^{-1}$, is called a *binary transformation.*

In the next section we will consider two concrete examples of such a binary transformation.

Now suppose that the operator L has a constraint of the form

$$L^\dagger R^\dagger = RL,$$

where R is in some formally invertible (matrix differential) operator†. We wish to find binary transformations that preserve this constraint. That is—using the notation of the above definition—we want $\widehat{L}$ to satisfy the constraint whenever L does.

Examples for the choice of R include

- $R = I$. L is self-adjoint. An example of the application of this is quoted in the introduction.
- $R = iI$. L is skew-adjoint. This corresponds to the CKP reduction of the KP hierarchy[2] and the reduction of the Kuperschmidt "$k = 0$" non-standard hierarchy[3, 4].
- $R = \partial$. This corresponds to the BKP[5] or the Kuperschmidt "$k = 1$" reduction.

The binary transformation we will discuss in the next section will preserve generalizations of these three reductions.

Let the gauge transformation G, such that $\widehat{L} = GLG^{-1}$, preserve the constraint $L^\dagger = RLR^{\dagger^{-1}}$. Then

$$\widehat{L}^\dagger - R\widehat{L}R^{\dagger^{-1}} = 0$$

which means that

$$G^{\dagger^{-1}}L^\dagger G^\dagger - RGLG^{-1}R^{\dagger^{-1}} = G^{\dagger^{-1}}RLR^{\dagger^{-1}}G^\dagger - RGLG^{-1}R^{\dagger^{-1}} = 0.$$

This leads to the single condition $RG = G^{\dagger^{-1}}R$.

Note that the relation between L and its adjoint imposes a relationship between the solution sets S and $S^\dagger$. In particular, for each $\theta \in S$, $R^\dagger(\theta) \in S^\dagger$. Hence, in the case of a binary transformation $G = G_{\theta,\phi}$, we may make the choice $\phi = R^\dagger(\theta)$.

3. Darboux transformations for general operators

In this section we describe two classes of Darboux transformation for general classes of matrix differential operators of arbitrary order. The first is originally due to Matveev[6] and has also been considered recently by Oevel[7]. We will present a very simple proof of this result. The second was found by Oevel & Rogers[8] in the case of scalar operators in the context of Sato theory. We will derive a more general version here.

In both cases, the results are remarkably general. There is however a serious drawback. There is, in this general case, absolutely no guarantee that the transformed operator we have the same "form" as the original and so only in special cases does one get a transformation that induces an auto-Bäcklund transformation.

---

†It is tempting to look for a constraint of the form $L^\dagger S = RL$ but this in fact corresponds to two constraints since on taking adjoints $L^\dagger R^\dagger = S^\dagger L$.

First, consider

$$L = \partial_t + \sum_{i=0}^{n} u_i \partial^i \quad \text{and} \quad \widetilde{L} = \partial_t + \sum_{i=0}^{n} \widetilde{u}_i \partial^i,$$

where u_i and $\widetilde{u}_i$ are $N \times N$ (not necessarily constant) matrices. Let the operator G be such that

$$\widetilde{L} = GLG^{-1} = L + [G, L]G^{-1}.$$

Hence G must satisfy

$$[G, L]G^{-1} = \sum_{i=0}^{n} (\widetilde{u}_i - u_i)\partial^i.$$

Taking $G = \theta\partial\theta^{-1}$, where θ is a non-singular $N \times N$ matrix, and hence $G^{-1} = \theta\partial^{-1}\theta^{-1}$, we get

$$[G, L]G^{-1} = [G, L]\theta\partial^{-1}\theta^{-1} = \sum_{i=0}^{n} a_i \partial^{i-1}\theta^{-1},$$

for some matrices a_i. For $i = 1, \ldots, n$, $a_i = \widetilde{u}_{i-1} - u_{i-1}$ and in order that G define a Darboux transformation we must have

$$a_0 = 0.$$

This condition gives $[G, L](\theta) = 0$ i.e. $G(L(\theta)) = 0$ since $G(\theta) = 0$. Hence we only need require that $L(\theta) = \theta C$, for some x-independent matrix C. Note that if $L(\theta) = 0$ then for $\theta' = \theta \exp(\partial_t^{-1}(C))$, $L(\theta') = \theta' C$. Also, $G_{\theta'} = G_\theta$ and so we may suppose, without loss of generality, that $C = 0$‡. Thus we find that $\theta \in S$.

The second case we consider is

$$L = \partial_t + \sum_{i=1}^{n} u_i \partial^i \quad \text{and} \quad \widetilde{L} = \partial_t + \sum_{i=1}^{n} \widetilde{u}_i \partial^i,$$

where u_i and $\widetilde{u}_i$ are again $N \times N$ matrices. Note that the multiplicative term in L and $\widetilde{L}$ is omitted.

As in the first case, a gauge operator G must satisfy

$$[G, L]G^{-1} = \sum_{i=1}^{n} \widetilde{u}_i - u_i.$$

There are now two simple choices. First, let $G = G_\theta^{(1)} = \theta^{-1}$, an (invertible) $N \times N$ matrix. Then

$$[G, L]G^{-1} = \sum_{i=0}^{n} a_i \partial^i,$$

‡Note that if L is an ordinary differential operator, then taking $C \neq 0$ is a genuine generalization. For example, this is exploited in the classical "discrete eigenvalue adding" Darboux transformation for the time-independent Schrödinger operator.

and so $a_0 = [G, L](\theta) = G(L(\theta)) = 0$, i.e. $\theta \in S$.

Second, let $G = G^{(2)}_\rho = \rho_x^{-1}\partial$, where ρ_x is an invertible $N \times N$ matrix. Now

$$[G, L]G^{-1} = [G, L]\partial^{-1}\rho_x = \sum_{i=1}^{n+1} a_i \partial^{i-1}\rho_x,$$

and we must have $a_1 = 0$, i.e. $[G, L](\partial^{-1}(\rho_x)) = G(L(\rho)) = 0$. Thus $L(\rho) = C$, an x-independent matrix. Again, we may take $C = 0$ without loss of generality and so $\rho \in S$.

As in the scalar case[8], it is the composition of the two gauge transformations which is of most interest, and we take $G_\theta = G^{(2)}_{G^{(1)}_\theta(1)} G^{(1)}_\theta = (\theta^{-1})_x^{-1}\partial\theta^{-1}$.

4. Binary transformations and reductions

To determine the binary transformations $G_{\theta,\phi}$ corresponding to the two Darboux transformations found above we must determine two additional things: the mapping $i: \hat{S} \to \tilde{S}^\dagger$ and then the element $\hat{\theta} \in \hat{S}$ in terms of θ and ϕ.

First consider $L = \partial_t + \sum_{i=0}^{n} u_i \partial^i$, $G_\theta = \theta\partial\theta^{-1}$. Here the condition $G^\dagger_\theta(i(\theta)) = -\theta^{\dagger^{-1}}\partial(\theta^\dagger i(\theta)) = 0$ is satisfied by the choice $i(\theta) = \theta^{\dagger^{-1}}$. Further,

$$\begin{aligned} \hat{\theta} &= \left(G_\theta^{\dagger^{-1}}(\phi)\right)^{\dagger^{-1}} \\ &= -\left(\theta^{\dagger^{-1}}\partial^{-1}(\theta^\dagger\phi)\right)^{\dagger^{-1}} \\ &= -\theta\Omega^{-1}, \end{aligned}$$

where $\Omega = \partial^{-1}(\phi^\dagger\theta)$. It may be shown that for all operators $L = \sum_{i=0}^{n} u_i \partial^i$, Ω is exact in the sense that $d\Omega = \phi^\dagger\theta dx + A(u_1, \ldots, u_n, \theta, \phi)dt$[7].

In this case the binary transformation is

$$\begin{aligned} G_{\theta,\phi} = G_{\hat{\theta}}^{-1} G_\theta &= \theta\Omega^{-1}\partial^{-1}\Omega\partial\theta^{-1} \\ &= \theta\Omega^{-1}\partial^{-1}(\partial\Omega - \Omega_x)\theta^{-1} \\ &= 1 - \theta\Omega^{-1}\partial^{-1}\phi^\dagger. \end{aligned}$$

For discussion of the reduction we will also need

$$G^{\dagger^{-1}} = 1 - \phi\Omega^{\dagger^{-1}}\partial^{-1}\theta^\dagger.$$

Now suppose that L satisfies the constraint $L^\dagger R^\dagger = RL$ where $R = A$, a (not necessarily constant) matrix. Then we may choose $\phi = R^\dagger(\theta) = A^\dagger\theta$. The condition $RG_{\theta,\phi} = G^{\dagger^{-1}}_{\theta,\phi} R$ now gives

$$\begin{aligned} A - A\theta\Omega^{-1}\partial^{-1}\phi^\dagger = A - \phi\Omega^{\dagger^{-1}}\partial^{-1}\theta^\dagger A \quad &\Longleftrightarrow \quad A\theta\Omega^{-1}\partial^{-1}\theta^\dagger A = A^\dagger\theta\Omega^{\dagger^{-1}}\partial^{-1}\theta^\dagger A \\ &\Longleftrightarrow \quad A^\dagger = \pm A. \end{aligned}$$

This establishes the following theorem.

Theorem 1 *Let the matrix operator* $L = \sum_{i=0}^{n} u_i \partial^i$ *satisfy the constraint*

$$L^\dagger A = AL,$$

where A is an Hermitian or skew-Hermitian matrix. Then the binary transformation

$$G = 1 - \theta \Omega^{-1} \partial^{-1} \theta^\dagger A$$

where $\Omega = \partial^{-1}(\theta^\dagger A \theta)$, *preserves the above constraint, i.e.* $\widehat{L} = GLG^{-1}$ *satisfies* $\widehat{L}^\dagger A = A\widehat{L}$.

For the second case, $L = \sum_{i=1}^{n} u_i \partial^i$, $G_\theta = (\theta^{-1})_x^{-1} \partial \theta^{-1}$ and hence $i(\theta) = (\theta^{\dagger^{-1}})_x$.

To determine the binary transformation it is notationally convenient to write an element of $S^\dagger$ as ϕ_x rather than ϕ as we did above. Also, it is necessary to introduce two integrals

$$\Omega = \partial^{-1}(\phi^\dagger \theta_x) \quad \text{and} \quad \Omega' = \partial^{-1}(\phi_x^\dagger \theta),$$

where

$$\Omega + \Omega' = \phi^\dagger \theta.$$

Now

$$\begin{aligned} i(\widehat{\theta}) = (\widehat{\theta}^{\dagger^{-1}})_x &= G_\theta^{\dagger^{-1}}(\phi_x) \\ &= -(\theta^{\dagger^{-1}})_x \partial^{-1}(\theta^\dagger \phi_x) \end{aligned}$$

and so

$$(\widehat{\theta}^{-1})_x = -\partial^{-1}(\phi_x^\dagger \theta)(\theta^{-1})_x = -\Omega'(\theta^{-1})_x.$$

Integrating by parts and taking inverses, we get

$$\begin{aligned} \widehat{\theta} &= \left(-\Omega' \theta^{-1} + \partial^{-1}(\Omega'_x \theta^{-1})\right)^{-1} \\ &= \left(\phi^\dagger - \Omega' \theta^{-1}\right)^{-1} \\ &= \theta \Omega^{-1}. \end{aligned}$$

We may now obtain

$$\begin{aligned} G_{\theta,\phi_x} &= \widehat{\theta} \partial^{-1} (\widehat{\theta}^{-1})_x (\theta^{-1})_x^{-1} \partial \theta^{-1} \\ &= -\theta \Omega^{-1} \partial^{-1} \Omega' \partial \theta^{-1} \\ &= 1 - \theta \Omega^{-1} \partial^{-1} \phi^\dagger \partial, \end{aligned}$$

and in a similar way

$$G_{\theta,\phi_x}^{\dagger^{-1}} = 1 - \partial \phi \Omega'^{\dagger^{-1}} \partial^{-1} \theta^\dagger.$$

Suppose that L satisfies the constraint $L^\dagger R^\dagger = RL$ where $R = A\partial$, A a matrix, and choose $\phi_x = R^\dagger(\theta) = -(A^\dagger\theta)_x$, i.e. $\phi = -A^\dagger\theta$. The condition $RG_{\theta,\phi_x} = G^{\dagger^{-1}}_{\theta,\phi_x}R$ is

$$A\partial - A\partial\theta\Omega^{-1}\partial^{-1}\theta^\dagger A\partial = A\partial - \partial A^\dagger\theta\Omega'^{\dagger^{-1}}\partial^{-1}\theta^\dagger A\partial \iff A^\dagger = \pm A \text{ and } A_x = 0.$$

With these conditions on A, $\Omega = \pm\Omega'^\dagger$.

This establishes a second theorem.

Theorem 2 *Let the matrix operator* $L = \sum_{i=1}^n u_i\partial^i$ *satisfy the constraint*

$$L^\dagger A\partial + A\partial L = 0,$$

where A *is an* x*-independent Hermitian or skew-Hermitian matrix. Then the binary transformation*

$$G = 1 - \theta\Omega^{-1}\partial^{-1}\theta^\dagger A\partial$$

where[§] $\Omega = \partial^{-1}(\theta^\dagger A\theta_x)$, *preserves the above constraint, i.e.* $\widehat{L} = GLG^{-1}$ *satisfies* $\widehat{L}^\dagger A\partial + A\partial\widehat{L} = 0$.

5. Examples

5.1. *Davey-Stewartson I*

This system has Lax pair

$$L = \partial_y + \alpha\partial + Q, \quad M = i\partial_t + \alpha\partial^2 + \frac{1}{2}(Q\partial + \partial Q + \alpha Q_y) + D,$$

where $Q = \begin{pmatrix} 0 & u \\ \epsilon\bar{u} & 0 \end{pmatrix}$ and $D = \begin{pmatrix} U & 0 \\ 0 & V \end{pmatrix}$ is real.

Let $A = \begin{pmatrix} 1 & 0 \\ 0 & -\epsilon \end{pmatrix}$. Then

$$L^\dagger(iA)^\dagger = (iA)L, \quad M^\dagger A^\dagger = AL.$$

Hence we may use Theorem 1 (with $A = I$) to obtain a binary transformation. This transformation has been used to obtain a wide class of solutions including dromions[9].

5.2. *Sawada-Kotera equation*

The equation is a reduction of the BKP equation and so has Lax pair admitting the BKP reduction:

$$L = (\partial^2 + 3u)\partial, \quad M = \partial_t + (9\partial^5 - 15\partial u\partial + 30(\partial^2 u + u\partial^2) + 15u^2)\partial,$$

[§] For a better notation we have replaced Ω with $-\Omega$ in the statement of the theorem

where

$$L^{\dagger}\partial + \partial L = M^{\dagger}\partial + \partial M = 0.$$

We may use Theorem 2 (with $A = I$) to obtain the binary transformation. Note that this is given by $G = \theta^{-1}\partial^{-1}\theta^{2}\partial\theta^{-1}$ and coincides the with the well-known "Darboux" transformation[10, 11].

5.3. Modified Novikov-Veselov equation

This system[12, 13, 14] belongs to the two component BKP hierarchy and has a Lax pair

$$L = \partial_y + S\partial, \quad M = \partial_t + (S\partial^2 + T\partial + \partial T + U)\partial,$$

where $S = \begin{pmatrix} \cos u & \sin u \\ \sin u & -\cos u \end{pmatrix}$, and T and U are give skew-symmetric and symmetric real matrices respectively. Again

$$L^{\dagger}\partial + \partial L = M^{\dagger}\partial + \partial M = 0$$

and Theorem 2 gives the binary transformation.

6. Conclusions

We have discussed the general construction of binary transformations from Darboux transformations. In particular we have carried this out for two classes of operators. More importantly, we have shown that the well studied reductions of these classes (the multi-component BKP and CKP reductions) are among those that the binary transformations preserve.

Iteration of the binary transformation is, of course, possible and leads to closed-form expressions for solutions—of the linear problems and for the integrable systems that are their compatibility conditions—in terms of (multi-component) Grammian determinants. In the case of the reduction described in Theorem 2, one may see how these Grammians are transformed into Pfaffians by the reduction process. These features will be discussed in more details elsewhere[15].

7. Acknowledgements

I wish to thank Ralph Willox and Walter Oevel for useful discussions relating to this work and the Royal Society of London for financial support enabling my attendance at this workshop.

References

1. Matveev, V.B. and Salle, M.A. (1991) *"Darboux Transformations and Solitons"*, Nonlinear Dynamics: Springer, Berlin.

2. Date, E., Jimbo, M., Kashiwara, M. and Miwa, T. (1981) *J. Phys. Soc Japan*, **50**, 3813.

3. Kuperschmidt, B. (1985) *Commun. Math. Phys.*, **99**, 51.

4. Konopelchenko, B.G. and Oevel, W. (1993) *Publ. RIMS, Kyoto Univ.*, **29**, 581.

5. Date, E., Jimbo, M., Kashiwara, M. and Miwa, T. (1982) *Physica*, **4D**, 343.

6. Matveev, V.B. (1979) *Lett. Math. Phys.*, **3**, 217.

7. Oevel, W. (1993) *Physica A*, **195**, 533.

8. Oevel, W. and Rogers, C. (1993) *Rev. Math. Phys*, **5**, 299.

9. Nimmo, J.J.C. (1992) *Inverse Problems*, **8**, 219.

10. Aiyer, R.N., Fuchsteiner, B. and Oevel, W. (1986) J. Math. Phys., **19**, 3755.

11. Nimmo, J.J.C. (1990) Proc. R. Soc. Lond. A, **431**, 361.

12. Konopelchenko, B.G. (1992) *"Introduction to Multi-Dimensional Integrable Equations: The Inverse Spectral Transform in* $2+1$ *Dimensions"*, Plenum Press, New York.

13. Nimmo, J.J.C. (1993) in *Applications of Analytic and Geometric Methods to Nonlinear Differential Equations"* ed. P.A. Clarkson, Kluwer, Dordrecht.

14. Oevel, W and Schief, W. (1993) Preprint (University of New South Wales, Applied Mathematics Report AMR93/2).

15. Nimmo, J.J.C. (1994) In preparation.

INERTIAL MANIFOLDS AND THE STRONG SQUEEZING PROPERTY

J. C. ROBINSON
Department of Applied Mathematics & Theoretical Physics
University of Cambridge
Silver Street, Cambridge, CB3 9EW, U.K.

ABSTRACT

A concise geometric proof of the existence of an inertial manifold is given: the strong squeezing property alone is sufficient. The theorem can be applied directly, or general conditions found to ensure its applicability. The latter approach leads to the familiar spectral gap condition.

To the Reynolds family & Schola Cantorum of Atlanta

1. Introduction

1.1 Inertial manifolds

Many partial differential equations seem to exhibit finite-dimensional behaviour: the existence of a finite-dimensional global attractor is known for many examples,[1−3] and numerical investigation of the Kuramoto-Sivashinksy equation[4] has produced a bifurcation diagram consistent with a small number of degrees of freedom.

For a large class of these equations it is possible to show rigorously that the original infinite-dimensional system is asymptotically equivalent to a finite set of ordinary differential equations using the theory of inertial manifolds.[5] An inertial manifold is a finite-dimensional Lipschitz manifold $\mathcal{M}$ which attracts all the orbits exponentially - thus trajectories "eventually" lie on this manifold and are determined by the ambient flow restricted to $\mathcal{M}$.

There are many existence proofs in the literature,[5−12] and most are based on analytical approaches which, although successful, are somewhat uninformative about the mechanisms that conspire to produce this simplifying behaviour. The present paper presents a proof[13] based solely on the geometric strong squeezing property,[14] often used indirectly in the existing literature.[3,6,12] To show that this property is sufficient to ensure the existence of inertial manifold is the object of the following analysis. Related approaches can be found in references 6, 11, and 15.

1.2 The equation and the associated projection operators

The following evolution equation on a Hilbert space H (usually L^2) will be considered,

$$du/dt + Au + f(u) = 0, \tag{1}$$

where A is a positive self-adjoint linear operator and f is a globally bounded Lipschitz map from $D(A^{\alpha})$ into $D(A^{\beta})$ $(0 \leq \alpha - \beta < 1)$ with support contained in some ball Ω_ρ in $D(A^\alpha)$,

$$\begin{gathered}
\operatorname{supp}(f) \subset \Omega_\rho \equiv \{u : |u|_\alpha \leq \rho\}, \\
\sup_{u \in \Omega_\rho} |f(u)|_\beta \leq C_0 \ \forall\, u \in D(A^\alpha), \text{ and} \\
|f(u_1) - f(u_2)|_\beta \leq C_1 |u_1 - u_2|_\alpha \ \forall\, u_1, u_2 \in D(A^\alpha),
\end{gathered}$$

where $|v|_\theta = |A^\theta v|$ and $|\cdot|$ is the norm on H. Many dissipative PDEs can be expressed in this form via a suitable truncation of the nonlinear term.[3,5,11] Standard existence and uniqueness results[16,17] provide a semigroup $S(t)$ on $D(A^\alpha)$ which gives the solution $u(t)$ with $u(0) = u_0$ as $S(t)u_0$: in fact, $u(t) \in D(A^{1+\beta})$ for $t > 0$, and this increased regularity will be useful later.

The linear operator A has an orthonormal basis of eigenvectors w_j with corresponding eigenvectors λ_j,

$$A w_j = \lambda_j w_j, \qquad \lambda_{j+1} \geq \lambda_j,$$

and these generate a natural set of finite-dimensional projection operators P_N and their orthogonal complements Q_N according to

$$P_N u = \sum_{j=1}^{N} (u, w_j) w_j \qquad Q_N u = (I - P_N) u = \sum_{j=N+1}^{\infty} (u, w_j) w_j,$$

where $(\cdot,\cdot)$ is the inner product on H. The inertial manifold will be searched for as the graph of a Lipschitz function over PH (choosing some fixed value for N), i.e.

$$\mathcal{M} = \mathcal{G}[\phi] \equiv \{p + \phi(p) : p \in PH\}, \text{ with}$$

$$|\phi(p_1) - \phi(p_2)|_\alpha \leq l |p_1 - p_2|_\alpha. \tag{2}$$

1.3 The Strong Squeezing Property

The strong squeezing property[14] is made up from two components, the first of which is the "cone condition". Consider two trajectories $u_i(t)$ with $u_i(0) \in \mathcal{M}$. Then they lie on the manifold for all time and so from Eq. (2)

$$|Qu_1(t) - Qu_2(t)|_\alpha \leq l |Pu_1(t) - Pu_2(t)|_\alpha \tag{3}$$

for all $t \geq 0$. The cone condition generalises this to certain trajectories that lie off the manifold: if Eq. (3) holds for two trajectories for some t_0 then it holds for all $t \geq t_0$. Put another way, one can say that the cone

$$\mathcal{C}_l = \{w : w = u_1 - u_2, |Qw|_\alpha \leq l |Pw|_\alpha\}$$

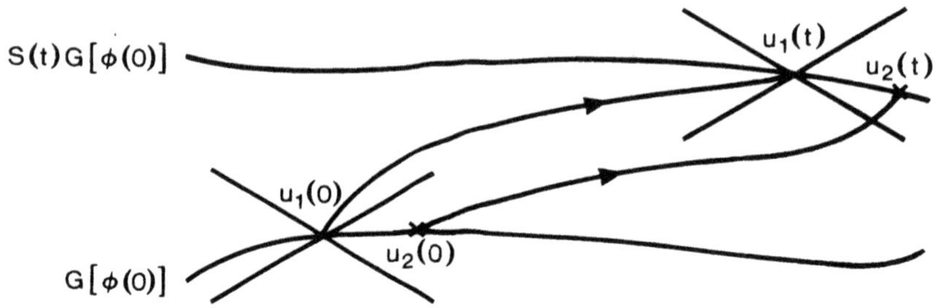

Figure 1 The cone condition

is invariant under $S(t)$ (illustrated in figure 1). This geometric condition turns out to be sufficient (given the form of Eq. (1)) to guarantee the existence of an *invariant* manifold.

The exponential attraction is given by the "squeezing" part of the property, which says that the difference of trajectories that do not lie in the interior the cone decays exponentially: if $w(t_0) \notin \text{int}(\mathcal{C}_l)$ then

$$|Qw(t)|_\alpha \leq |Qw(0)|_\alpha \, e^{-kt} \tag{4}$$

for all $0 \leq t \leq t_0$ and some $k > 0$. In fact this decay can be used to prove the stronger property of asymptotic completeness, i.e. the existence for any $u(t)$ of an asymptotically equivalent trajectory on $\mathcal{M}$.

2. The existence theorem

The essential idea of the theorem is to use the cone condition to convert the semiflow $S(t)$ on $D(A^\alpha)$ to a semiflow $T(t)$ on a space of Lipschitz functions ϕ whose graphs could be an inertial manifold for $S(t)$. The appropriate space of functions to consider is $\mathcal{F}_l$, functions $\phi : PH \to QH \cap D(A^\alpha)$ with

$$\|\phi\| \equiv \sup_{p \in PH} |\phi(p)|_\alpha < \infty,$$
$$|\phi(p_1) - \phi(p_2)|_\alpha \leq l|p_1 - p_2|_\alpha, \text{ and}$$
$$\text{supp}(\phi) \in P\Omega_\rho.$$

Theorem 1. *If the cone condition holds for Eq. (1) then there exists an invariant manifold given as the graph of a Lipschitz function $\phi \in \mathcal{F}_l$. If the strong squeezing property holds then for any solution $u(t)$ of Eq. (1) there exists an initial condition u_0 on this invariant manifold such that*

$$|S(t)u_0 - u(t)|_\alpha \leq C(X)e^{-kt},$$

where $C(X)$ is uniform for $u(0) \in X$, a bounded set in $D(A^\alpha)$; i.e. there exists an asymptotically complete inertial manifold for Eq. (1).

The proof uses a theorem of Hale[2] which is a generalisation to infinite dimensional spaces of the result that, for the flow generated by a set of ordinary differential equations, an absorbing region in $\mathbb{R}^n$ will contain a fixed point:

Theorem 2 (Hale). *Consider a semigroup $T(t)$ defined on a Banach space or a closed convex subset of a Banach space. If the semigroup $T(t)$ is dissipative, C^0, and for any bounded set B the closure of $T(t)B$ is compact for $t > 0$, then $T(t)$ has a fixed point.*

Proof (theorem 1). Because of the cone condition, the flow $S(t)$ on $D(A^\alpha)$ induces a flow on $\mathcal{F}_l$ (see figure 1) which will be denoted $T(t)$ and is given by

$$S(t)\mathcal{G}[\phi(\tau)] = \mathcal{G}[\phi(t+\tau)] = \mathcal{G}[T(t)\phi(\tau)].$$

Indeed, any two points on $\mathcal{M}_t = S(t)\mathcal{G}[\phi(0)]$ must satisfy $|q_1 - q_2|_\alpha \le l|p_1 - p_2|_\alpha$, and so for each value of p in $P\mathcal{M}_t$ there exists a unique value of q, denoted $\psi(p)$, for which $p + \psi(p) \in \mathcal{M}_t$. Over its domain of definition, ψ satisfies the Lipschitz bound necessary for functions in $\mathcal{F}_l$. Since $\text{supp}(f) \subset \Omega_\rho$ it is clear that $\psi(p) = 0$ for $|p|_\alpha > \rho$. It remains to show that there are "no holes" in $\mathcal{M}_t$, i.e. that $P\mathcal{M}_t = PH$.

This follows by adapting the standard "transversality" proof of the Brouwer fixed point theorem[18] since the map $\pi_\phi(t) : PH \to PH$ defined by

$$p \mapsto P(S(t)(p + \phi(p))),$$

is equal to $e^{-At}p$ for $|p|_\alpha > e^{At}\rho$. Thus the map $e^{At}\pi_\phi(t)$ is continuous and equal to the identity for $|p|_\alpha > e^{At}\rho$. If there is some point $y \in PH$ that is not contained in the image of this map then this allows one to construct a retraction r of the unit ball onto its boundary (by defining $r(p)$ to be the point on the line from p through y which lies on the boundary) which is impossible (thanks to Toby Hall for suggesting this approach). This implies that $\pi_\phi(t)$ itself is onto. Consequently, $S(t)\mathcal{G}[\phi(0)] = \mathcal{G}[\phi(t)]$ with $\phi(t) = \psi$ and $\phi(t) \in \mathcal{F}_l$.

The operator $T(t)$ defined by $T(t)\phi(0) = \phi(t)$ clearly inherits the semigroup properties of $S(t)$ (thanks are due here to an anonymous referee who supplied the following argument). Indeed, consider $\pi_\phi^{-1}(t) \equiv \gamma_\phi(t) : PH \to PH$ (possible since the cone condition ensures that π_ϕ is injective), and set

$$[T(t)\phi](p) = QS(t)(\gamma_\phi(t)p + \phi(\gamma_\phi(t)p)).$$

Then

$$S(t)(p + \phi(p)) = p(t) + (T(t)\phi)(p(t)),$$

where $p(t) = \pi_\phi(t)p$. Setting $u_\phi = p + \phi(p)$, one has

$$u_{T(t)\phi}(P(S(t)u_\phi)) = S(t)u_\phi. \qquad (5)$$

To show that $T(t)$ is indeed a semigroup, use Eq. (5) repeatedly:

$$\begin{aligned} u_{T(t+s)\phi}(P(S(t+s)u_\phi)) &= S(t+s)u_\phi \\ &= S(t)(S(s)u_\phi) \\ &= S(t)[u_{T(s)\phi}(P(S(s)u_\phi))] \\ &= u_{T(t)T(s)\phi}[P(S(t)u_{T(s)\phi}(P(S(s)u_\phi)))] \\ &= u_{T(t)T(s)\phi}(P(S(t+s)u_\phi)). \end{aligned}$$

If $T(t)$ has a fixed point ϕ, then the manifold $\mathcal{G}[\phi]$ must be invariant under the flow $S(t)$. Theorem 2 is now shown to ensure the existence of such a fixed point.

To show that the semiflow $T(t)$ is dissipative and possesses the compactness property needed for theorem 2 requires a bound on $\|\phi(t)\|$ and $\|A^\epsilon\phi(t)\|$ for some $\epsilon > 0$. This can be obtained by considering the evolution equation for the Q component of u,

$$dq/dt + Aq + Qf(u) = 0,$$

which using the variation of constants formula gives

$$q(t) = e^{-At}q(0) - \int_0^t e^{-A(t-s)}Qf(u(s))\,ds.$$

Thus $|q(t)|_{\alpha+\epsilon}$ (choosing ϵ so that $\alpha - \beta + \epsilon < 1$) is bounded by

$$\begin{aligned} |q(t)|_{\alpha+\epsilon} &\le |q(0)|_\alpha\, b_\epsilon(t) + C_0 \int_0^t \|A^{\alpha-\beta+\epsilon}e^{-AQ(t-s)}\|\,ds \\ &\le |q(0)|_\alpha\, b_\epsilon(t) + C_0 I_{\alpha-\beta+\epsilon}, \end{aligned}$$

using standard expressions[3,5] to bound the integral and to obtain $b_\epsilon(t) \sim e^{-\Lambda t}$. Therefore

$$\|A^\epsilon\phi(t)\| \le \|\phi(0)\| b_\epsilon(t) + C_0 I_{\alpha-\beta+\epsilon}. \tag{6}$$

With $\epsilon = 0$ it is clear that $\{\phi \in \mathcal{F}_l^n : \|\phi\| \le 2C_0 I_{\alpha-\beta}\}$ is absorbing for $T(t)$, i.e. $T(t)$ is dissipative.

To show that a bounded set in $\mathcal{F}_l$ is mapped by $T(t)$ into a compact set X it is necessary to show that X is equicontinuous and pointwise compact (i.e. the closure of $\cup_{\phi\in X}\phi(p)$ is compact in $QH \cap D(A^\alpha)$ for each $p \in P\Omega_\rho$) and then apply the Arzelà-Ascoli theorem.[19] Clearly X consists of equicontinuous functions since they all have Lipschitz constant at most l, and X is pointwise compact since Eq. (6) with $\epsilon > 0$ gives a uniform bound on $\phi(p)$ in $D(A^{\alpha+\epsilon})$ for all $\phi \in X$, and a closed bounded subset of $D(A^{\alpha+\epsilon})$ is compact in $D(A^\alpha)$. Thus X *is* compact.

Therefore $T(t)$ satisfies all the conditions of theorem 2 and there exists an invariant manifold for the flow $S(t)$ given as the graph of a function $\phi \in \mathcal{F}_l$.

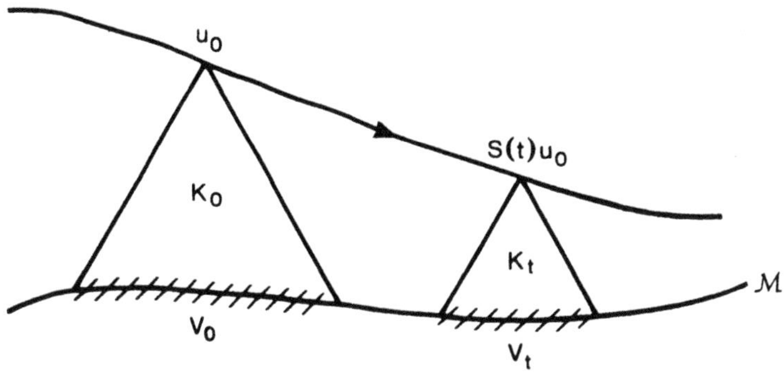

Figure 2 The shadow construction

The proof of the asymptotic completeness property is based on that in ref. 20 with considerable simplifications (see also ref. 6). Consider the complement of $\mathcal{C}_l$ centred at the point $u(t)$ of a trajectory,

$$\mathcal{C}'(u(t)) \equiv \{v \in D(A^\alpha) : |Qw|_\alpha \geq l|Pw|_\alpha, w = v - u(t)\}.$$

In the language of ref. 20, this cone casts a "shadow" $\mathcal{V}(u(t))$ on the inertial manifold, $\mathcal{V}(u) = \mathcal{C}'(u) \cap \mathcal{M}$. This construction is illustrated in figure 2.

Denote by $V(u)$ the closure of $\mathcal{V}(u)$ so that

$$V(u) = \{v \in \mathcal{M} : |Qw|_\alpha \geq l|Pw|_\alpha, w = v - u\}.$$

As the distance of $u(t)$ to the manifold decreases so does the distance between u and all points in $V(u)$. The idea of the proof is to find a point in V_0 that remains in V_t for all time.

The first stage is to show that $V(u)$ is compact. Since $V(u)$ is by definition closed and contained in $\mathcal{M}$, it suffices to show that $V(u)$ is bounded. Because $V(u) \subset \mathcal{M}$, this is equivalent to $PV(u)$ bounded in $PD(A^\alpha)$. Indeed, if $u = p + q$ and $v \in V(u)$ with $v = \pi + \phi(\pi)$, $|p - \pi|_\alpha \leq l|q - \phi(\pi)|_\alpha \leq l(|q|_\alpha + \|\phi\|)$, and as $|p - \pi|_\alpha \geq |\pi|_\alpha - |p|_\alpha$, compactness of $V(u)$ is ensured by the bound $|\pi|_\alpha \leq |p|_\alpha + l(|q|_\alpha + \|\phi\|)$.

Since both $\mathcal{M}$ and the complement of $\mathcal{C}'(u(t))$ are positively invariant, trajectories can only leave the shadows $\mathcal{V}_t$. Thus the sequence of points $u(n) = S(-t_n)v(t_n)$ with $v(t_n) \in \mathcal{V}_t$ are all contained in V_0. Since V_0 is compact there exists a subsequence of $u(n)$ which converges to a point $u_0 \in V_0$. This point must be contained in V_t for all $t \geq 0$, for if not there exists a neighbourhood N of u_0 in $\mathcal{M}$ and a time t_0 such that for all $t \geq t_0$, $S(t)N$ is disjoint from V_t, contradicting the definition of u_0.

Now the squeezing property of Eq. (4) is used: since $S(t)u_0 \in V_t$ for all t, the difference $w(t) = S(t)u_0 - u(t)$ lies outside $\mathrm{int}(\mathcal{C})$ for all $t \geq 0$, and so

$$\begin{aligned}|u(t) - S(t)u_0|_\alpha &= |w(t)|_\alpha \\ &\leq (1+l^{-1})|Qw(t)|_\alpha \\ &\leq (1+l^{-1})|Qw(0)|_\alpha\, e^{-kt} \\ &\leq (1+l^{-1})|Q(u(0)-u_0)|_\alpha\, e^{-kt} \\ &\leq (1+l^{-1})(|Qu(0)|_\alpha + \|\phi\|)e^{-kt},\end{aligned}$$

which is the asymptotic completeness property. In particular the manifold is exponentially attractive,

$$\mathrm{dist}_\alpha(S(t)u(0), \mathcal{M}) \leq C(X)e^{-kt} \qquad C(X) = (1+l^{-1})(\varrho + \|\phi\|),$$

uniform for $u(0)$ contained in any bounded set $X \subset \Omega_\varrho$.

3. Applications

Two applications of theorem 1 are now given. The first is a direct use in the case of the one-dimensional Kuramoto-Sivashinsky equation, and the second a general condition (the familiar spectral gap condition) is derived to ensure the strong squeezing property and hence an inertial manifold.

3.1 The Kuramoto-Sivashinsky equation

There has been much research on inertial manifolds for the Kuramoto Sivashinsky equation

$$u_t + u_{xxxx} + u_{xx} + uu_x = 0 \qquad u(x+L,t) = u(x,t). \tag{7}$$

The analytical work of ref. 21 culminated in an existence proof of inertial manifolds,[22] the dimension N being bounded by $cL^{3.75}$. Since the first result, many different methods of proof have been advocated, and the dimension estimates have improved slowly. The equation can be written in the form of (1) with $Au = u_{xxxx}$.[3]

The analysis of ref. 14 proceeds via the strong squeezing property in both $H = L^2$ and $V = D(A^{1/4})$ and a fixed point argument. It is proved that there exist

$$N_H \sim L^{2.46} \qquad \text{and} \qquad N_V \sim L^{2.89}$$

such that for $N > N_H$ the strong squeezing property holds in H, and for $N > N_V$ the strong squeezing property holds in V. The dimension estimates follow from a recent result[23] on the size of the absorbing set for Eq. (7). Clearly these results can be applied immediately to deduce the existence of an asymptotically complete inertial manifold in H and V with the same bounds on the dimension.[24]

3.2 General conditions for the Strong Squeezing Property

Attempting to find general conditions for the strong squeezing property leads to the "spectral gap condition", familiar in the literature. The proof of the following proposition is standard,[3,12] but is given here for completeness.

Proposition 3. *Suppose that* $0 \leq \alpha - \beta \leq \frac{1}{2}$. *If there exists an* n *such that the eigenvalues* $\lambda = \lambda_n$ *and* $\Lambda = \lambda_{n+1}$ *satisfy*

$$\Lambda - \lambda > C_1\Big((1+l)\lambda^{\alpha-\beta} + (1+l^{-1})\Lambda^{\alpha-\beta}\Big) \tag{8}$$

then the strong squeezing property holds with $k = \Lambda - C_1\Lambda^{\alpha-\beta}(1+l^{-1})$, *i.e. there exists an asymptotically complete inertial manifold which attracts at rate* k.

Proof. The idea is simply to consider

$$\frac{d}{dt}(|Qw|_\alpha - l|Pw|_\alpha),$$

and to show that this is negative when $|Qw|_\alpha = l|Pw|_\alpha$ by bounding $d|Qw|_\alpha/dt$ above and $d|Pw|_\alpha/dt$ below.

The difference w of two solutions satisfies the equation

$$dw/dt + Aw + f(u_1) - f(u_2) = 0. \tag{9}$$

Set $p = Pw$ and $q = Qw$. It is straightforward to obtain a lower bound on $d|Pw|_\alpha/dt$ by taking the scalar product of Eq. (9) with $A^{2\alpha}p$ to give

$$\left(\frac{d}{dt}|p|_\alpha\right)_{t=0} \geq -(\lambda + C_1\lambda^{\alpha-\beta}(1+l))|q(0)|_\alpha/l.$$

when $|q(0)|_\alpha = l|p(0)|_\alpha$.

For the Q equation,

$$dq/dt + Aq + QR(u_1) - QR(u_2) = 0 \tag{10}$$

the natural approach is to multiply Eq. (10) by A^α and then take the inner product with $A^\alpha q$, since

$$\frac{d}{dt}|q|_\alpha^2 = 2(A^\alpha dq/dt, A^\alpha q), \tag{11}$$

when q and dq/dt are in $D(A^\alpha)$. However, this need not be the case: from standard existence theorems[16,17] the guaranteed regularity is only $dq/dt \in D(A^\beta)$. All is not lost however, as Eq. (11) could be rewritten

$$(A^{2\alpha-\beta}q, A^\beta dq/dt), \tag{12}$$

using the self-adjointness of A, and since it is assumed that $\alpha-\beta \leq \frac{1}{2}$ it follows that $2\alpha-\beta \leq 1+\beta$, so Eq. (12) makes sense as it is guaranteed[16,17] that $q(t) \in D(A^{1+\beta})$.

So if one multiplies Eq. (11) by A^β and takes the scalar product with $A^{2\alpha-\beta}q$ then when $|q(0)|_\alpha = l|p(0)|_\alpha$ the equation becomes

$$\left(\frac{1}{2}\frac{d}{dt}|q|_\alpha^2\right)_{t=0} \leq -|q|_{\alpha+1/2}^2 + C_1(1+l^{-1})|q|_\alpha|q|_{2\alpha-\beta}$$
$$\leq -|q|_{\alpha+1/2}^2 + C_1\Lambda^{\alpha-\beta-1/2}(1+l^{-1})|q|_\alpha|q|_{\alpha+1/2},$$

since $\alpha+\frac{1}{2} \geq 2\alpha-\beta$. Now, observing that $-X^2+cX$ is decreasing[3] for $X > c/2$, provided that

$$|q|_{\alpha+1/2} > C_1\Lambda^{\alpha-\beta-1/2}(1+l^{-1})|q|_\alpha/2$$

which reduces to

$$\Lambda^{1-\alpha+\beta} > C_1(1+l^{-1})/2 \tag{13}$$

the following expression is valid :

$$\left(\frac{d}{dt}|q|_\alpha\right)_{t=0} \leq -(\Lambda - C_1\Lambda^{\alpha-\beta}(1+l^{-1}))|q(0)|_\alpha.$$

Then at $t=0$

$$\frac{d}{dt}(|q|_\alpha - l|p|_\alpha)_{t=0} \leq -\left(\Lambda - \lambda - C_1(1+l)\lambda^{\alpha-\beta} - C_1(1+l^{-1})\Lambda^{\alpha-\beta}\right)|q(0)|_\alpha,$$

which is negative provided that Eq. (8) holds, and so C_l is invariant. Note also that Eq. (8) implies Eq. (13). The exponential decay outside the cone follows directly from the Q component of the equation with $|q|_\alpha \geq l|p|_\alpha$.

4. Conclusion

This concise proof demonstrates explicitly the central rôle played by the strong squeezing property in the existence of inertial manifolds, and that this is the important property to try to prove given any particular equation. The derivation of the spectral gap condition in proposition 3 should come as no surprise since the cone condition can be closely related to the standard Lyapunov-Perron fixed point method.[13]

Acknowledgements

This work was supported by a grant from the SERC and by Trinity College, Cambridge. Many thanks to my PhD supervisor, Paul Glendinning, who has been exceedingly helpful throughout. Most of this paper was written while I was at the Georgia Institute of Technology, where Jack Hale, Shui-Nee Chow, and John Mallet-Paret not only made me feel welcome but provided me with many helpful discussions. Special thanks to the Reynolds family and the other members of Schola Cantorum of Atlanta who helped me feel sufficiently at home to work productively.

References

1. J. Mallet-Paret, *J. Diff. Eq.* **22** (1976) 331.
2. J.K. Hale, *Asymptotic Behavior of Dissipative Systems* (Math. Surveys and Monographs Vol 25, Amer. Math. Soc., Providence, RI, 1988).
3. R. Temam, *Infinite Dimensional Dynamical Systems in Mechanics and Physics* (Springer AMS 68; 1988).
4. J.M. Hyman and B. Nicolaenko, *Physica* **18D** (1986) 113.
5. C. Foias, G.R. Sell, and R. Temam, *J. Diff. Eq.* **73** (1988) 309.
6. H. Ninomiya, *J. Math. Kyoto Univ.* **32** (1992) 678.
7. S.-N. Chow, K. Lu, and G.R. Sell, *J. Math. Anal. Appl.* **169** (1992) 283.
8. P. Constantin, C. Foias, B. Nicolaenko, and R. Temam, *Integral Manifolds and Inertial Manifolds for Dissipative Partial Differential Equations* (Springer AMS 70, 1988).
9. P. Constantin, C. Foias, B. Nicolaenko, and R. Temam, *J. Dyn. Diff. Eq.* **1** (1989) 45.
10. A. Debussche, *Diff. Integral. Eq.* **3** (1990) 467–486.
11. J. Mallet-Paret and G.R. Sell, *J. Amer. Math. Soc.* **1** (1988) 805.
12. A. Rodriguez Bernal, *Appl. Anal.* **37** (1990) 95.
13. J.C. Robinson, *Dyn. Sys. Appl.* **2** (1993) 311 and **4** (1994) 145.
14. C. Foias, B. Nicolaenko, G.R. Sell, and R. Temam, *J. Math. Pures Appl.* **67** (1988) 197.
15. D.A. Jones and E.S. Titi, "C^1 Approximations of Inertial Manifolds for Dissipative Nonlinear Equations", *J. Diff. Eq.* (submitted).
16. D. Henry, *Geometric Theory of Semilinear Parabolic Equations*, (Lecture Notes in Math. Vol 840, Springer-Verlag, New York, 1981).
17. M. Miklavčič, *Pacific J. Math.* **118** (1985) 199.
18. V. Guillemin and A. Pollack, *Differential Topology* (Prentice-Hall, Englewood Cliffs, New Jersey, 1974).
19. A.W. Naylor and G.R. Sell, *Linear Operator Theory in Engineering and Science* (Springer AMS 40, 1982).
20. C. Foias, G.R. Sell, and E.S. Titi, *J. Dyn. Diff. Eq.* **1** (1989) 199.
21. B. Nicolaenko, B. Scheurer, and R. Temam, *Physica* **D16** (1985) 155.
22. C. Foias, B. Nicolaenko, G.R. Sell, and R. Temam, *C. R. Acad. Sci. Paris I* **301** (1985) 285.
23. P. Collet, J.-P. Eckmann , H. Epstein, and J. Stubbe, *Comm. Math. Phys.* **152** (1993) 203.
24. J.C. Robinson, *Phys. Lett. A* **184** (1994) 190.

D-BAR METHOD FOR FINITE-ZONE REFLECTIVE KdV POTENTIALS

YURY RODIN
Dept. of Mathematics, Wayne State University.
Detroit MI 48202 ,USA

ABSTRACT

The Cauchy problem for KdV equation with initial conditions determining finite-zone continued spectrum is reduced to a scalar boundary value problem and then to an elliptic differential system on a Riemann surface.Solvability conditions have been found.

1.Generalized Jost functions

Let

$$L\psi \equiv -\psi'' + u(x,t)\psi = \lambda\psi, \qquad -\infty < x < \infty \tag{1}$$

be the one-dimensional Schrödinger operator with a real potential $u(x,t)$ satisfying the KdV equatiuon

$$u_t = 6uu_x - u_{xxx}. \tag{2}$$

We will consider the Cauchy problem $u(x,0) = f(x)$ with such an initial function that the continuous spectrum of the operator L in the space $L_2(-\infty,\infty)$ at the moment $t = 0$ and hence for any t consists of a finite number of zones

$$E = \bigcup_{k=0}^{g-1}(\lambda_{2k},\lambda_{2k+1})\bigcup(\lambda_{2g},\infty) = \bigcup_{k=0}^{g} E_k \tag{3}$$

Let $\theta(x,\lambda)$ and $\phi(x,\lambda)$ be solutions of (1) determining by the conditions

$$\theta(x_0,t) = \phi'_x(x_0,t) = 1, \qquad \theta'_x(x_0,t) = \phi(x_0,t) = 0 \tag{4}$$

and $\psi_j(x,t)$ be the Weyl solutions of (1)

$$\psi(x.\lambda) = \theta(x,t) + m_j(\lambda)\phi(x,t), \qquad j = 1,2 \tag{5}$$

determined by the conditions

$$\int_{-\infty}^{x_0} |\psi_1(x,\lambda)|^2 dx < \infty, \qquad \int_{x_0}^{\infty} |\psi_2(x,\lambda)|^2 dx < \infty$$

for any non-real λ(see [5]). The Weyl functions $m_j(\lambda)$ satisfy the relations

$$m_j(\lambda) = (-1)^j \int_E [\frac{1}{\tau-\lambda} + \frac{\tau}{1+\tau^2}] d\rho_j(\tau) + a_j, \qquad j = 1,2 \tag{6}$$

where $\rho_j(\lambda)$ are the spectral densities of the operator L in $L_2(-\infty, x_0)$ for $j = 1$ and $L_2(x_0, \infty)$ for $j = 2$.

Let us denote by $\tilde{E}_j$ the cuts along the spectral zones E_j and by $\lambda_\pm$ the points belonging to the upper (+) and lower (-) sides of the cut $\tilde{E}_j$. Then, as it follows from (6)

$$\Im m_j(\lambda_\pm) = \pm(-1)^j \rho'_j(\lambda). \tag{7}$$

We will consider (see [2]) $f_j(x,\lambda)$ as analytic functions on different sheets of the surface $M = M_1 \cup M_2$ of the function

$$w^2 = R(\lambda) \equiv (\lambda-\lambda_0)(\lambda-\lambda_1)...(\lambda-\lambda_{2g}).$$

Note that the signs of $\Im m_1(\lambda)$ and $\Im m_2(\lambda^*)$ on opposite sides of the cut $\tilde{E}_j$ coincide.

There is a linear relation between ψ_j and $\bar{\psi}_j$ at any point λ of the continuous spectrum E.. Coefficients of this relation depend on the side of $\tilde{E}_j$ and the values of ψ_j have to be cosidered on different sheets M_j of M. Taking into account the rule of glueing of sheets we have the relation

$$\psi_1(x,\lambda) = a(\lambda) f_2(x,\lambda_*) + b(\lambda)\overline{f_2(x,\lambda^*)}, \qquad \lambda \in \tilde{E} = \cup \tilde{E}_j \tag{8}$$

where λ is a point on one of sides of the cut (λ_+ or λ_-) and λ^* is the point on the opposite side,i.e.λ_- or λ_+, correspondingly.

Because of (4) and (5) we have

$$W(\psi_j, \bar{\psi}_j) = -2\Im m_j(\lambda), \qquad \lambda \in \tilde{E}, \tag{9}$$

$$a(\lambda) + b(\lambda) = 1, \qquad a(\lambda) m_2(\lambda^*) + b(\lambda)\overline{m_2(\lambda^*)} = m_1(\lambda). \tag{10}$$

The following inequality is valid

$$|a(\lambda| > |b(\lambda)|, \qquad \lambda \in \tilde{E}. \tag{11}$$

Indeed,it follows from Eq.(10) that $\Im a(\lambda) = -\Im b(\lambda)$. Further,

$$\Im m_1(\lambda) = \Re a(\lambda)\Im m_2(\lambda^*) - \Re b(\lambda)\Im m_2(\lambda^*)$$
$$+\Im a(\lambda)\Re m_2(\lambda^*) + \Im b(\lambda)\Re m_2(\lambda^*) = [\Re a(\lambda) - \Re b(\lambda)]\Im m_2(\lambda^*).$$

Because of Eq.(7) the values $\Im m_1(\lambda)$ and $\Im m_2(\lambda^*)$ have the same sign that implies (11).

Let

$$\Psi_j(x,\lambda) = \alpha_j(\lambda)\psi_j(x,\lambda), \qquad j = 1,2, \qquad \lambda \in M. \tag{12}$$

Then

$$\Psi_1(x,\lambda) = \hat{a}(\lambda)\Psi_2(x,\lambda^*) + \hat{b}(\lambda)\overline{\Psi_2(x,\lambda^*)}, \qquad \lambda \in \tilde{E} \tag{13}$$

where

$$\hat{a}(\lambda) = a(\lambda)\alpha_1(\lambda)/\alpha_2(\lambda^*), \quad \hat{b}(\lambda) = b(\lambda)\alpha(\lambda)/\overline{\alpha_2(\lambda^*)}$$

. We will choose the multipliers satisfying the condition for $\lambda \in \tilde{E}$

$$\alpha_1(\lambda) = \Delta(\lambda)\alpha_2(\lambda^*), \quad \Delta(\lambda) = \sqrt{|a(\lambda)|^2 - |b(\lambda)|^2}, \quad \alpha(\infty) = 1. \tag{14}$$

Considering Eq.(14) as a Riemann boundary value problem on M we conclude that the problem is solvable in the class of functions holomorphic in M_1 and M_2 if and only if [2] the conditions

$$l_\mu = \int_{\tilde{E}} \ln \Delta(\tau)dw_\mu(\tau) \equiv 2\pi i(\text{mod periods of } dw_\mu), \ \mu = 1,2,...,g \tag{15}$$

are satisfied. Here

$$dw_\mu(\lambda) = \frac{P_{\mu-1}(\lambda)}{\sqrt{R(\lambda)}}, \qquad \mu = 1,2,...,g \tag{16}$$

are normalized Abelian differentials of M ($P_{\mu-1}$ are polynomials of the degree $\mu - 1$).

If conditions (15) do not valid, the problem (14) has no analytic single-valued solution but it is possible to choose a non-analytic solution. Let $A_0(s)d\overline{z(s)}$ where $z(s)$ is a local coordinate of the point $s \in M$ be a differential form of the type (0,1)piece-wise continuous on M and such that

$$\int_M A_0(s)d\overline{z(s)} \wedge dw_\mu(s) = l_\mu, \qquad \mu = 1,2,...,g. \tag{17}$$

Then there exists [2] a single-valued solution of the equation

$$\overline{\partial}\alpha = A_0(s)\alpha \tag{18}$$

satisfying the boundary condition (14)

$$\alpha_j(s) = \exp[\frac{1}{2\pi i}\int_{\tilde{E}} \ln \Delta(\tau) M(\tau, s) dz(\tau)$$

$$+\frac{1}{2\pi i}\int\int_m A_0(p) M(p, s) d\overline{z(p)} \wedge dz(p)], \qquad j = 1, 2. \tag{19}$$

Here $M(\tau,, s)$ is a Cauchy kernel of the surface M. Functions (12) with multipliers satisfying (13) will be called generalized Jost functions because the coefficients of the generalized scatterimg relation (8) satisfies the unitarity condition

$$|a(\lambda)|^2 - |b(\lambda)|^2 = 1 \tag{20}$$

For a big λ the Jost solutions have a behavior

$$\Psi_j(x, \lambda) = \exp[-i\sqrt{\lambda}x] + o(1), \qquad \lambda \in M_j, \qquad j = 1, 2 \tag{21}$$

where the branch of $\sqrt{\lambda}$ on M_j is determined by the relation $\text{sign}\{\Im\sqrt{\lambda}\} = (-1)^{j+1}$.

Let the branches of the function $\sqrt{R(\lambda)}$ be determined by the same rule $\text{sign}\{\Im\sqrt{R(\lambda)} = (-1)^{j+1}$ on M_j and $(\sqrt{R(\lambda)})_j$ mean a corresponding branch. The differential form

$$-\frac{\sqrt{R(\lambda)} + (\sqrt{R(a)})_j}{(z-a)\sqrt{R(\lambda)}} d\lambda, \qquad a\bar{\in}E$$

has a pole of the first order at the point a of the sheet M_j with a residue -1 and a pole with residue +1 at infinity. Let

$$d\omega_{a,j} = \left[-\frac{\sqrt{R(\lambda)} + (\sqrt{R(a)})_j}{(z-a)} + P_{g-1}(\lambda)\right] \frac{d\lambda}{\sqrt{R(\lambda)}} \tag{22}$$

where $P_{g-1}(\lambda)$ is a polynomial of degree $g-1$ with *real* coefficients such that

$$\int_{k_{2l}} d\omega_{a,j} = 0, \qquad l = 1, 2, ..., g. \tag{23}$$

Here $k_{2l}, ; (l = 1, 2, ..., g$ are cycles of canonic homology basis of M homotopic to the corresponding cuts $\tilde{E}_l$. The forms $d\omega_{a,j}$ are the normalized Abelian differentials of the third kind on M. We consider the functions

$$\beta_j(\lambda) = \exp\int_{a_0}^{\lambda} d\omega_{a,\hat{j}}, \quad \hat{0} = 1, \quad \hat{1} = 0, \quad \Im a_0 = 0, \quad a_0 < \lambda_0. \tag{24}$$

The function $\beta_j(\lambda)$ is single-valued on the correspondimg sheet M_j , has no zeros , a pole of the first order at infinity and real on $\tilde{E}$. Therefore, we have a representation

$$\Psi_j(x,\lambda) = \exp[-i\beta_j(\lambda)x]\Psi_j^0(x,\lambda), \qquad \Psi_j^0(x,\infty) = 1, \qquad \lambda \in M_j.(25)$$

The relation (13) can be rewritten in the form

$$\Psi_1^0(x,\lambda) = a_0(\lambda)\Psi_2^0(x,\lambda) + b_0(\lambda)\overline{\Psi_2^0(x,\lambda)}, \qquad \lambda \in \tilde{E},$$

$$a_0(\lambda) = \hat{a}(\lambda)\frac{\exp -i\beta_2(\lambda)}{\exp -i\beta_1(\lambda)}, b_0(\lambda) = \hat{b}(\lambda)\exp i\beta_2(\lambda)\exp_- ibeta_1(\lambda)] \qquad (26)$$

2.D-bar method

We consider the boundary problem (27 in the form

$$\Psi_1^0(\lambda) = a_0(\lambda)\Psi_2^0(\lambda) + b_0(\lambda)\overline{\Psi_2^0(\lambda)}, \qquad \bar{\partial}\Psi^0 = A_0\Psi^0, \qquad \Psi^0(\infty) = 1 \quad (27)$$

Because of Eqs.(20),(27)

$$0 < 1 + \frac{b_0(\lambda)\overline{\Psi_2^0(\lambda)}}{a_0(\lambda)\Psi_2^0(\lambda)} < 2$$

and hence the total number of zeros and poles of a solution reckoning multiplicity is equal to the Cauchy index of $a_0(\lambda)$

$$\kappa = \frac{1}{2\pi}\int_{\tilde{E}} d\arg a_(\lambda).$$

The function $Imm_2(\lambda)$ has no zeros outside of the real axis. Therefore, because of the relation

$$a(\lambda) = \frac{W(\psi_1,\bar{\psi_2})}{\Im m_2(\lambda)} \qquad (28)$$

where $W(\psi_1,\bar{\psi_2})$ is a Wronskyan , zeros of $a(\lambda)$ coincide with zeros of W and consequently with a discrete spectrum which is located in M_1. Let this spectrum be $a_1, a_2, ..., a_\kappa$. To separate the soliton component we can use the Abelian differential of the third kind

$$d\omega = \left[\sum_{k=1}^{\kappa} -\frac{2\Re\sqrt{R(a_k)}}{z-a_k} + P_{g-1}(\lambda)\right]\frac{d\lambda}{\sqrt{R(\lambda)}}$$

satisfying the condition (23) and the single-valued in M_1 function

$$\delta(\lambda) = \exp \int_{\infty}^{\lambda} d\omega. \tag{29}$$

Letting $f_1(\lambda) = \delta(\lambda)\Psi_1^0(\lambda)$, $f_2(\lambda) = \Psi_2^0(\lambda)$ we will get the problem

$$f_1(\lambda) = a_1(\lambda) f_(\lambda) + b_1(\lambda)\overline{f_2(\lambda)}, \qquad \mathrm{ind} a_1(\lambda) = 0. \tag{30}$$

Following [1] we continue functions $a_1(\lambda)$ and $b_1(\lambda)$ in M_1 such that $|b_1(\lambda)| < |a_1(\lambda)|$ for $\lambda \in M_1$(see[3]) and let

$$\Delta_1(\lambda) = \sqrt{|a_1(/lb)|^2 + |b(\lambda)|^2}.$$

We introduce functions

$$w(\lambda) = \begin{cases} \frac{\overline{a_1(\lambda)}}{\Delta^2(\lambda)} f_1(\lambda) - \frac{b(\lambda)}{\Delta^2(\lambda)} \overline{f_1(\lambda)}, & \lambda \in M_1 \\ f_2(\lambda) & \lambda \in M_2. \end{cases} \tag{31}$$

This function is a regular solution of the elliptic equation

$$\overline{\partial} w = q\overline{\partial w} + Aw + B\bar{w},$$

$$q(\lambda) == -\frac{b_1}{a_1}, \quad A(\lambda) = \frac{a_1 A_0 - \overline{\partial} a_1}{a_1}, \quad B(\lambda) = \frac{b_1 A_0 - \overline{\partial} b}{a}. \quad , \lambda \in M_1,$$

$$q = A = B = 0, \lambda \in M_2. \tag{32}$$

Let $q0(\lambda)$ be a smooth function on M such that $|q(\lambda) - q_0(\lambda)| < \epsilon$ a.e..We will look for a solution in the form

$$w(\lambda) = \exp T\phi + q_0(\lambda) \exp \overline{T\phi}, \tag{33}$$

$$T\phi = \frac{1}{2\pi i} \int \int_M \phi(s) M(s, \lambda) \overline{dz(s)} \wedge dz(s) \tag{34}$$

where $M(s\lambda)$ is the Cauchy kernel with *real* periods (see Appendix), a soution is looked in the space $L_r(M)$, $2 < r < 2 + \beta$ and

$$\|.\|_{L_r(M)} = \|.\|_{L_r(M_1)} + \|.\|_{L_r(M_2)}$$

We get the singular integral equation

$$\phi(\lambda) = K\phi, \quad K\phi = \frac{\exp \overline{T\phi}}{\exp T\psi} [\mu \overline{S\phi} + b] + a$$

$$a = A + B\overline{q}_0 + q\overline{\partial q_0}, \quad \mu = \frac{q - q_0}{1 - q\overline{q_0}}, \quad |\mu| < 1.$$

$$b = Aq_0 + b - \overline{\partial} q_0. \tag{35}$$

Here

$$S\phi = \overline{\partial} T\phi = \frac{1}{2\pi i} \int\int_M \phi(s) M(s, \lambda \overline{dz(s)} \wedge dz(s). \tag{36}$$

The norm of the operator S in $L_r(M)$ for a small β is close to 1, $\|S\|_{L_r(M)} < 1 + \eta$ (see3). Let

$$\Pi\phi = \frac{\exp \overline{T\phi}}{\exp T\phi}, \qquad H\phi = \mu(\Pi\phi)\overline{S\phi}. \tag{37}$$

Because of $|\pi\phi| = 1$ we have the estimation

$$\|H\phi\|_{L_r(M)} \leq \sup \|\overline{\phi}\|_{L_r(M)} < \epsilon(1+\eta) < 1$$

for a small β, $2 \leq r < 2 + \beta$. herefore, the operator $1 - H$ is convertable and

$$\phi = H_0\phi = (1 - H)^{-1}(b\Pi\phi + a). \tag{38}$$

The operator H_0 is completely continuous and

$$R = \|H_0\phi\|_{L_2(M)} \leq \|(1 - H)^{-1}\|_{L_2(M)}(\|a\| + \|b\|) + L_2(M).$$

Therefore, the operator H_0 in the space $L_2(M)$ maps the sphere S_R of radius R into itself and hence the Eq.(35)is solvable.

It is possible to show that this point is unique.[4]

Let ϕ be the fixed point of the operator K. The function $w(\lambda)$ determined by the Eq.(33) is, generally speaking, multi-valued. It can be shown [3] that in order the function $w(/lb)$ to be single-valued it is necessary and sufficient that

$$\int_M \phi\overline{dz(s)} \wedge dw_k(s) \equiv 0(\text{mod periods of } dw_k), \quad k = 1, 2, ..., g \tag{39}$$

The function ϕ ,of course ,depends on q_0 chosen . But the conditions (37) are valid or not valid for all q_0 simultaneously (for small enought ϵ).

Dynamics of the coefficients can be found from the second equation of the L-A pair. The case $b \equiv 0$ gives the classical theory of finite-zone potentials[7].

APPENDIX. Cauchy Kernel

There are different analogs of the Cauchy kernel on a Riemann surfaces(see[2]).The simplest variant can be constructed by the formula

$$M(p,q)dz(p) = \partial_{z(p)}[\omega_{p_0p}(q) - \omega_{p_0p}(q_0)]. \tag{40}$$

Here $\omega_{p_0p}(q)$ is an Abelian integral of the third kind with poles $q = p_0, p$ with residues -1 and +1 ,correspondingly. Points p_0, q_0 are arbitrary selected points of the surface and $z(p)$ is a local coordinate of p. The integral ω_{p_0p} is detrrmined up to an arbitrary combination of Abelian integrals of the first kind (16). If $p_0 = \infty$, we can let

$$\omega_{\infty,p}(q) = \int_{q_0}^{q} d\omega_{aj} \tag{41}$$

where differential (22) corresponds to the point $p = (a, j)$ of the surface M . We will get a kernel with real periods if instead of Eq.(23) we will use conditions

$$\int_{k_l} d\omega_{a,j} = 0, \qquad l = 1,2,...2g. \tag{42}$$

Coefficients of P_{g-1} are complex. The cycle k_{2l-1} consists of two segments $[\lambda_{2l-1}, \lambda_{2l}]$ passed twice,on each of sheets of M.

References

1. Bojarsky *On Stability of the Hilbert Problem for Holomorphic Vector* Bull. of Acad. of Sci, Georgia SSR **21**,391-398 (1958)

2. Rodin Yu.L. *The Riemann Boundary Problem on Riemann Surfaces.*D.Reidel Publ. Co. Dordrecht (1987)

3. Rodin Yu.L.*Singular Integral Operators on Riemann Surfaces*,pp.1-12 (submitted to "Jorn. Geometric Analysis")

4. Rodin Yu. L. *Single-valuedness Classes for Elliptic Differential Equations on Riemann Surfaces and Boundary Value Problems*(submitted in "AMS Transactions").

5. Titchmarsh E.S. *Eigenfunction Expansions Associated with Second-order Differential Equations.* Clarendon Press Oxford (1946)

6. Vekua I.N. *Generalized Analytic Functions.*Pergamon Press London(1962)

7. Zakharov V.E., Manakov S.V., Novikov S.P. Pitaevsky L.P. *Soliton Theory.Inverse Problem Method.*Pergamon Press NY(1984)

ELEMENTS OF CRITICAL POINTS THEORY FOR (3+1)-DIMENSIONAL TOPOLOGICAL SOLITONS: STRUCTURE OF TORONS IN THE FADDEEV MODEL

YURII P. RYBAKOV and VALERII I. SANYUK
Theoretical Physics Department
Peoples' Friendship University of Russia
Miklukho-Maklay Str. 6, Moscow 117198, Russia
e-mail: sanyuk@udn.msk.su

In this report we mostly deal with the theory of critical points for the (3+1)-dimensional $\vec{n}$-field (or Faddeev) model. It is shown here that the energy functional of the model is bounded from below through the Hopf index, attains its minimum in all homotopy classes on axisymmetric configurations, and that there exist G_2-invariant critical points of the toron type.

1. Introduction

The main goal of the report is to present results on the critical points theory for (3+1)-dimensional nonlinear σ-models, which prove to be attractive both from physical and mathematical points of view. From the physical point one might consider these models as universal ones due to wide range of their applications. In Particle Physics, where σ-models originated, they provide with a non-point-like description of baryons (the Skyrme model) and heavy mesons (the Faddeev model), reproduce the Current Algebra results for hadronic sector of QCD (see [1-3] and references therein), if not to speak about perspectives related with the Topological σ-model approach (see, e.g. [4]) as a part of the 'Theory of Everything'. Among other fields of σ-model applications here should be mentioned models of nematic liquid crystals and ferromagnets, nonlinear elasticity and fluid dynamics, and more. We are not going to discuss here the latter topic at any length, but rather concentrate on some mathematical problems, related with the structure of critical points in the Faddeev model.

In our recent papers [5, 6] we formulated the problem of construction a critical points theory for (3+1) dimensional nonlinear σ-models. Therefore here we just reproduce the main questions (with respect to a descriptive functional $H[\phi]$), which one has to answer in order to construct a theory of critical points, analogous to the finite dimensional Morse theory or Lusternik-Schnirelman theory (see e.g. [7, 8]). The questions are the following:

1. Whether $H[\phi]$ is bounded from below?
2. Does $H[\phi]$ attains its minimum?

3. Does $H[\phi]$ have critical points?

4. What is the structure of critical points?

Let us remind, that the role of fundamental variables (elements of the configuration space) for topologically nontrivial σ-models in (3+1) dimensions, is played by mappings

$$\phi(\vec{x}, t) : R^3 \otimes R^1 \to \Phi, \tag{1}$$

from the Minkowskian space-time $R^3 \otimes R^1$ into *a field manifold* Φ. In the Skyrme model case we have $\Phi = 3DS^3 \simeq SU(2))$, and in the case of the Faddeev model $\Phi = 3DS^2$. We regard as *topologically nontrivial σ-models* those, where fields (1) are endowed with a nontrivial topological index (a degree of mapping, a Hopf index, a Chern-Pontryagin number etc.) For this type of models the positive answer to the first question (from the above list) is almost straightforward, since the descriptive functionals, say $H[\phi]$, are bounded from below through the corresponding topological index (or charge) Q:

$$H[\phi] > f(|Q|), \tag{2}$$

where $f(|Q|)$ is a monotonously increasing function.

Actually, this is the case for the class of $G-$invariant models (where $G-$ is a compact group), which admit existence of topological charges, and, in particular, for chiral models. In chiral models vacuum states correspond to fixed points, like $\varphi_\infty \in \Phi$, and therefore field configurations, reasonable from the point of physics, are to satisfy natural boundary conditions $\varphi \to \varphi_\infty$ as $x \to \infty$. In this way we come to the effective compactification of the domain $\boldsymbol{R}^3$ into S^3, and this allows one to classify physical states of the model by means of the third homotopy group $\pi_3(M)$.

To get answers to the second question, and partially to the forth one from the list, we consider special classes of equivariant field configurations $\phi_0(\vec{x})$, defined by the condition

$$\phi_0(\vec{x}) = 3D\hat{T}_g\phi_0(g^{-1}\vec{x}), \quad g \in G, \tag{3}$$

with $\hat{T}_g$ being the representation operator of the group G. For more details see e.g. [1], [2]. In accord with the Coleman-Palais principle extremals of a G-invariant functional in the class of invariant fields are in fact true extremals.

For (3+1)-dimensional chiral models it was found that two groups G_i are of special interest, namely

$$\begin{aligned} G_1 &= 3D\mathrm{diag}\,[SO(3)_I \otimes SO(3)_S], \\ G_2 &= 3D\mathrm{diag}\,[SO(2)_I \otimes SO(2)_S], \end{aligned}$$

with $SO(n)_I$, $n = 3D2, 3$ being the group of isospin rotations and $SO(n)_S$ denoting the group of spatial rotations. We claim that these are the only maximal compact continuous groups which possess the equivariant fields with nontrivial topological indices in (3+1)-dimensional models.

The special role of G_i-invariant fields together with a subsequent strategy in searching for minima of G-invariant functionals comes out from the following

Theorem. *Let a G_i-invariant field ϕ_0, $i = 3D1, 2$, realizes the minimum of the Hamiltonian H, restricted to the class of G_i-invariant fields. Assume the functional $H[\phi]$ to be convex with respect to the derivatives $\partial_i\phi$ at the point $\phi = 3D\phi_0$. Then the field ϕ_0 realizes* the true minimum *of $H[\phi]$.*

In other words, the theorem allows to reduce the initial variational problem $\delta H[\phi] = 3D0$ to that for G-invariant fields only: $\delta H[\phi_{\text{inv}}] = 3D0$. The obtained results might be supported or checked up by *Direct Minimization in Extended Phase Space Procedure*, combined with *Spherical Rearrangement Method*, and at the end of this procedure one needs to check whether the obtained minimal configurations would satisfy the equation of motion.

A detailed description of these methods in application to the Skyrme model one can find in Ref. 1, Ch. 5. In our previous publications [5, 6] we already discussed thus obtained results in the light of critical points theory. Here we will try to answer the questions listed above for another popular (3+1)-dimensional model, known as n-field or Faddeev's model.

2. G-Invariant Critical Points for the Faddeev model

First, this model was suggested by L.D. Faddeev [9] as a model of leptons, but later on it was found to be applicable as well for magnets, liquid crystals and so on. The fundamental variables of the model is the set of three scalar fields $n^a(t,\vec{r}) : \boldsymbol{R}^1 \times \boldsymbol{R}^3 \to S^2$, $|\,\vec{n}\,| = 3D1$, subject to the boundary conditions $n^a(t,\infty) = 3D\delta_3^a$. Introducing an auxiliary 4-vector a_μ, we can relate it with the n-field through

$$f_{\mu\nu} = 3D\partial_\mu a_\nu - \partial_\nu a_\mu = 3D2\,\varepsilon_{abc}\partial_\mu n^a \partial_\nu n^b\, n^c. \tag{4}$$

Therewith the n-fields are classified with the help of the homotopy group $\pi_3(S^2) = 3D\boldsymbol{z}$, and characterized by the Hopf index Q_H. The latter is defined as the number of connectedness of $\vec{b}$−lines, related with the n-field through the relation

$$f_{ik} = 3D\varepsilon_{iks}b_s = 3D2\varepsilon_{abc}\partial_i n^a \partial_k n^b n^c.$$

Then the Hopf index might be expressed as

$$Q_H = 3D - \frac{1}{(8\pi)^2}\int d^3x(\vec{a}\,\vec{b}), \quad \vec{b} = 3D\mathrm{rot}\,\vec{a}. \tag{5}$$

From (5) and the structure of Faddeev's model Lagrangian

$$\mathcal{L}_F = 3D - \frac{\varepsilon^2}{4} f_{\mu\nu}^2 + \lambda^2(\partial_\mu n^a)^2 \tag{6}$$

one can derive the following estimate

$$E > \mu|\,Q_H\,|^{3/4}, \quad \mu = 3D\mathrm{const}. \tag{7}$$

The details one can find, for example, in Ref. 10. In this way we have got the positive answer to the first question, as listed in Introduction.

Next we have to study the structure of critical points in the Faddeev model. To this end let us write down the static Hamiltonian with an additional mass term $M^2(1 - n_3)$, which provides the n-field with the fast enough fall down at spatial infinity

$$H = 3D \int dx \, [\varepsilon^2 \vec{b}^2/2 + \lambda^2 (\partial_i \vec{n})^2 + M^2 (1 - n_3)], \tag{8}$$

and perform the so-called "lifting procedure", which consists in passing to an intermediate S^3-sphere manifold, determined by angles A, B, C. This "lifting" can be done by introducing a "spinor" field

$$\psi = 3D\mathrm{col}\,[\cos A + i \sin A \cos B, \; sin A \sin B \, e^{iC}]$$

and putting

$$\vec{n} = 3D\psi^+ \vec{\tau} \psi. \tag{9}$$

Then it can be easily checked, that

$$\vec{b} = 3D - 4i \, [\nabla\psi^+ \nabla\psi] = 3D\mathrm{rot}\, \vec{a}, \quad \vec{a} = 3D - 2i \, \psi^+ \overleftrightarrow{\nabla} \psi. \tag{10}$$

Next we substitute (9) and (10) into Eq. 8 and find

$$\begin{aligned} H - \mu |Q_H|^{3/4} &= 3D \textstyle\int dx \, \{2\varepsilon^2 \sin^2 \phi [\nabla\phi \nabla(C - \varphi)]^2 + \\ &+ M^2 (1 - \cos\phi) + \lambda^2 [(\nabla\phi)^2 + \sin^2\phi (\nabla C - \nabla\varphi)^2]\} - \\ &- \mu \left| \tfrac{1}{8\pi^2} \textstyle\int dx \, \sin\phi \, ([\nabla\phi \nabla\varphi] \nabla C) \right|^{3/4} =, \end{aligned} \tag{11}$$

where it is denoted

$$\sin A \sin B = 3D \sin(\phi/2), \quad \mathrm{tg}\, A \cos B = 3D\mathrm{tg}\, \varphi. \tag{12}$$

The substitution (9), which is known as *the Hopf mapping*, brings the Hopf index Q_H to the form identical to that of the degree of mapping $\mathbf{R}^3 \to S^3$, or, after transformations (12) to the form

$$Q_H = 3D \frac{1}{8\pi^2} \int dx \, \sin\phi \, ([\nabla\phi \nabla \, v] \nabla C), \tag{13}$$

which one can consider as the equivalence relation between homotopy groups $\pi_3(S^2) = 3D\pi_3(S^3)$.

Now we are in the position to apply the minimization in an extended configuration space. By changing the directions of vectors $[\nabla\phi \nabla(C - \varphi)]$, $\nabla\phi$, $\nabla(C - \varphi)$, we get convinced that the minimum of (11), which corresponds to the minimal density of topological charge, is realized either when $[\,[\,\nabla\phi\,\nabla\varphi\,]\,\nabla C\,] = 3D0$, or if

$$(\nabla C\,\nabla\phi) = 3D(\nabla C\,\nabla\varphi) = 3D0.$$

Taking into account the latter condition, one finds that the Hamiltonian will contain the term $(\nabla C)^2$, and the situation becomes analogous to that of the Skyrme Hamiltonian for higher homotopy classes (see Ref. 1, Ch. 7). Therefore the very same argumentation (which we are not going to repeat here) leads to the G_2 –invariant minimal configuration, with $C = 3Dk\alpha$, $k \in \boldsymbol{z}$. The corresponding ansatz for the $\vec{n}$ –field would be axisymmetric one :

$$\beta = 3D\beta(\rho, z); \quad \gamma = 3Dk\alpha + v(\rho, z), \tag{14}$$

where ρ, α, z are cylindrical coordinates.

The expression for the topological charge Q_H on the ansatz (13) becomes simpler. In fact, from Eqs. 10, 13 and 14 one can deduce

$$\phi = 3D\beta, \quad \varphi = 3D - v, \quad \gamma = 3DC - \varphi, \quad C = 3Dk\alpha. \tag{15}$$

Substitution of (15) into the formula (13) leads to

$$Q_H = 3D\frac{k}{4\pi}\int\limits_0^\infty d\rho \int\limits_{-\infty}^{+\infty} dz\, \sin\beta\,(\beta_\rho v_z - \beta_z v_\rho). \tag{16}$$

Further, by noting that $\vec{e}_\alpha\, \sin\beta\,(\beta_\rho v_z - \beta_z v_\rho) = 3D\mathrm{rot}\,(v\nabla(1-w))$, where $w = 3D\cos\beta$, the integral in (16) via Stokes' theorem can be transformed into

$$Q_H = 3D\frac{k}{4\pi}\oint\limits_l v\,(\vec{dl}\,\nabla)(1-w). \tag{17}$$

From=20(17) it follows that $Q_H \neq 3D0$, if the contour l in the ρ, z-plane envelops the lines C_i, where the angle function v breaks. In accord with the Eq. 15 the jumps $[v]_i$ of this function on the lines C_i should be multiple of 2π, i.e. $[v]_i = 3D2n_i\pi$, $n_i \in \boldsymbol{z}$. Therewith from (17) we get the following expression for the Hopf index:

$$Q_H = 3D - \frac{k}{2}\sum_i n_i \int\limits_{C_i} d(1-w). \tag{18}$$

At the same time from the $\vec{n}$ –field single-valuedness it follows that the lines C_i might end at the points, which are mapped into poles of the S^2-sphere where $w = 3D \pm 1$. The finiteness of the vector $\vec{b}$ on the Z-axis implies $w(\infty) = 3D1$, and $w(\rho = 3D0) = 3D1$, therefore the lines C_i either have their origin at infinity, or on the Z-axis, and that corresponds to the north pole of the sphere S^2. If $Q_H \neq 3D0$, then one has to assume that lines C_i are ended at the points, which are mapped into the south pole of S^2, where $w = 3D - 1$. Thus the formula (18) can be presented in the form

$$Q_H = 3Dk \sum_i n_i. \tag{19}$$

Further simplification comes under the assumption that there is the only line C_i and that it corresponds to the jump $[v] = 3D2n\pi$, then from (19) one gets the simplest expression for the Hopf index

$$Q_H = 3Dk\, n. \tag{20}$$

As it is evident from (15) the surface $\beta = 3D\mathrm{const}$ is homeomorphic to the torus T^2, and an image of a point on the S^2-sphere is a closed $\vec{b}-$ line, which is situated on the torus T^2, and which winds it k times lengthwise and n times in the transverse direction. Then it proves convenient to choose the toroidal coordinates ξ, η, putting

$$\rho = 3Da \triangle \operatorname{sh} \eta, \quad z = 3Da \triangle \sin \xi, \quad \triangle = 3D(\operatorname{ch} \eta - \cos \xi)^{-1},$$

where $a-$ is a length scale and it is assumed that $\xi \in [-\pi, \pi]$, $\eta \in [0, \infty)$.

The structure of the configuration (14) is determined by the break-line C of the function v. In terms of toroidal coordinates this line is naturally identified with the interval $z = 3D0$, $\rho \in [0, a]$. Then the function v has to satisfy the boundary condition $v(\xi = 3D\pi) - v(\xi = 3D - \pi) = 3D2n\pi$, which is fulfilled on the ansatz of the form

$$v = 3Dn\,\xi + u, \quad n \in \mathbf{z}, \tag{21}$$

where u is a function periodic in ξ with the period 2π. One can further note that the extremal value of the functional

$$I_1[\beta, v] = 3D\frac{1}{2}\int_0^{\infty} \rho d\rho \int_{-\infty}^{+\infty} dz \, \sin^2 \beta \{(\nabla v)^2 + [\nabla \beta \nabla v]^2\} \tag{22}$$

does not depend on a particular choice of the curve C. The functional (22) is related with the Hamiltonian of the model as $H = 3D4\pi\varepsilon\lambda\sqrt{2}\,I$, $I = 3DI_0 + I_1$, where

$$I_0[\beta] = 3D\tfrac{1}{2}\int_0^{\infty} \rho d\rho \int_{-\infty}^{+\infty} dz \Big[(\nabla\beta)^2 \; + \tfrac{k^2}{\rho^2} \sin^2 \beta \times \\ \times (1 + (\nabla\beta)^2) + \nu\,(1 - \cos\beta)\Big], \tag{23}$$

and the following notation $2M^2\varepsilon^2/\lambda^4 = 3D\nu$, and the scale transformation $\vec{r} \to \vec{r}\sqrt{2}\,\varepsilon/\lambda$ where used. In fact, let us denote by $\vec{n}$ and $\vec{t}-$ the normal and the tangent vectors of the curve C, respectively, and let $P = 3D\{z = 3D0, \ \rho = 3Da\}-$ be the end point of this curve. Then, taking into account the equation for v, and performing an integration by parts, one can convert the functional I_1 into the form

$$I_1 = 3Dn\,\pi \int_C dl \, \rho \, \sin^2 \beta \, (J \nabla_t \beta - \nabla_n v)$$

where $J = 3D\nabla_n\beta\,\nabla_t v - \nabla_t\beta\,\nabla_n v = 3D\beta_z v_\rho - \beta_\rho v_z$, $\nabla_n = 3D(\vec{n}\,\nabla)$, $\nabla_t = 3D(\vec{t}\,\nabla)$. But the function v is subject to the equation

$$\partial_z\left[\,\rho\sin^2\beta\,(v_z - \beta_\rho\,J)\right] + \partial_\rho\left[\,\rho\sin^2\beta\,(v_\rho + \beta_z\,J)\right] = 3D0,$$

therewith one can put

$$[\,\rho\sin^2\beta\,(v_\rho + \beta_z\,J)] = 3DU_z, \quad [\,\rho\sin^2\beta\,(v_z - \beta_\rho\,J)] = 3D - U_\rho \tag{24}$$

and can bring the functional I_1 to the form

$$I_1 = 3Dn\,\pi\int\limits_C dU = 3Dn\,\pi\,U(P), \tag{25}$$

where it is chosen $U(\infty) = 3DU(0) = 3D0$.

We can also consider the transformations (24) as a change of variables $\beta,\, v \rightarrow \beta,\, U$, then one is able to perform a symmetrization of functions β, U with respect to reflections $z \rightarrow -z$. In accord with Eq. 22 the extremal value of the functional I_1 remains unchanged under these transformations, while the value of $I_0[\beta]$ will decrease on the decreasing rearrangement of the function β (see Ref. 1, §5.2). In view of this the optimal configuration corresponds to the even function β with respect to z. Next we make a decomposition $v = 3Dv_+ + v_-$ into even and odd in z parts, and, using (22), find that

$$I_1[\beta,\, v] = 3DI_1[\beta,\, v_+] + I_1[\beta,\, v_-].$$

Thus the minimum of I_1 corresponds to the choice $v_+ = 3D0$, and the sought for configuration will obey the following symmetry:

$$\beta(\rho,\, -z) = 3D\beta(\rho,\, z), \quad v(\rho,\, -z) = 3D - v(\rho,\, z). \tag{26}$$

In terms of toroidal coordinates this means that the function β is even with respect to ξ and also periodic with the period 2π.

The structure of the functionals I_0 and I_1 tells us, that $\sin\beta \rightarrow 0$ for $\eta \rightarrow \infty$ and also for $\eta \rightarrow 0$. Noting that the point P corresponds to the north pole of S^2-sphere, we convince ourselves that the function $\beta(\xi,\, \eta)$ has to satisfy the boundary conditions:

$$\beta(\eta = 3D0) = 3D0, \quad \beta(\eta = 3D\infty) = 3D\pi. \tag{27}$$

Thus we got the answer to the forth question from the list in Introduction that the critical points of the Faddeev functional in all homotopy classes will have toroidal structure. Next we have to prove the existence of these torons and with this in mind it proves to be useful to rewrite our functionals in toroidal coordinates:

$$\begin{aligned} I_0 &= 3D\tfrac{1}{2}\int\limits_0^\infty d\eta\int\limits_{-\pi}^{\pi} d\xi\,\Big[(\beta_\eta^2 + \beta_\xi^2)\left(a\Delta\,\mathrm{sh}\,\eta + k^2\tfrac{\sin^2\beta}{a\Delta\,\mathrm{sh}\,\eta}\right) + \\ &\quad + k^2\sin^2\beta\tfrac{a\Delta}{\mathrm{sh}\,\eta} + \nu a^3(1-\cos\beta)\Delta^3\mathrm{sh}\,\eta\Big], \\ I_1 &= 3D\tfrac{1}{2}\int\limits_0^\infty d\eta\int\limits_{-\pi}^{\pi} d\xi\,\sin^2\beta\,\mathrm{sh}\,\eta\,\times \\ &\quad\times\left[(v_\eta^2 + v_\xi^2)a\Delta + \tfrac{1}{a\Delta}(\beta_\eta v_\xi - \beta_\xi v_\eta)^2\right]. \end{aligned} \tag{28}$$

3. The Existence of Torons in the Faddeev Model

First let us obtain some *a priori estimates.* In notations $\vec{X} = 3D\nabla\beta$, $\vec{Y} = 3D\sin\beta\,\nabla v$, and assuming that $k \geq 1$ we can rewrite the inequality (9.18) from Ref. 10 in the following form

$$\begin{aligned}\int_0^\infty \rho d\rho \int_{-\infty}^{+\infty} dz \Big\{\vec{X}^2\left(1+\tfrac{k^2}{\rho^2}\sin^2\beta\right) + \vec{Y}^2 + [\vec{X}\,\vec{Y}]^2 + \\ + \tfrac{k^2}{\rho^2}\sin^2\beta + 2\nu\,\sin^2\tfrac{\beta}{2} - \tfrac{4k}{\rho}\sin\beta\,[\vec{X}\,\vec{Y}]_\alpha\Big\} > 0.\end{aligned} \tag{29}$$

Taking the second variation of the integrand in the above expression with respect to increments $\delta\vec{X} = 3D\vec{x}$, $\delta\vec{Y} = 3D\vec{y}$, when imposing the condition of local convexity of the functional with respect to derivatives, as the necessary condition for minimum, we get

$$\begin{aligned}F = 3D \quad \vec{x}^2\left(1+\tfrac{k^2}{\rho^2}\sin^2\beta\right) + \vec{y}^2 + ([\vec{X}\,\vec{y}] + [\vec{x}\,\vec{Y}])^2 + \\ +2\,[\vec{X}\,\vec{Y}]\cdot[\vec{x}\,\vec{y}] - \tfrac{4k}{\rho}\sin\beta\,[\vec{x}\,\vec{y}]_\alpha > 0.\end{aligned} \tag{30}$$

The quadratic form F would be positive if the following inequality holds

$$1 + \frac{k^2}{\rho^2}\sin^2\beta > \left(\frac{2k}{\rho}\sin\beta - [\vec{X}\,\vec{Y}]_\alpha\right)^2, \tag{31}$$

which is equivalent to the boundedness of derivatives $\nabla\beta$ and ∇v in the domain Ω, where $\sin\beta > \delta$, $\rho > \delta_1$. It is almost evident that $\Omega = 3D\boldsymbol{R}^3\backslash(\Omega_1 \cup \Omega_2 \cup \Omega_3)$, where it is denoted

$$\Omega_1 = 3D\{\eta > T \to \infty\},\ \Omega_2 = 3D\{\rho < \delta_1 \to 0\},\ \Omega_3 = 3D\{r > R \to \infty\}.$$

Thus we are in the situation similar to that in higher homotopy classes of the Skyrme model (see e.g. [1], Ch. 7; [5], [10]) and have to examine the functional $I(\Omega)$ with almost the same structure. Therefore we can construct minimizing sequences of functions β_n, $v_n \in W^1_\infty(\Omega)$, which would be *-weakly convergent and the functional $I(\Omega)$ would be lower weakly semicontinuous.

Now we are going to enlarge the domain Ω to $\boldsymbol{R}^3$ and to get convinced that the functional $I(\Omega)$ depends continuously on its domain. In order to see it we will examine the behavior of functions β, v, treating them as solutions of corresponding Euler-Lagrange equation in the domains of degeneracy Ω_i.

1) In the domain Ω_1 we put $\eta \to \infty$, $\pi - \beta = 3Dy \to 0$, $u \to 0$, $v \to n\,\xi$, which leads to the functional

$$\begin{aligned}I(\Omega_1) &= 3D\tfrac{a}{2}\int_T^\infty d\eta \int_{-\pi}^{\pi} d\xi\,\Big\{y_\xi^2 + y_\eta^2 + y^2\times \\ &\times\left[v_\xi^2 + v_\eta^2 + (y_\xi\,v_\eta - y_\eta\,v_\xi)^2\,e^{2\eta}/a^2\right]\Big\}.\end{aligned} \tag{32}$$

Separating the variables in Euler-Lagrange equations and assuming $k,\,n \geq 1$, we find

$$y = 3DA\,e^{-n\eta}, \quad u = 3DO[\sin\xi\,\exp(-\frac{B_n}{2n}e^{2n\eta})], \tag{33}$$

where $A = 3D\mathrm{const}$, $B_1 = 3DA^2/a^2$, $B_{n>1} = 3D1$.

The substitution of (33) into $I(\Omega_1)$ gives $I(\Omega_1) = 3DO(e^{-2nT}) = 3DO(\delta^2)$, where $\delta = 3De^{-nT}$ is determined from the condition $\delta \geq \pi - \beta = 3DO(e^{-n\eta})$ =AF=E0=A8 $\eta \geq = T$.

2) It proves convenient to use cylindrical coordinates in the domain Ω_2 putting $\beta \to 0$ and bringing the functional $I(\Omega_2)$ to the form

$$\begin{aligned} I(\Omega_2) &= 3D\tfrac{1}{2}\int_0^{\delta_1}\rho d\rho \int_{-\infty}^{+\infty} dz \left\{(\beta_\rho^2 + \beta_z^2)\left(1 + \tfrac{k^2}{\rho^2}\beta^2\right) + \right. \\ &\left. + \beta^2\left(\tfrac{k^2}{\rho^2} + \tfrac{\nu}{2}\right) + \beta^2[v_\rho^2 + v_z^2 + (\beta_\rho v_z - \beta_z v_\rho)^2]\right\}. \end{aligned} \tag{34}$$

Observing that $v = 3Dn\xi + u$, where for $\rho \to 0$

$$\xi = 3D\mathrm{arctg}\,\frac{2za}{z^2 + \rho^2 - a^2}, \quad (\nabla\xi)^2 = 3D\frac{4a^2}{(z^2 + \rho^2 - a^2)^2 + 4z^2a^2} = 3DO(1),$$

and taking $k \geq 1$, it is possible to separate variables in corresponding Euler-Lagrange equations asymptotically, and to get

$$\beta = 3DO[\rho^k/\mathrm{ch}\,(\text{æ} z)], \quad \text{æ} = 3D(\nu/2)^{1/2}, \quad u = 3DO(\rho^2 z). \tag{35}$$

Substituting (35) into $I(\Omega_2)$, we find $I(\Omega_2) = 3DO(\delta_1^{2k}) = 3DO(\delta^2)$, where $\delta = 3D\delta_1^k$ is determined from the condition $\delta \geq \beta = 3DO(\rho^k)$.

3) Finally, in the region Ω_3 we use spherical coordinates. Noting that for $r \to \infty$ $\xi = 3D2\cos\vartheta\,(a/r)$, $\beta \to 0$, and putting $k \geq 1$, we can write down the functional $I(\Omega_3)$

$$\begin{aligned} I(\Omega_3) = 3D\tfrac{1}{2}\int_R^{\infty} r^2 dr \int_0^{\pi} \sin\vartheta\, d\vartheta \left\{(\nabla\beta)^2 + \beta^2 \times \right. \\ \left. \times \left[\tfrac{k^2}{r^2\sin^2\vartheta} + \tfrac{\nu}{2} + (\nabla v)^2\right]\right\}. \end{aligned} \tag{36}$$

Introducing new functions $A = 3D\beta\,\cos v$, $B = 3D\beta\,\sin v$ and separating the variables in Euler-Lagrange equations, we get

$$A = 3DO[\sin^k\vartheta\, e^{-\text{æ} r}/r], \quad B = 3D\cos\vartheta\, O[\sin^k\vartheta\, e^{-\text{æ} r}/r]. \tag{37}$$

After substitution of (37) into $I(\Omega_3)$ we come to the estimate $I(\Omega_3) = 3DO(e^{-2\text{æ} R})$, combining which with the previous one, we find that $I(\Omega) \to I$ as $\delta \to 0$, $\delta_1 \to 0$, $R \to \infty$, $T \to \infty$.

In this way we managed to prove the existence of G_2-invariant configurations of the $\vec{n}$ –field with nontrivial Hopf invariant. Let us note, that contrary to the Skyrme model, where the absolute minimum of the energy is realized on G_1-invariant (hedgehog) configurations, known as *skyrmions* for $|Q| = 3D1$, and only minimizers in higher homotopy classes ($|Q| > 1$) have G_2-invariant structure, in the Faddeev model for all values of Q_H we have G_2-invariant minimizers. Since the configuration with $|\,Q_H\,| = 3D1$ is the unique one, then it realizes the absolute minimum of energy in Faddeev model and therefore is globally stable.

In view of possible Particle Physics applications of the model it might be instructive to estimate the mass of toron with $Q_H = 3D1$. For this purpose using the Hopf map (9) we can construct a trial function

$$\begin{aligned} \cos\beta &= 3D1 - 2\sin^2\Theta\,\sin^2\vartheta, \\ v &= 3D\mathrm{arctg}\,(\mathrm{tg}\,\Theta\,\cos\vartheta) + \tfrac{\pi}{2}\,\mathrm{sgn}\,\cos\vartheta\,(1 - \mathrm{sgn}\,\mathrm{tg}\,\Theta), \end{aligned} \tag{38}$$

where $\Theta = 3D\Theta(r)$ is a radial function subject to boundary conditions $\Theta(0) = 3D\pi$, $\Theta(\infty) = 3D0$. Substituting these functions into the model functional for $\nu = 3D0$, $k = 3D1$ and integrating it with respect to angle variable ϑ, we come up to the expression

$$I[\Theta] = 3D\frac{16}{3}\int_0^\infty dr\,[\,\Theta'^2(r^2/2 + 2\sin^2\Theta) + \sin^2\Theta + \sin^4\Theta/r^2\,]. \tag{39}$$

The obtained functional (39) differs from the Skyrmion energy functional (see, for example, formula (6.1) in Ref. 1) by the factor 16/3. Therefore the upper bound for the toron energy (as expressed in units $4\pi\varepsilon\lambda\sqrt{2}$ and for $\nu << 1$) would be:

$$I_{tor} \leq (16/3)\cdot 8,206749 \approx 43,769328. \tag{40}$$

Aside from the obtained upper estimate the performed calculations disclose once again the interrelations between the Skyrme and Faddeev models, which have been already discussed in some length in [2], [10], and earlier in [11, 12]. For example, if in the Skyrme model Hamiltonian (Eq. 5.2 in Ref. 1) one fixes the value of the angle variable $\beta = 3D\pi/2$, then, besides the corresponding configurations become electrically neutral, one will obtain the static Faddeev Hamiltonian. The rest angle variables Θ, γ would determine a unit vector on the sphere S^2, which might form configurations with nontrivial Hopf index, but with trivial topological charge of the degree of mapping type. In the Skyrme model such states correspond to mesons and are unstable [13]. To get the lower bound of the toron energy it proves convenient to use the so-called string (or vortex) approximation [14, 15] of torons, obtained by putting

$$\beta = 3D\beta(\rho), \quad v = 3D\ae z, \quad \beta(0) = 3D\pi.$$

Taking a segment of the string of length l in such a way that being closed up this configuration will obey a nontrivial Hopf index $Q_H = 3D\ae l/2\pi = 3DN$. This approximation looks reasonable for $N >> 1$, when the radius of the closure is big enough. Then the lower bound of energy for a toron with the Hopf index equal to N (in units $8\pi\varepsilon\lambda$) can be expressed as

$$E_N > 23,65\,N.$$

Acknowledgments

One of us (V.I.S) is very grateful to International Science Foundation for the Travel Grant, which made the participation in NEEDS-94 possible. It is also a great pleasure to express the deepest gratitude to Prof. V.G. Makhankov, for numerous discussions of the subject of this report, and for the tremendous hospitality provided by him and his family during the stay in Los Alamos.

References

1. V.G. Makhankov, Yu.P. Rybakov, and V.I. Sanyuk , *The Skyrme Model: Fundamentals, Methods, Applications* (Springer-Verlag, Heidelberg-Berlin-New York, 1993) 265 pp.

2. V.G. Makhankov, Yu.P. Rybakov, and V.I. Sanyuk, *Usp.Fiz.Nauk* **162** (1992) 1-62

3. V.A. Novikov, M.A. Shifman, A.I. Vainshtein and V.I. Zakharov, *Phys. Rep.* **116** (1984) 103-171

4. D. Birmingham, M. Blau, M. Rakowski, and G. Thompson, *Phys.Reports,* **209** (1991) 129-340

5. Yu.P. Rybakov, and V.I. Sanyuk, *Physica* **D** (1995), to be published

6. V.I. Sanyuk, *Lect. Notes in Phys.* (1995) Proc. of Gürsey Memorial Conference on Strings and Symmetries, to be published

7. S.P. Novikov and I. Schmelzer *Funk. Anal. Appl.* **15** 3 (1981) 54-66; S.P. Novikov, *ibid.* **15** 4 (1981) 37-52 (in Russian)

8. C.H. Taubes, in *Progress in Gauge Field Theory* (Plenum Publ., New York, 1984) pp. 563-587; *Commun. Math.Phys.* **95** (1984) 345-391; *ibid.* **97** (1985) 473-540

9. L.D. Faddeev, *Dokl. AN SSSR* **210** 4 (1973) 807-810; *Trudy MIAN* **135** (1975) 218-223 (in Russian)

10. Yu.P. Rybakov, in *Itogui Nauki i Tekhniki. Classical Field Theory and Theory of Gravitation* Vol. 2, (VINITI, Moscow, 1991) 56-111 (in Russian)

11. V.I. Sanyuk, in *Topological Phases in Quantum Theory,* (World Scientific, Singapore, 1989), pp. 316-332

12. V.I. Sanyuk, in *Problems of High Energy Physics and Field Theory,* Proc. of the XII Workshop, Protvino, 1989, V.A. Petrov, Ed., (Nauka, Moscow, 1990), pp. 254-259

13. J.M. Gipson, and H.Ch. Tze *Nucl. Phys.* **B183** 3 (1981) 524-546

14. A. Kundu and Yu.P.Rybakov, J.Phys. A: Math. Gen. **15** 1 (1982) 269-275

15. Yu.P. Rybakov, in *Problems of the Theory of Gravitation and Elementary Particles,* (Moscow, Energoatomizdat, 1982) Vol. 13, 187-192 (in Russian)

ON ALGEBRA OF DIRECT LINEARIZATION

Pierre C. SABATIER
Laboratoire de Physique Mathématique et Théorique
Université Montpellier II
34095 MONTPELLIER Cedex 05 - FRANCE

1 Introduction

In a form which is more transparent than in previous papers [1,2,3] we give a method of direct linearization that yields all the integrable equations obtained as well by D-bar methods with polynomial dispersion relations. The method can also be extended as much as D-bar methods can (e.g singular dispersion relations) and in some other ways, but we do not present these extensions. Also, the method gets at classes of solutions larger than those reached by D-bar methods but here we do not study them either. Our aim is to show that this direct linearization works in problems of arbitrary number of space variables without any need of advanced mathematical analysis or advanced algebra and hence is conceptually much more simple than all other methods. Beyond its pedagogical interest, this fact may be important for solving Cauchy problems.

2 Starting the method

In this method, the starting point is the integral equation

$$\Psi(k,y) = I + \int\int \frac{d\sigma(\lambda,\Lambda)}{-k+\Lambda} T(\lambda,\Lambda,y)\Psi(\lambda,y) \tag{2.1}$$

where $k, \lambda, \Lambda \in \mathcal{C}$, the integration is on the product $\mathcal{C} \times \mathcal{C}$ called the spectral plane, $y \in \mathcal{C}^{M+M'}$ as long as we consider only formal derivations, Ψ is wanted in a class of sufficiently smooth functions of k, y, with values in $\mathcal{M}^N, d\sigma(\lambda,\Lambda)$ is a measure in the spectral plane and T a function with values in $\mathcal{M}^N$, such that the product $Td\sigma$ makes sense. For each couple $d\sigma, T$, which is introduced, we assume that Eq.(2.1) has at most one solution, which can be expanded in powers of k^{-1} up to any order we shall need as so :

$$\Psi - I = \sum_{i=1}^{L} \frac{\Psi^{(i)}}{k^i} + O(k^{-L-1}) \tag{2.2}$$

For a linear differential operator L, acting on matrix-valued functions of y, we readily get the identity

$$\begin{aligned} L(k)\Psi(k,y) &= L(k)I + \oint [L(k)-L(\Lambda)](T(\lambda,\Lambda,y)\Psi(\lambda,y)) \\ &+ \oint L(\Lambda)(T(\lambda,\Lambda,y)\Psi(\lambda,y)) - T(\lambda,\Lambda,y)L(\lambda)\Psi(\lambda,y) \\ &+ \oint T(\lambda,\Lambda,y)L(\lambda)\Psi(\Lambda,y) \end{aligned} \tag{2.3}$$

where we used the short notation $\oint$ for $\int\int \frac{d\sigma}{-k+\Lambda}$. If conditions on T, called "dispersion relations", are such that the third term, called "parasite term", does vanish, we say that L is compatible with these dispersion relations. The case with independent λ and Λ is called nonlocal. If $d\sigma$ contains $\delta(\lambda-\Lambda)$ (or any other measure which reduces integrations on λ and Λ to that on Λ only), we call it a local case. In either case, the vanishing of parasite terms usually determines the possible structure of acceptable dispersion relations and the correlated structure of compatible operators.

3 Properties of compatible operators

3.1 Linear differential operators acting on Ψ

The most general one, say, $\mathbf{E}$, is made out of endomorphisms of the space C^{N^2}, written with the help of elements of $\mathcal{M}^N$:

$$\mathbf{E}\Psi = \sum_{n,i,j} E^n_{ij}(k,y)\partial^n\Psi\tau_{ij} \tag{3.1}$$

where $\partial^n\Psi$ stands for any linear combination of derivatives of order n, $\tau_{ij} \in \mathcal{M}_n$ is defined by

$$[\tau_{ij}]_{k\ell} = \delta_{ik}\delta_{j\ell} \tag{3.2}$$

and we use boldface to tell operators from elements of $\mathcal{M}^N$. Two operators $\mathbf{E}$ and $\mathbf{F}$ can be composed together by using standard properties of matrix multiplication and of derivatives. The following formula is readily derived for the "components" G_{ij} of $\mathbf{G} = \mathbf{E}o\mathbf{F}$:

$$G_{ij} = \sum_{\ell,n,n'} E^n_{\ell j}\partial^n F^{n'}_{i\ell}\partial^{n'} \tag{3.3}$$

This composition law is associative, but not commutative. It is distributive for the addition in $\mathcal{M}^N$, and commutes with the multiplication by a scalar. The identity operator $\mathbf{I}$ has components $\delta_{ij}I$, where I is the unit matrix in $\mathcal{M}^N$. Other simple operators are the right multiplicative one and the left multiplicative one :

$$\mathbf{M}^R : \mathbf{M}^R\Psi = M\Psi = \sum_{i,j} M\delta_{ij}\Psi\tau_{ij} \tag{3.4}$$

$$\mathbf{M}^L : \mathbf{M}^L\Psi = \Psi M = \sum_{i,j}[M]_{ij}\Psi\tau_{ij} \tag{3.5}$$

In the following we use lower case letters for numbers-say m_{ij} instead of $[M]_{ij}$. Using Eq.(3.4) we can write $T\Psi$ with the operator notation $\mathbf{T}\Psi$.

3.2 Compatibility with a given dispersion relation : DR

The operators which are compatible with T satisfying DR satisfy for any of these T the relation
(non local case)

$$\mathbf{E}(\Lambda)o\mathbf{T}(\Lambda,\lambda) = \mathbf{T}(\Lambda,\lambda)o\mathbf{E}(\lambda) \tag{3.6}$$

(local case)

$$\mathbf{E}(\Lambda)o\mathbf{T}(\Lambda) = \mathbf{T}(\Lambda)o\mathbf{E}(\Lambda) \tag{3.7}$$

We did not set explicitly the dependences on y.

Theorem

If $\mathbf{E}$ and $\mathbf{F}$ are compatible with DR, so is $\mathbf{E}o\mathbf{F}$. (Proof : use associativity and Eq.(3.6), Eq.(3.7)). Hence, the set $\mathcal{C}$ of operators which are compatible with DR is an algebra for the composition law in the set of all linear differential operators (LDO) on $\mathcal{M}^N$.

3.3 Structure of possible DR and their compatible operators

We first notice that parasite terms for the general zeroth order LDO read as

$$\sum_{ij}[E^0_{ij}(\Lambda)T - TE^0_{ij}(\lambda)]\Psi\tau_{ij} \tag{3.8}$$

Assuming they vanish for several independent $\Psi's$ yields the condition $E^0_{ij} = e^0_{ij}I$, where e^0_{ij} may depend on y and, in the local case only, on Λ. *Hence a left multiplicative operator is compatible with any DR.*

At order 1, the most general LDO reads

$$\mathbf{E}\Psi = \sum_{i,j,\ell}(E^0_{ij}(k)\Psi + E^{1\ell}_{ij}(k)\frac{\partial\Psi}{\partial y_\ell})\tau_{ij} \tag{3.9}$$

In the generic case, which is the only one studied here[2], the dispersion relations enabling cancellation of parasite terms are

$$\frac{\partial T}{\partial y_\ell} + \Omega^\ell(\Lambda)T - T\Omega^\ell(\lambda) = 0 \tag{3.10}$$

Cancelling P.T. yields $E_{ij}^{1\ell} = e_{ij}^{1\ell}$ and by using Eq.(3.10) :

$$E_{ij}^0(\Lambda) = e_{ij}^0 + \sum_\ell e_{ij}^{1\ell}\Omega^\ell(\Lambda) \tag{3.11}$$

where e_{ij}^0 and $e_{ij}^{1\ell}$ may depend on y and, in the local case only, on Λ. Hence the generic first order operator compatible with the D.R. Eq.(3.10) is E :

$$\mathbf{E}\Psi = \sum_\ell\left(\frac{\partial \Psi}{\partial y_\ell} + \Omega^\ell(k)\Psi\right)E^{1\ell} + \Psi E^0 \tag{3.12}$$

where $E^{1\ell}$ and E^0 take arbitrary values in $\mathcal{M}^N$. It follows from Eq.(3.12) and from the structure of Eq.(3.2) that the algebra of operators compatible with Eq.(3.10) can be generated by composition of the left multiplicative operators and of the 1st order generator $\dfrac{\partial \Psi}{\partial y_\ell} + \Omega^\ell(k)\Psi$. This is easy to see by considering a compatible LDO whose leading order differential term is, say, $(\dfrac{\partial}{\partial y_1})^{m_1}...(\dfrac{\partial}{\partial y_\ell})^{m-1}$. Compatibility is possible only if the coefficient $E_{i,j}^{m_1,...,m_\ell}$ reduces to numbers, and thus defines a left multiplicative operator $E^{m_1,...,m_\ell}$. Hence we can reproduce this term by composing first order operators and multiplying the result on its right by $E^{m_1,...,m_\ell}$. Subtracting it from the original LDO again gives a compatible LDO with one leading order differential term less and we can continue.

Assume now we are in the local case, with a given DR.

We can prove the following theorem :

Polynomial Theorem

Let R be any LDO compatible with DR and polynomial of d^0n in k. The action of R on Ψ is equal to that of a left multiplicative operator which is polynomial of d^0n in k and whose coefficients can be exactly expressed in terms of $\Psi^{(1)}, \Psi^{(2)}, ..., \Psi^{(n-1)}$ and derivatives.

Proof. For a compatible operator $R = \sum_{m=0}^n R_m k^m$, the two first terms of Eq.(2.3), called together the inhomogeous term of R (say, ITR), read

$$(ITR) = RI - \int \frac{R(k) - R(\Lambda)}{k - \Lambda} T\Psi d\sigma \tag{3.13}$$

From Eq.(2.1) and Eq.(2.2), we readily derive

$$\Psi^{(n)} = -\int T\Psi\Lambda^{n-1}d\sigma \tag{3.14}$$

and using it in Eq.(3.13)

$$(ITR) = \sum_{m=0}^{n} k^m[R_m I + \sum_{\ell=m+1}^{n} R_\ell(\Psi^{(\ell-m)})] \tag{3.15}$$

By the same token, we show that if $S = \sum_{m=0}^{n} S_m k^m$ is a left multiplicative operator

$$(ITS) = \sum_{m=0}^{n} k^m[S_m + \sum_{\ell=m+1}^{n} \Psi^{(\ell-m)}S_\ell] \tag{3.16}$$

It is easy to make Eq.(3.15) and Eq.(3.16) equal by solving for $\{S_m\}$ the triangular system :

$$S_n = R_n I$$

$$S_{n-1} + \Psi^{(1)}S_n = R_{n-1}I + R_n(\Psi^{(1)})$$

$$S_m + \sum_{\ell=m+1}^{n} \Psi^{(\ell-m)}S_\ell = R_m I + \sum_{\ell=m+1}^{n} R_\ell(\Psi^{(\ell-n)}) \tag{3.17}$$

Thus the inhomogeneous term of the compatible operator which takes Ψ into $R\Psi - \Psi S$ vanishes, and thanks to the uniqueness assumption, so does the operator. Q.E.D.

4 Derivation of linearizable nonlinear equations

We only study the case of polynomials $\Omega_\ell(k)$. It follows from this assumption that all the operators compatible with the corresponding DR are polynomials of k.

4.1 The local case

Assume that y stands for one space variable, say, x, and one time variable, say, t. The two first order generators of the compatible operators algebra are polynomials in k and hence

$$A\Psi = \frac{\partial \Psi}{\partial x} + \Gamma\Psi = \Psi U \tag{4.1}$$

$$B\Psi = \frac{\partial \Psi}{\partial t} + \Theta\Psi = \Psi V \tag{4.2}$$

The same is true for their commutator

$$[AB]\Psi = \{-\frac{\partial \Theta}{\partial x} + \frac{\partial \Gamma}{\partial t} + [\Theta\Gamma]\}\Psi = \Psi W \tag{4.3}$$

Since A and B generate the algebra, relations between operators in the algebra can be obtained from that between U, V, W. We can calculate Eq.(4.3) by applying B to the last r.h.s. of Eq.(4.1), and using Eq.(4.2) in the result, then applying A to the last (r.h.s.) of Eq.(4.2), and using Eq.(4.3) in the result. Hence we get

$$W = -\frac{\partial V}{\partial x} + \frac{\partial V}{\partial t} + [V, U] \tag{4.4}$$

We can also calculate W by using Eq.(3.17), and thus obtain a set of nonlinear equations for $\Psi^{(1)}, \Psi^{(2)} ... \Psi^{(n)}$ (number according to the degree of polynomes Γ and Θ). These equations are solved by the solutions of (1.1) when the D.R. involving Γ and Θ are satisfied by T, and this is why we can say that they have been "linearized". However, some of the equations, or combinations of them, may be linear equations (we called them ghosts) or may be independent of t, or may be together redundant or contradictory ; we talk of "constraints"). This system is interesting because it can always be used when Γ and Θ are given. It is identical to that obtained by $\bar{D}$ methods.[4]

4.2 Another method

So as to construct integrable nonlinear equations, it is often easier to seek, first in the class of operators generated by A, then in the whole algebra, operators E which are both linear in k and such that $EI = 0$. If we can construct two operators F and C with these properties, it follows from Eq.(3.17) that their effect on Ψ can be calculated exactly in terms of $\Psi^{(1)}$, and comparing the asymptotic expansions into powers of k^{-1} of two different values of their commutator produces an equation for $\Psi^{(1)}$:

$$[F_0 + kF_1, G_0 + kG_1]\Psi$$

$$= [F_0, G_0]\Psi + k[F_1, G_0] + [F_0, G_1]\Psi + k^2[F_1, G_1]\Psi \tag{4.5}$$

$$= (F_0 + kF_1)\Psi G_1\Psi^{(1)} - (G_0 + kG_1)\Psi F_1\Psi^{(1)} \tag{4.6}$$

Indeed, the terms $0(k^2)$ vanish, the term $0(k)$ are trivially equal in Eq.(4.5) and Eq.(4.6), and the term $0(1)$ yield

$$(F_1G_0 - G_1F_0)\Psi^{(1)} = F_1(\Psi^{(1)}G_1\Psi^{(1)}) - G_1(\Psi^{(1)}F_!\Psi^{(1)}) \tag{4.7}$$

whereas the following orders calculate $\Psi^{(2)}...\Psi^{(n)}$ from $\Psi(1)$. It would be equivalent to use the polynomial theorem and the expressions of E and F obtained from the algebra generators to derive and equate the left multiplicative operators corresponding to Eq.(4.5) and Eq.(4.6). Hazards of course exist as in the method given in Eq.(4.1), and may produce ghosts or/and constraints.

4.3 Nonlocal case

A cheap form of the polynomial theorem remains. *A* glance at Eq.(3.13) and Eq.(3.14) shows that if E is linear in k and $EI = 0$, then

$$E\Psi = (E_0 + kE_1)\Psi = \Psi E_1\Psi^{(1)} \tag{4.8}$$

Hence the method using Eq.(4.5) and Eq.(4.6) asymptotic expansions still applies, and nonlinear evolution equations are obtained thanks to a clever use of algebra generators to derive operators as F and G. Ghosts [1,2] appear if there exist operators of this type and which are zeroth degree polynomials in k. Constraints usually appear[1] as soon as the number M of space variables and/or N exceed 2. Their effect is generally to confine the classes of solutions obtained from (1.2) to manifolds of dimension ≤ 2.

4.4 Examples

Let us give an example, for which the derivation of results (but not the results themselves[3]) is new. The case is nonlocal, with two space variables x_1 and x_2, and time x_3. The algebra generators are $A_i = \dfrac{\partial\Psi}{\partial x_i} + \Omega_i\Psi, i = 1, 2, 3$. For the sake of simplicity, we choose Ω_i independent of x_1, x_2, x_3. Now if we choose

$$\Omega_1(k) = \Omega(k) \quad , \quad \Omega_2 = \Omega^2(k) \quad , \quad \Omega_3 = \Omega^3(k) \tag{4.9}$$

We easily construct from $A_1, A_1^2, A_1^3, A_2, A_1A_2, A_3$, two compatible operators that cancel I :

$$F\Psi = \frac{\partial^2\Psi}{\partial x_1^2} + 2\Omega\frac{\partial\Psi}{\partial x_1} - \frac{\partial\Psi}{\partial x_2} \tag{4.10}$$

$$G\Psi = \frac{\partial^3\Psi}{\partial x_1^3} - 3\frac{\partial^2\Psi}{\partial x_1\partial x_2} + 3\Omega(\frac{\partial^2\Psi}{\partial x_1^2} - \frac{\partial\Psi}{\partial x_2}) - 6\frac{\partial\Psi}{\partial x_3} \tag{4.11}$$

If $\Omega = k\Gamma, F$ and G are linear in k and the method of Eq.(4.2) applies to give

$$\frac{\partial^2\Psi^{(1)}}{\partial x_2^2} + \frac{1}{3}\frac{\partial^4\Psi^{(1)}}{\partial x_1^4} - 4\frac{\partial^2\Psi^{(1)}}{\partial x_1 \partial x_3} + 2\frac{\partial}{\partial x_1}[\frac{\partial\Psi^{(1)}}{\partial x_1}\Gamma\frac{\partial\Psi^{(1)}}{\partial x_1}]$$

$$= 2\frac{\partial\Psi^{(1)}}{\partial x_1}\Gamma\frac{\partial\Psi^{(1)}}{\partial x_2} - \frac{\partial\Psi^{(1)}}{\partial x_2}\Gamma\frac{\partial\Psi^{(1)}}{\partial x_1}) \tag{4.12}$$

There are no ghost and no constraint if $N = 2$, constraints if $N > 2$. The equations can reduce to KP.

5 Miscellaneous topics

This direct linearization method has been extended to difference equations[5]. We studied also various generalizations, where $T\Psi$ is replaced by $\mathbf{T}\Psi$, T being an endomorphism of C^{N^2} more general than a right multiplicative operator. However, when we succeeded to make the method fully workable, it produced at best new solutions of known integrable equations. Other generalizations can be introduced by using a kernel which is not $(-k+\Lambda)^{-1}$ or using convolution instead of multiplication, or starting from a free term in Eq.(2.1) which is not I. Again, it seems difficult to obtain better than new solutions. The extension non singular dispersion relations as in Léon[4] is also possible but complicated.

Managing the method to deal with ordinary differential equations and with general Cauchy-Goursat problems for 1+1 partial differential equations is in our current study.

6 References

1. P.C. Sabatier, *Spectral transform for nonlinear evolution equations with N-space dimensions. Physics Letters* **A 161**, (1992) 345-351.

2. M. Boiti, F. Pempinelli and P.C. Sabatier, *First and second order nonlinear partial differential evolution equations from an inverse spectral problem. Inverse Problems* **9**, (1993) 1-37.

3. P.C. Sabatier, *Constructing nonlinear integrable equations by direct linearization. Teoreticheskaia i matematicheskaia fisica* **99**, (1994) 517-527.

4. J. Leon, *Nonlinear Systems related to an arbitrary space-time dependence of the spectral transform. J. Math. Phys.* **35** (1994) 3504-3524.

5. P.C. Sabatier, *Spectral transform for nonlinear evolution equations with N-discrete space dimensions. Inverse Problems* (1992) L29-L33.

CONTROL OF COHERENCE BY VARYING NONLINEARITY

V.I.Yukalov
Department of Mathematics and Statistics
Queen's University, Kingston
Ontario K7L 3N6, Canada

Abstract

Nonlinear evolution equations for an ensemble of radiating spins are analysed. The nonlinearity is due to direct dipole interactions between spins and to the feedback electromagnetic field induced by a resonator. The amplitude of this feedback field depends on the strength of coupling of the spin system with a resonator, which can be characterized by a coupling constant. The value of the latter can be changed by varying an external magnetic field and by tuning the parameters of the resonator. Varying in this way the coupling constant, one can get different radiation regimes. The main attention is paid to the coherent regimes such as collective induction, pure superradiance and triggered superradiance. The onset of coherence is characterized by the time - dependent coherence coefficients. The analytical results are compared with computer simulations.

Self - organized coherence can arise in radiating systems only when the equations of motion are nonlinear. The nonlinearity can be due to the existence of direct interactions between radiating objects or it can be caused by effective correlations induced by a resonator feedback field. Before coherence can develop, there should pass time, called the delay time. The life - time of a coherent effect is short as compared to the dephasing time. This is why coherent phenomena are called transient. The solutions of nonlinear equations describing these transient phenomena are somewhat similar to quasisolitons [1], also having finite life - time, and to some extent they remind instanton solutions [2].

For practical purpose it is important to know how the strength of nonlinearity influences the features of transient coherent phenomena. This study is the aim of the present communication.

As an example of a radiating system let us consider nuclear ferromagnet prepared in a nonequilibrium state and placed inside a resonator. The choice of this spin system is reasoned by the fact that self - organized coherent phenomena in spin systems are studied much less than analogous phenomena in atomic systems.

The radiation process in nonequilibrium ferromagnets can be described by the Bloch equations which, for a dimensionless magnetization $\vec{s} = \{s_x, s_y, s_z\}$, are

$$\frac{ds_x}{dt} = -\gamma_2 s_x + \omega_0 s_y,$$

$$\frac{ds_y}{dt} = -\omega_0 s_x - \gamma_2 s_y + \gamma_3 h s_z, \tag{1}$$
$$\frac{ds_z}{dt} = -\gamma_3 h s_y - \gamma_1 (s_z - \xi).$$

Here γ_1 and γ_2 are the longitudinal and transverse widths, respectively; ω_0 is a Larmor frequency of spins in an external magnetic field; ξ is a stationary solution for s_z ; and h is a dimensionless magnetic field induced by a resonance electric circuit. The resonator field obeys the Kirchhoff equation

$$\frac{dh}{dt} + 2\gamma_3 h + \omega^2 \int_0^t h dt = -2g \frac{ds_x}{dt} + \gamma_3 f, \tag{2}$$

in which ω and γ_3 are the resonator natural frequency and the attenuation, respectively; g is the coupling parameter characterizing the strength of coupling between moving spins and resonator; and f is the resonator Nyquist noise plus an external electromotive force, if any.

In the system of equations (1),(2) the role of direct spin interactions is reduced to the existence of the linewidth γ_2 . Thus, the sole origin of nonlinearity here is the coupling between spins and the resonator.

Eqs.(1),(2) can be solved by a method suggested in ref. [3]. Thus method is a combination of three main approaches: parametric separation of slow and fast variables [3], Krylov - Bogolubov averaging method of first order [4], and a generalized asymptotic expansion [5] above the found first - order approximation.

The solution of Eqs.(1),(2) shows several coherent regimes, depending on the value of the coupling parameter g , the sign of $z_0 \equiv s_z(0)$, and on the existence of initial coherence imposed by nonzero value of $v_0 \equiv \sqrt{s_x^2(0) + s_y^2(0)}$. These regimes are:

(i) *free induction*,

$$g < g_c \qquad (g_c \approx \frac{\gamma_2}{\gamma_3}), \qquad v_0 \neq 0. \tag{3}$$

(ii) *collective induction*,

$$g > g_c, \qquad z_0 \geq 0, \qquad v_0 \neq 0. \tag{4}$$

(iii) *triggered superradiance*,

$$g > g_c, \qquad z_0 < 0, \qquad v_0 \neq 0. \tag{5}$$

(iv) *coherent luminescence*, when there is a powerful external pumping $f = f_0 \cos \omega t$ such that

$$f_0^2 \gg \frac{16\gamma_1\omega^2}{\pi^2\gamma_2^3}.$$

It is worth stressing that the value of the critical coupling parameter g_c strongly depends on the retardation time, γ_3^{-1} , of the resonator feedback field. Without taking this retardation into account [6,7], the value of g_c becomes unrealistically small.

Note that pure superradiance cannot be described by Eqs.(1),(2) since the Nyquist noise is too weak to be able to induce noticeable correlations between spins. In order to obtain the regime of pure superradiance, one has to take into account the nonlinearity caused by direct spin interactions. This can be done by considering a microscopic model with the Hamiltonian

$$\hat{H} = \mu H_0 \sum_{i=1}^{N} S_i^z - \mu H \sum_{i=1}^{N} S_i^x + \hat{H}_d,$$

$$\hat{H}_d = \frac{1}{2} \sum_{i \neq j}^{N} \frac{\mu^2}{r_{ij}^3} \left[\vec{S}_i \vec{S}_j - 3 \left(\vec{S}_i \vec{n}_{ij} \right) \left(\vec{S}_j \vec{n}_{ij} \right) \right],$$

in which μ is the magnetic moment of a particle; $\vec{S}$ is a spin vector; H is the magnetic field induced by a resonator; and

$$\vec{n}_{ij} \equiv \frac{\vec{r}_{ij}}{r_{ij}}, \qquad \vec{r}_{ij} \equiv \vec{r}_i - \vec{r}_j, \qquad r_{ij} \equiv \left| \vec{r}_{ij} \right| .$$

The equations of motion for this microscopic model have been solved numerically [8,9] for the number of spins $N = 125$ and for different values of the coupling parameter g. It is interesting to note that the behaviour of the system is almost symmetric with respect to time inversion. The results of numerical simulations for positive time have been presented earlier [8,9]. Here the figures corresponding to negative time are given. The following notation is used: K_{coh} is the coherence coefficient [6,10]; P is the current power in arbitrary units; and

$$p_z \equiv -\frac{1}{N} \sum_{i=1}^{N} \langle S_i^z \rangle .$$

The time is measured in units of γ_2^{-1}. Different curves correspond to different initial conditions of $p_z(0)$ and to different coupling parameters: the larger is g, the longer is the delay time of superradiant pulse. The first three figures represent pure superradiance; the fourth, triggered superradiance.

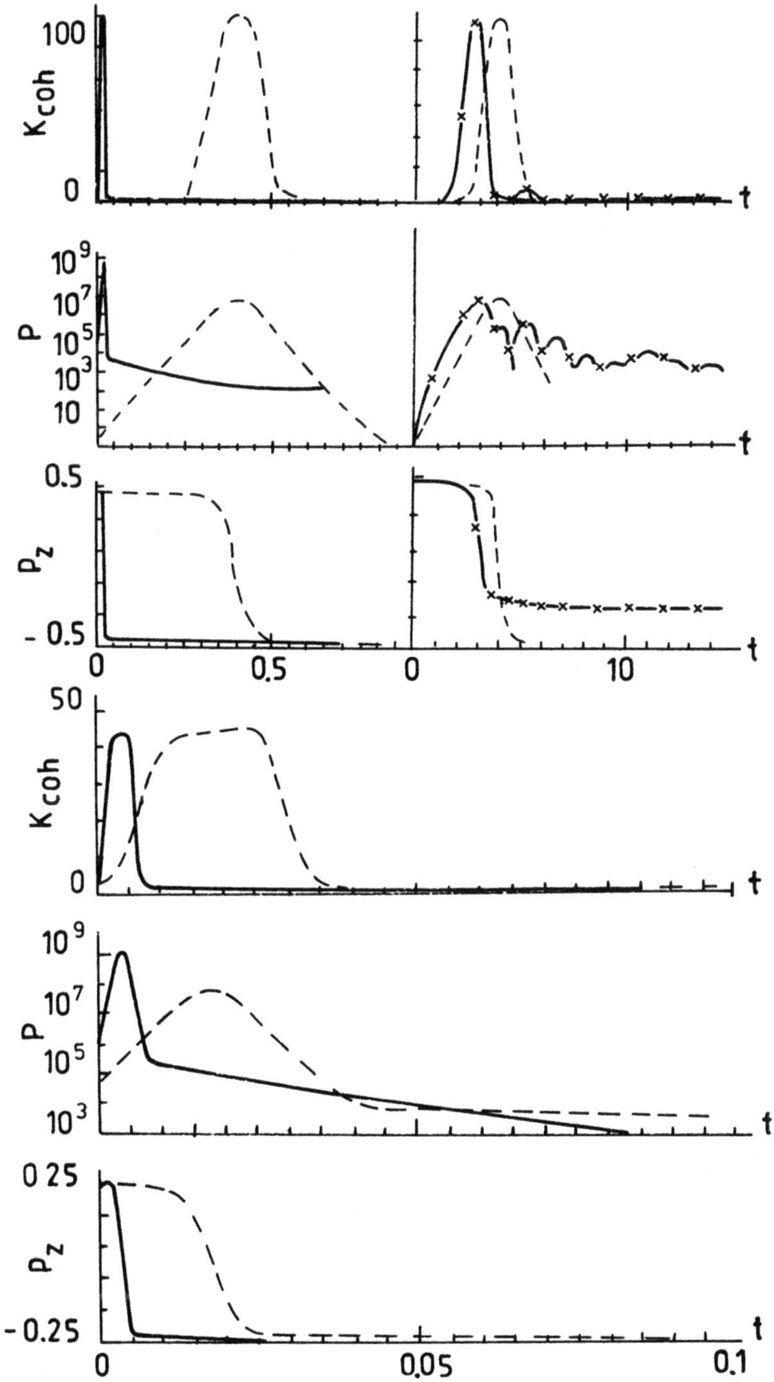

100
K_{coh}
0
t
10^9
10^7
P
10^5
10^3
10
t
0.5
p_z
- 0.5
t
0
0.5
0
10
50
K_{coh}
0
t
10^9
10^7
P
10^5
10^3
t
0 25
p_z
-0.25
t
0
0.05
0.1

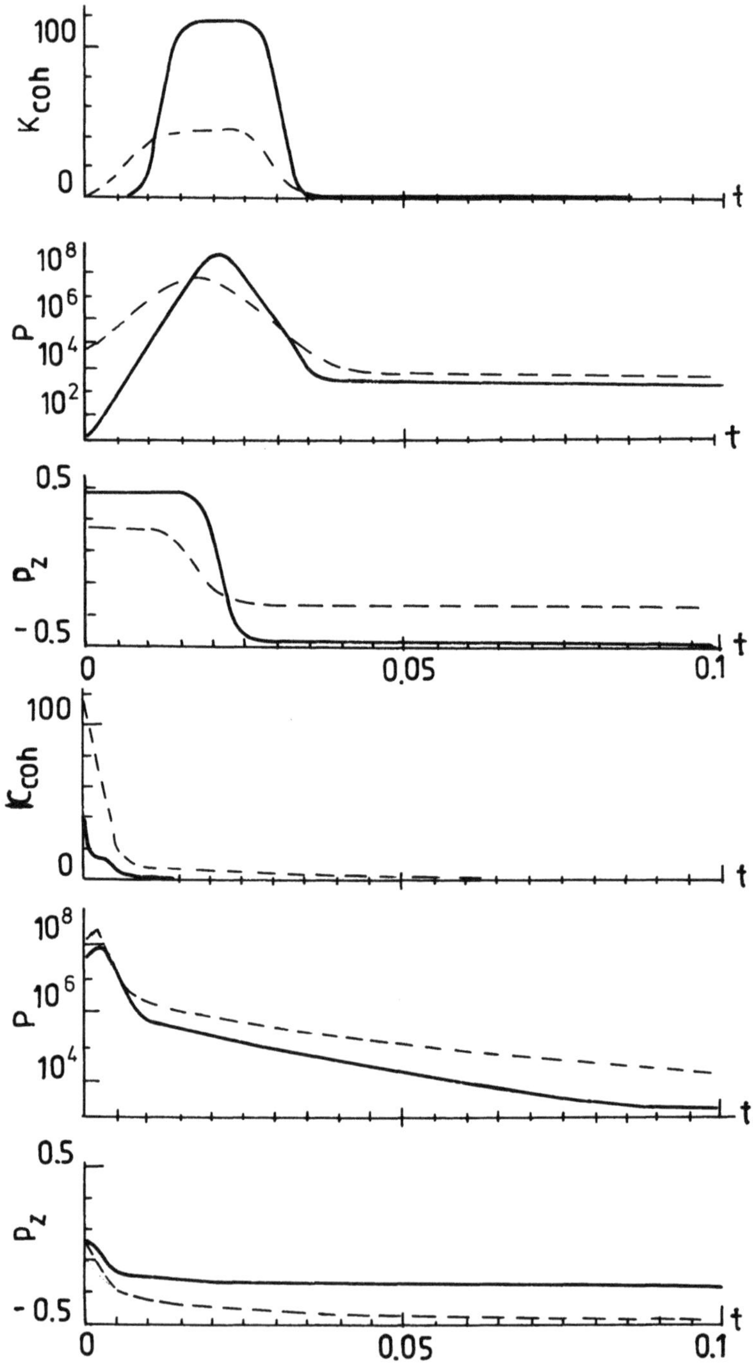
100
K_{coh}
0
t
10^8
10^6
P
10^4
10^2
t
0.5
p_z
- 0.5
t
0
0.05
0.1
100
K_{coh}
0
t
10^8
10^6
P
10^4
t
0.5
p_z
- 0.5
t
0
0.05
0.1

References

[1] V.G.Makhankov, *Phys. Part. Nucl.* **14** (1983) 123.

[2] M.K.Prasad, *Physica* **D1** (1980) 167.

[3] V.I.Yukalov, *Laser Phys.* **3** (1993) 870.

[4] N.N.Bogolubov and Y.A.Mitropolsky, *Asymptotic Methods in the Theory of Non - Linear Oscillations* (Gordon and Breach, New York, 1961).

[5] A.H.Nayfeh, *Problems in Perturbation* (Wiley, New York, 1985).

[6] V.I.Yukalov, *Laser Phys.* **2** (1992) 559.

[7] N.P.Fokina, K.O.Khutsishvili, and S.G.Chkhaidze, *J. Eksp. Teor. Fiz.* **100** (1992) 1013.

[8] T.S.Belozerova, V.K.Henner, and V.I.Yukalov, *Phys. Rev.* **B46** (1992) 682.

[9] T.S.Belozerova, V.K.Henner, and V.I.Yukalov, *Laser Phys.* **2** (1992) 545.

[10] V.I.Yukalov, *J. Mod. Opt.* **37** (1990) 1361.

Multi-phase periodic solutions of KdV-type functional equations

J. Zagrodziński
Institute of Physics, Polish Academy of Sciences
al. Lotników 32/46, 02-668 Warszawa, Poland
email: zagro@ifpan.edu.pl

Abstract

The semi-discrete (ode) system and doubly discrete i.e. functional system of KdV-type equations are discussed. It appears that in addition to the soliton solutions and simple periodic solutions found by K.W.Chow, there exist also multi-phase quasi-periodic solutions which cannot be expressed in terms of elliptic functions. The general Riemann theta functions are the proper objects here and results are very close to the finite gap solutions known for KdV pde, although the dispersion equation technique was used to determine them. Moreover the applied formalism allows to find the solutions in form of solitons on the quasi-periodic background, also in full analogy to KdV equation considered as pde case.

PACS number 3.40.Kf

A direct motivation of this work was a paper of K. Chow [2], where there has been shown that some of the Korteweg de Vries (KdV) type ordinary differential functional equations (ode) and functional equations (FE) with shifted arguments have periodic (i.e. single-phase) solutions. The multisoliton solutions were known in the past [5]. Thus the situation is quite analogous as for partial differential equations (pde).

Since a growing interest can be observed in the analysis of a certain class of nonlinear functional equations, whose solutions have soliton-type properties, e.g. [3,4], we intend to present that this analogy goes further. We shall show that in this case also multiphase quasi-periodic solutions exist and moreover, also the solutions in form of solitons on the background of quasi-periodic process can be found. Thus the presented contribution can be considered as the generalization of results announced in [4].

The reader is referred also to the paper [12], where some problems are clarified in details.

The traditional terminology used in [2], after Hirota [5] is rather not adequate. The systems of ode's and FE's were called originally as semi-discrete (sd-KdV) and doubly-discrete KdV (dd-KdV) equations. It must be stressed however, that in both cases the independent variables x and t can be always considered as continuous ones. In this context the term "discrete" is not correct. Nevertheless, we conserve this primary terminology of sd-KdV and dd-KdV equations.

We shall consider the system of equations

$$u(x,t)_{,t} = [1+u(x,t)]^2[u(x-1/2,t) - u(x+1/2,t)], \tag{1}$$

called sd-KdV, with a substitution

$$u(x,t) := \tau(x-1/2,t)\tau(x+1/2,t)/\tau(x,t)^2 - 1, \tag{2}$$

and dd-KdV equation with a suitable substitution, (δ – *const.*)

$$-[1/u(x,t+\delta/2) - 1/u(x,t-\delta/2)]/\delta = u(x-1/2,t) - u(x+1/2,t), \tag{3}$$

$$u(x,t) := [\tau(x-1/2,t)\tau(x+1/2,t)]/[\tau(x,t+\delta/2)\tau(x,t-\delta/2)] - 1. \tag{4}$$

It is obvious that as $\delta \to 0$, then (3)$\Rightarrow$(1) and (4)$\Rightarrow$(2). Less obvious is the reason that (2) relates to KdV equation. Indeed, by substitution $v = \ln(1+u)$, equation (1) can be transformed into

$$v(x,t)_{,t} = \exp[v(x,t) + v(x-1/2,t)] - \exp[v(x,t) + v(x+1/2,t)], \tag{5}$$

and by a standard procedure [1], substituting $v(x,t) = h^2 w(x+h/2,t)$, with a space step h, taking the truncated Taylor series up to h^5, one obtains $w_{,t} = h^3 w_{,x} + h^5(2ww_{,x} + w_{,xxx}/24)$, which with variables $X = 8^{1/2}(x+h^3t)$, and $T := h^5 8^{1/2} t/3$ takes a standard KdV form.

There are at least three approaches to the problem:

1. The inverse scattering transform in a quasi-periodic version for functional equations as it has been done for the generalization of the Calogero-Bruschi system in paper [4];

2. The Hirota, bilinear operator approach, very effective for soliton solutions, but less effective for quasi-periodic ones;

3. The dispersion equations approach [10,11], following the addition theorem, formally a very close bilinear technique, but more effective for quasi-periodic processes and which will be applied bellow.

The technique of dispersion equations developed earlier for pde's, allows one to determine the multi-periodic (multi-gap) solutions, not referring them however to the structure of the Riemann surfaces being a natural scenery when IST is applied. The most surprising conclusion is, that for the functional equations the formalism of dispersion equations and IST also holds.

Before we shall pass to the concrete equations, let us consider the properties of functions $F : C^g \to C$ having the following factorization property:

$$F(z+w)F(z-w) = \sum_{\epsilon} W_\epsilon(w) Z_\epsilon(z), \tag{6}$$

where the sum is over a finite set (here always Z_2^g) and the functions Z_ϵ and W_ϵ also map $C^g \to C$. The relation is known as the "addition property" and there are a few classes of functions which fulfill (6). Among others, the exponential functions occurring in solutions of soliton type equations, and the Riemann theta functions have this property.

It is obvious that if Z_ϵ are linearly independent, W_ϵ is symmetric, i.e. $W_\epsilon(w) = W_\epsilon(-w)$.

The fundamental consequence of the property (6) is that for the functions F having this property one can immediately write down the subsequent derivatives.

For example, if the property (6) holds, we have for the first derivatives of the shifted F functions,

$$\left[\ln[F(z+w)/F(z-w)]\right]_{,j} = \sum_{\epsilon} W_{\epsilon,j}(w) Z_{\epsilon}(z)/[F(z+w)F(z-w)], \tag{7}$$

where $\partial/\partial z_j := (.)_{,j}$. Denoting $L := \ln F(z)$, for the second and fourth derivatives we have

$$L(z)_{,ij} = 1/2 \sum_{\epsilon} W_{\epsilon,ij}(0) Z_{\epsilon}(z)/F(z)^2, \tag{8}$$

$$L_{,ijkl} + 2(L_{,ij}L_{,kl} + L_{,ik}L_{,jl} + L_{,il}L_{,jk}) = \sum_{\epsilon} W_{\epsilon,ijkl}(0) Z_{\epsilon}(z)/F(z)^2. \tag{9}$$

Note that in (8) and (9) the set of functions of argument z on the right hand side, which forms a basis, is the same. Moreover the differential properties of F functions satisfying (6) determine a certain, known before, so-called second hierarchy of KdV equations. The first member of this hierarchy, related to (9), is the KdV equation and the next one would be the Kotera-Sawada equation, etc. [10].

This fact allows one to write easily the dispersion equation for a standard KdV equation [10,11], and the same procedure can be applied to other commonly known soliton type equations.

Starting from the KdV equation in the form of $u_t + 6uu_x + u_{xxx} = 0$, and using the substitution $u(x,t) = 2(\ln F)_{xx}$, where $F = F(\kappa x + \omega t)$, $\kappa, \omega \in C^g$ and c is a constant, we obtain

$$\sum_{i,j=1}^{g} \omega_i \kappa_j L_{,ij} + \sum_{i,j,k,l=1}^{g} \kappa_i \kappa_j \kappa_k \kappa_l (L_{,ijkl} + 2(L_{,ij}L_{,kl} + L_{,ik}L_{,jl} + L_{,il}L_{,jk})) = c. \tag{10}$$

Considering $Z_\epsilon(z)$ as linearly independent, for any ϵ specified by (6), equation (10) reduces to the system of algebraic equations

$$\sum_{i,j=1}^{g} \omega_i \kappa_j W_{\epsilon,ij}(0) + \sum_{i,j,k,l=1}^{g} \kappa_i \kappa_j \kappa_k \kappa_l W_{\epsilon,ijkl}(0) - cW_\epsilon(0) = 0. \tag{11}$$

Since (11) involves the propagation vectors κ_j and angular frequencies ω_j, it is natural to call this system of equations as a system of dispersion equations (de). It is a system of 2^g equations for $2g+1$ unknown quantities $\kappa_j/c^{1/4}, \omega_j/c^{3/4}, (j = 1, \ldots g)$ and thus for $g > 2$ it is overdetermined.

Hence for $g > 2$ the system of de's supplies the conditions for the elements of B matrix being a parameter of the solution. In the language of algebraic geometry this is equivalent for B matrix to be a period matrix of a suitable Riemann surface. Further details can be found in [3,6-8,13].

For $g = 1$, all functions appearing in (1)-(11) are functions of only one variable and de's (11) reduce then to a simple formula

$$\omega = -\kappa^3 [W_1(0) W_0^{(4)}(0) - W_0(0) W_1^{(4)}(0)]/[W_1(0) W_0^{(2)}(0) - W_0(0) W_1^{(2)}(0)]. \tag{12}$$

For $g = 2$, the relations are slightly more complicated. For $g \geq 3$ the situation becomes qualitatively different. We get also the additional conditions for the elements of B matrix.

Now, let us consider the sd-KdV equation, (1) with the substitution (2). Observe that (2) represents in fact the discretized version of $(\ln f)_{,xx}$. Using a shorthand notation

$$\tilde{\tau}(\alpha) := \tau(\kappa(x+\alpha)+\omega t)), \tag{13}$$

where $\kappa \in C^g, \omega \in C^g$, sd-KdV takes a form

$$\sum_{i=1}^{g} \omega_i \left\{ \ln\left[\frac{\tilde{\tau}(1/4+1/4)}{\tilde{\tau}(1/4-1/4)} \cdot \frac{\tilde{\tau}(-1/4-1/4)}{\tilde{\tau}(-1/4+1/4)} \right] \right\}_{,z_i} =$$

$$= \frac{\tilde{\tau}(-1/4+3/4)\tilde{\tau}(-1/4-3/4)}{\tilde{\tau}(-1/4+1/4)\tilde{\tau}(-1/4-1/4)} - \frac{\tilde{\tau}(1/4+3/4)\tilde{\tau}(1/4-3/4)}{\tilde{\tau}(1/4+1/4)\tilde{\tau}(1/4-1/4)}, \tag{14}$$

which is convenient in order to make use of property (6). Through (6)-(13), similarly as in the case of a standard KdV equation, we obtain from (14) a system of equations (for each ϵ specified by addition property)

$$\sum_{i=1}^{g} \omega_i W_{\epsilon,i}(\kappa/4) + C_0 W_\epsilon(\kappa/4) + W_\epsilon(3\kappa/4) = 0, \tag{15}$$

where C_o is a constant and $W_{\epsilon,i}(\kappa/4) := \left.\frac{\partial W_\epsilon}{\partial w_i}\right|_{w=\kappa/4}$.

Since in this equation, the quantities to be determined are $\kappa \in C^g, \omega \in C^g$ and C_o, it is natural to call these equations also a system of dispersion equations.

In particular, for $g=1$ we have

$$\omega = -[W_0(3\lambda)W_1(\lambda) - W_1(3\lambda)W_0(\lambda)]/[W_0'(\lambda)W_1(\lambda) - W_1'(\lambda)W_0(\lambda)], \tag{16}$$

where $\lambda = \kappa/4$, and W' denotes the derivative with respect to argument.

Repeating the same trick with respect to the dd-KdV (3) with the substitution (4) we have

$$W_\epsilon(^1w)/\delta + W_\epsilon(^2w) + CW_\epsilon(^3w) = 0, \tag{17}$$

i.e. a system of de's, where

$$^1w := (3\omega\delta+\kappa)/4,\ ^2w := (\omega\delta+3\kappa)/4,\ ^3w := (\omega\delta-\kappa a)/4;\ ^1w, ^2w, ^3w \in C^g. \tag{18}$$

and C is again a constant with respect to x and t, dependent however on $\delta\omega$ and κ. Now, in contrast to sd-KdV, both $\kappa \in C^g$ and $\omega \in C^g$ appear as the arguments of W functions.

As it was mentioned above sd-KdV⇒dd-KdV when $\delta \to 0$. This property should concern also the dispersion equations. Indeed since for $\delta \to 0$, $C \sim -1/\delta + C_o + 0(\delta)$, also (17)⇒(15).

For $g=1$ the dispersion relation reduces to

$$\delta = -[W_0(^1w)W_1(^3w) - W_1(^1w)W_0(^3w)]/[W_0(^2w)W_1(^3w) - W_1(^2w)W_0(^3w)]. \tag{19}$$

Till now, a class of functions that our considerations remain valid, was not specified precisely. There was only one restriction: the functions have the addition property (6). It is clear that the exponential functions belong to this class and thus the presented previously dispersion equations determine the commonly known conditions for the existence of a multisoliton solution. More interesting is that the same conditions lead to the quasi-periodic solution if the Riemann theta function is considered as the function F in (6). We precede the discussion of this type solutions giving elementary information concerning the Riemann theta functions.

Let B be a Riemann matrix, (i.e. $B \in C^{g\times g}$ is symmetric and with a positively definite imaginary part), $x \in C^g$. The Riemann theta function with characteristics $[\alpha,\beta]$, $\alpha,\beta \in \Re^g$, is defined as

$$\Theta[\alpha,\beta](z|B) := \sum_{n\in Z^g} \exp\{i\pi[2\langle(z+\beta),(n+a)\rangle + \langle(n+\alpha),B(n+\alpha)\rangle]\}, \tag{20}$$

$$\text{with} \qquad < z,n > := \sum_{i=1}^{g} z_i n_i, \tag{21}$$

The Riemann theta function (ordinary or generic) is defined then as

$$\Theta(z|B) := \Theta[0,0](z|B) = \sum_{n\in Z^g} \exp\{i\pi[2\langle z,n\rangle + \langle n,Bn\rangle]\}. \tag{22}$$

If n and m are integer, ($m,n \in Z^g$) then

$$\Theta(z+n+Bm|B) = \exp[-i\pi(2\langle z,m\rangle + \langle m,Bm\rangle)]\Theta(z|B). \tag{23}$$

In the one-dimensional case, when $x \in C^g$ and $B \in C^{gxg}$, the multiple sums in (20) reduce to a single one and then to the Riemann functions of a single variable. The four functions $\Theta[0,0](z|B)$, $\Theta[1/2,0](z|B)$, $\Theta[0,1/2](z|B)$ and $\Theta[1/2,1/2](z|B)$ are also known as the four basic Jacobi functions. Depending on how the functions F, W and Z in (6) are interpreted, the different equivalent forms of addition theorem .(modulo modular transformation) can be presented. We confine ourselves here to one, which can be useful for a further generalization [9-12].

Let F-functions be the Riemann theta functions

$$F(z+w) := \Theta(z+w|B) \quad \text{and} \quad F(z-w) := \Theta(z-w|B), \tag{24}$$

then in (6) the sum is over $\epsilon \in Z_2^g$, i.e. over g-dimensional hypercube and

$$Z_\epsilon(z) = \Theta(z+\epsilon/2|B)^2, \tag{25}$$

$$W(w) = 2^{-g} \sum_{\mu\in Z_2^g} (-1)^{\langle\epsilon,\mu\rangle} \exp(i2\pi\langle w,\epsilon\rangle)\Theta(2w+B\epsilon|2B)/\Theta(B\epsilon|2B). \tag{26}$$

Propagation vectors $\kappa \in C^g$, and angular frequencies $\omega \in C^g$, obtained from a relevant dispersion equation system, determine completely the quasi-periodic solution. As it was mentioned for $g \leq 2$ it can always be done, although for $g > 2$, the conditions for elements of B matrix have to be taken into account. In the table, below, we report the number of dispersion equations, number of wave parameters to be determined + constant, in dependence on the number of phases in a multiphase process, which is equal to the genus of the relevant Riemann surface.

$$\begin{array}{lll} \textit{Genus} & g & = 1\ 2\ 3\ 4\ 5 \\ \textit{No.eqns} & 2^g & = 2\ 4\ 8\ 16\ 32 \\ \textit{No.param.} & 2g+1 & = 3\ 5\ 7\ 9\ 11 \end{array}$$

It can be seen that for $g \geq 2$ the system is overdetermined, since the number of unknown quantities is greater than the number of equations.

As always, for $g = 1$ and a periodic solution, both dispersion equations become relatively simple and we obtain

$$\pi\omega\Theta'[1/2,1/2](0 \mid B) = \Theta[1/2,1/2](\pi\kappa \mid B), \text{ and } \Theta[1/2,1/2](\pi\omega\delta|B) = \delta\,\Theta[1/2,1/2](\pi\kappa|B), \tag{27}$$

for sd-KdV and dd-KdV, respectively, which up to the different definition of the Riemann theta, and thus the Jacobi functions, are equivalent to those given by K. Chow [2]. Looking for the dispersion equations for a single soliton solution we have next the equations

$$i\pi\omega = -\sinh(i\pi\kappa), \text{ and } (\pi\omega\delta) + \delta\sin(\pi\kappa) = 0, \tag{28}$$

for sd-KdV and dd-KdV, respectively.

For real δ the last equation has an imaginary solution ($\kappa, \omega \in iR$) — then (31) represents the soliton solution. There also exists a real solution ($\kappa, \omega \in R$) but then (31) represents a singular wavetrain solution [12].

As it was mentioned before, it is natural to consider the soliton processes as a limiting form of quasi-periodic ones and the multisoliton process as a limit of a multiphase quasi-periodic process. There is a class of "intermediate" solutions, where the limiting procedure does not relate to all the phases of a quasi-periodic solution, but only to a few of them. As a result, one obtains the solution in a form of multisoliton on a background of multiphase periodic process. Details of such procedure can be found in [10-12]. Denoting such limiting process as S-lim (soliton limit), in order to preserve the addition property, the operation S-lim should be

$$\text{S-lim}\,[F(z+w)]\ \text{S-lim}\,[F(z-w)] = \sum_{\epsilon} \text{S-lim}\,[W_\epsilon(w)]\ \text{S-lim}\,[Z_\epsilon(z)]. \tag{29}$$

For the Riemann theta functions, the operation S-lim can be now defined as follows. Let the Riemannian matrix $B \in C^{g\times g}$ be decomposed into blocks $B^{ss} \in C^{s\times s}, B^{pp} \in C^{p\times p}, B^{ps} \in C^{p\times s}$ and $B^{sp} = (B^{ps})^t$ where the superscript t denotes transposed matrix. The superscript s relates to the soliton subspace and p - a periodic subspace. Of course, $s+p = g$ and matrices B^{ss} and B^{pp} are also Riemannian. Let the argument $z \in C^g$ be similarly decomposed into $z_s \in C^s$ and $z_p \in C^p$. The matrix $D^{ss} := \text{Im}[\text{Diag}(B^{ss})]$ has s diagonal elements d_i. We define the S-lim operation as an s-fold limit

$$\text{S-lim}\,\Theta(z|B) := \lim_{d_i\to\infty} \Theta\left[\begin{array}{c} z^s - iD^{ss}e^s/2 \\ z^p \end{array}\middle|\begin{array}{cc} B^{ss} & B^{sp} \\ B^{ps} & B^{pp} \end{array}\right] =: T(z), \tag{30}$$

where all elements of diagonal matrix d_i tend to infinity and $e^s = (1, \dots, 1)^t \in Z^s$. We denote this limit as $T(z)$. It turns out that the limit for the fixed Riemannian matrix B always exists, and the relation (6) can be rewritten now as

$$T(z+w)T(z-w) = \sum_{\epsilon} W_\epsilon(w) T(z+\epsilon/2)^2, \tag{31}$$

where the sum is over g-dimensional hypercube $\epsilon \in Z_2^g$ and T takes a form

$$T(z) = \sum_{n \in (Z_2)^s} \exp\left[i\pi(2\langle n, z^s\rangle + \langle n, \tilde{B}^{ss} n\rangle)\right] \Theta(z^p + B^{ps} n | B^{pp}), \tag{32}$$

with $\tilde{B}^{ss} := B^{ss} - iD^{ss}$, i.e. having only real diagonal elements. W-function has a more complex form [10-12], which however does not cause the difficulties since the arguments w are fixed later and $W_\epsilon(w)$ appear as constants.

In application to the dd-KdV equation here we shall discuss only the simplest nontrivial case of a soliton on the background of a periodic wavetrain, $(s = p = 1)$.

Rewriting (19) now with the arguments ${}^i w \in C^2, (i = 1, 2, 3)$ for any $\epsilon \in Z^2$, (i.e. $\epsilon = (0,0), (1,0), (0,1), (1,1)$), we obtain the following de's

$$\sum_{i=1}^{3} \gamma_i \exp(i2\pi\, {}^i w_2) \Theta(2\, {}^i w_2 | 2B_{22}) = 0, \tag{33}$$

$$\sum_{i=1}^{3} \gamma_i \exp(i2\pi\, {}^i w_2) \Theta(2\, {}^i w_2 + B_{22} | 2B_{22}) = 0, \tag{34}$$

$$\begin{aligned} \sum_{i=1}^{3} \gamma_i \exp(i2\pi\, {}^i w_2) & \Big[\exp(i2\pi\, {}^i w_1) \Theta(2\, {}^i w_2 + B_{12} | 2B_{22}) \\ & + \exp(-i2\pi\, {}^i w_1) \Theta(2\, {}^i w_2 - B_{12} | 2B_{22}\Big] = 0, \end{aligned} \tag{35}$$

$$\begin{aligned} \sum_{i=1}^{3} \gamma_i \exp(i2\pi\, {}^i w_2) & \Big[\exp(i2\pi\, {}^i w_1) \Theta(2\, {}^i w_2 + B_{12} + B_{22} | 2B_{22}) \\ & + \exp(-i2\pi\, {}^i w_1) \Theta(2\, {}^i w_2 - B_{12} + B_{22} | 2B_{22}\Big] = 0, \end{aligned} \tag{36}$$

Ascribing the index 1 to the s-part and index 2 - to the p-part, according to [36], F is given now as

$$F(z) = T(z) = \Theta(z_2 | B_{22}) + \exp[i\pi(2z_1 + \tilde{B}_{11})] \Theta(z_2 + B_{12} | B_{22}), \tag{37}$$

where $\tilde{B}_{11}$ is real and without losing generality can be chosen as zero. It is seen that the solution $(\ln F)_{xx}$ asymptotically $(z_1 \to \pm i\infty)$ has a form of a periodic process, but there exists a shift between the left and right hand side asymptotics and the off-diagonal element of the starting B matrix (i.e. B_{12}) represents its measure.

The direct inspection shows a generic effect which we have denoted as a "dominance" of periodic processes for solutions in form of solitons on the background of the periodic process. The parameters of the periodic subprocess are determined by equations (33) and (34), which are of the same form either with or without a soliton. This means that for such solution the periodic process, is to some extent, primary.

It is interesting that the formalism originated by Hirota for pde, ode is effective also for functional equations.The essential seems to be that defining the discrete version of bilinear operator as

$$\widetilde{D}^n_a(f,g) := \sum_{k=0}^{n} \frac{(-1)^k n!}{k!(n-k)!a^n} f[x+(n/2-k)a]\, g[x-(n/2-k)a] = \frac{\sinh(D^n a/2)}{a/2}(f,g), \quad (38)$$

in the limit, one obtains

$$\lim_{a\to 0} \widetilde{D}^n_a(f,g) = D^n(f,g) = \frac{\partial^n}{\partial a^n} f(x+a)\, g(x-a)\Big|_{a=0} . \quad (39)$$

The definition of $\widetilde{D}_a$ is of course not unique and the definition proposed here differs from the definition of a discrete operator reported e.g. in [1,5]. But on the other hand, if the functions f have the addition property (6), all operators $\widetilde{D}^n_a(f,f)$ have the representation

$$\widetilde{D}^n_a(f,f) := \sum_{k=0}^{n} \frac{(-1)^k n!}{k!(n-k)!a^n} f[x+(n/2-k)a]\, f[x-(n/2-k)a] = \sum_{\epsilon} Y_\epsilon(n) Z_\epsilon(x), \quad (40)$$

$$\text{where } Y_\epsilon(n) := \sum_{k=0}^{n} \frac{(-1)^k n!}{k!(n-k)!a^n} W_\epsilon((n/2-k)a). \quad (41)$$

This means that $\widetilde{D}^n_a(f,f)$ of different n can be easily added, since the "basis" $Z_\epsilon(x)$ in (40) does not depend on the order of the operator and hence, as a final result we are able to obtain the system of dispersion equations.

The author is grateful to S. Lewandowski and M. Jaworski for stimulating discussions. This paper was supported by the Com. for Sci. Res. Grant No. 2-0480-91-01.

REFERENCES

[1] M. J. Ablowitz and H. Segur, *Solitons and the Inv. Scat. Tran.* (SIAM, Philad. 1981).
[2] K. W. Chow, Phys.Scripta (to be published).
[3] B. A. Dubrovin, Usp. Mat. Nauk, **36** :11 (1981), English transl.: Rus. Math. Surv.
[4] B. A. Dubrovin, A. S. Fokas and P. M. Santini, Integr.Funct.Eqns Alg.Geom.(to be publ.).
[5] R. Hirota, J. Phys. Soc. Japan, **43**, 1424 (1977).
[6] J. Igusa, *Theta-functions* (Springer, Berlin 1972).
[7] V. B. Matveev, *Abelian Funct.& Solitons*, Prep.373, Inst.Th.Phys., Wroclaw Univ. 1976.
[8] D. Mumford, *Tata Lectures on Theta Functions I,II* (Birkhauser, Boston, 1983, 1984).
[9] J. Zagrodziński, J. Math. Phys, **24**, 46 (1983).
[10] J. Zagrodziński, J. Phys. **A15**, 3109 (1982).
[11] J. Zagrodziński, in *Syst. dynam. nonlin.*, Univ. Montreal, 1986, ed. P. Winternitz.
[12] J. Zagrodziński, Phys. Rev. E (to be published 1994).
[13] V. E. Zakharov, S. V. Manakov, S. P. Novikov and L. P. Pitajevski, *Theory of Solitons* (Plenum, New York, 1984).

PART 2. APPLICATIONS

The Use of Generalized Coherent States in the Study of Spin Systems

V.G. Makhankov

Center for Nonlinear Studies, Los Alamos National Laboratory, P.O.Box 1663, Los Alamos, N.M. 87545, USA

and Joint Institute for Nuclear Research, LCTA, Head Post Office, P.O. Box 79, Dubna, Moscow 101000, Russia

Abstract

Generalized Coherent States (GCS) are constructed (and discussed) in order to study quasiclassical behaviour of quantum spin models of the Heisenberg type. Several such models are taken to their semiclassical limits, whose form depends on the spin value as well as the Hamiltonian symmetry. In the continuum approximation, SU(2)/U(1) GCS when applied give rise to the well-known Landau-Lifshitz classical phenomenology. For arbitrary spin values one obtains a lattice of coupled nonlinear oscillators. Corresponding classical continuum models are described as well.

1. In what follows our main aim is to provide a semiclassical description of quantum spin models of the Heisenberg type. For the sake of simplicity we confine ourselves to considering Heisenberg ferromagnets (and antiferromagnets) with uniaxial anisotropy of the following form:

$$\hat{H}_e = \mp J \sum_j (\hat{\vec{S}}_j \hat{\vec{S}}_{j+1} + \delta \hat{S}^z_j \hat{S}^z_{j+1}) \tag{1}$$

(exchange anisotropy) or

$$\hat{H}_s = \mp J \sum_j (\hat{\vec{S}}_j \hat{\vec{S}}_{j+1} + \delta \hat{S}^z_j \hat{S}^z_j) \tag{2}$$

(single-ion anisotropy).
Here $\hat{S}^x_j, \hat{S}^y_j, \hat{S}^z_j$ are the spin operators acting at site j, δ is the anisotropy coefficient.

In the one-dimensional case with spin $s = 1/2$, the study can be done completely [1]. Higher spin models (and more dimensional) require approximate treatment.

An approximation strategy we use later is the so-called trial function method. Based on minimization of the Hamiltonian with respect to a set of trial functions, this method is in fact very sensitive to the choice of these functions. There are certain thoughts (ideas) and even theorems aiding in the search for them. Usually these ideas are based on symmetry properties of the system under consideration (e.g., the Coleman-Palais theorem). Symmetry turns out to be the most powerful and effective tool here. This is why we choose *Generalized Coherent States* (GCS, for their definition and properties see, e.g. [2]) as the trial functions for models (1) and (2).

The paper is organized as follows. In the second part, we construct GCS defined on the homogeneous (complex projective) spaces $SU(2j+1)/SU(2j) \otimes U(1) \equiv \mathbf{CP}^{2j}$. The emphasis will be on the $SU(2)/U(1)$ GCS, reducing the quantum description to classical Landau-Lifshitz phenomenology. In the other cases, we shall only be interested in GCS in the faithful (fundamental) representation of correponding groups. Also GCS constructed on the noncompact group $SU(1,1)$, will be presented in order to treat the antiferromagnet models (+ sign in (1) and (2)).

In the third part, classical counterparts of the spin operators and their products are obtained and discussed.

The fourth part is devoted to deriving classical lattice Hamiltonians for various spin values, s, starting from (1) and (2). Here, we note that the dimensions of the spin phase space at a lattice site coincides with that of the corresponding coset space. Indeed, we have for an arbitrary spin state at a site,

$$| \Psi > = \sum_{j=1}^{2s+1} c_j \mid \psi_j > \tag{3}$$

where $\mid \psi_j >$ are pure spin states, e.g., $\mid \psi_j > = \mid s, m >$, and c_j are complex constants. The vector $\mid \Psi >$ is defined up to an arbitrary phase: $\mid \Psi > \simeq \mid \Psi > e^{i\theta}$ and should satisfy the normalization condition:

$$< \Psi \mid \Psi > = 1$$

These two real conditions reduce the dimension of the spin phase space, $\mathbf{S}$, by two:

$$dim\mathbf{S} = 2(2s+1) - 2 = 4s. \tag{4}$$

The dimension of $\mathbf{CP}^{2s}$ is

$$dim\mathbf{CP}^{2s} = (2s+1)^2 - 4s^2 - 1 = 4s, \tag{5}$$

so

$$dim\mathbf{CP}^{2s} = dim\mathbf{S}$$

and $\mathbf{CP}^{2s}$ GCS provide a complete description of the corresponding model.

The fifth part deals with classical lattice equations of motion derived by considering the quantum probability amplitude.

The sixth part considers continuum models resulting from the classical lattice models. Some properties of these continuum models are discussed.

2. GCS defined on the coset space $SU(2j+1)/SU(2j)\otimes U(1)$

Generalized coherent states (GCS), the extension of the well-known Glauber coherent states related to the Heisenberg-Weyl group, are defined and discussed in the book [3]. Here we choose them as the trial functions for at least three reasons:

1) Our aim is to obtain a semiclassical description of spin quantum models and GCS usually produce this because they minimize the uncertainty relation.

2) They provide a minimal description (no extra parameters) as we will see later.

3) They are chosen according to the Hamiltonian symmetry requirements.

Our concern in what follows will be with GCS defined on the complex projective spaces, $\mathbf{CP}^{2s}$. The simplest is the sphere, $\mathbf{S}^2 = \mathbf{CP}^1 = SU(2)/U(1)$. Points on the sphere are related to the *spin coherent states* discussed in detail in many papers and books (see, e.g. [2]and [3]). These states can be parametrized by the points of a real sphere, $\mathbf{S}^2$, or via the stereographic projection onto the complex plane, $\mathbf{C}$, by its points:

$$|\Psi>= e^{\alpha\hat{S}^+ - \bar{\alpha}\hat{S}^-}|0>= (1+|\psi|^2)^{-j}e^{\psi\hat{S}^+}|0> \qquad (6)$$

with

$$\hat{S}^{\pm} = \hat{S}^x \pm i\hat{S}^y \quad and \quad \psi = \frac{\alpha}{|\alpha|}tan|\alpha|,$$

where α and ψ are complex numbers; $|0>=|j,-j>$, the ground state and j defines the unitary representation of the group $SU(2)$, and hence the spin value. By stereographic projection, we have $\psi = -tan\frac{\theta}{2}e^{i\phi}, |\alpha| \leq \frac{\pi}{2}$. Thus, the set of GCS (6) has spherical symmetry and is (in principle) valid for any spin value. Hence, (6) can be used as the trial function basis when the symmetry of the quantum Hamiltonian is very close to spherical symmetry: $\delta = 0$ or $\delta \ll 1$.

GCS for other groups are constructed using their fundamental representation:

$$|\Psi>= exp\{\sum_i^{2s}(\zeta_i\hat{T}_i^+ - \bar{\zeta}_i\hat{T}_i^-)\}|0>= (1+\sum_i^{2s}|\psi_i|^2)^{-\frac{1}{2}}\{|0>+\sum_i^{2s}\psi_i|i>\} \quad (7)$$

where $\hat{T}_i^+$ and $\hat{T}_i^-$ are generators of the SU(2j+1) group in the fundamental representation and

$$\psi_i = \frac{\zeta_i}{|\zeta|}tan|\zeta|, \qquad |\zeta| = \sqrt{\sum_i^{2s}|\zeta_i|^2}$$

$$|0>=(0,...,0,1)^{tr}, \quad |i>=(0,...,0,1,\underbrace{0,...,0}_{i})^{tr}$$

For example, in the $\mathbf{CP}^2$ case we have $G = SU(3)$, the coset space $G/H = SU(3)/SU(2) \otimes U(1)$ and

$$|\,\Psi>=(1+|\psi_1|^2+|\psi_2|^2)^{-\frac{1}{2}}\{|0>+\psi_1|1>+\psi_2|2>\} \tag{8}$$

Let us consider GCS for $s=1$ systems. In this case, states (6) become

$$|\,\Psi>=\frac{1}{1+|\psi|^2}\{|0>+\sqrt{2}\psi|1>+\psi^2|2>\}. \tag{9}$$

These states are parametrized by a complex function, ψ, i.e. this system lives on a two-dimensional real manifold (the spin phase space). States (8) are parametrized by two complex functions, ψ_1 and ψ_2, so the system lives on a four-dimensional real manifold (the spin phase space). We saw earlier that the spin phase space in the $s=1$ case was four-dimensional and the second system in this regard was more appropriate.

One can easily check that under the conditions

$$\psi_1=\sqrt{2}\psi, \quad \psi_2=\psi_1^2/2$$

both systems coincide. It means that the first 2-d manifold is just a section of the second 4-d manifold, under the constraint

$$\psi_2=\frac{1}{2}\psi_1^2 \tag{10}$$

An analogous projection will occur for $s=\frac{3}{2}$, where

$$|\,\Psi>=(1+|\psi_1|^2+|\psi_2|^2+|\psi_3|^2)^{-\frac{1}{2}}(|0>+\psi_1|1>+\psi_2|2>+\psi_3|3>) \tag{11}$$

and the constraint determining the two-dimensional SU(2) section is

$$\psi_2=\frac{1}{\sqrt{3}}\psi_1^2, \quad \psi_3=\frac{1}{3\sqrt{3}}\psi_1^3 \tag{12}$$

Here it is of use to stress that, in the case $s=1$, the states (8) can be rewritten in the so-called angle parametrization (see, for example [4]):

$$|\Psi>=e^{-i\phi\hat{S}^z}e^{-i\theta\hat{S}^y}e^{i\gamma\hat{S}^z}e^{2ig\hat{Q}^{xy}}|0>=c_0|0>+c_1|1>+c_2|2> \tag{13}$$

where

$$\hat{Q}^{xy}=\frac{i}{2}\begin{pmatrix} 0 & 0 & -1 \\ 0 & 0 & 0 \\ -1 & 0 & 0 \end{pmatrix}$$

$$
\begin{aligned}
c_o &= e^{i\phi}(e^{-i\gamma}\sin^2\frac{\theta}{2}\cos g + e^{i\gamma}\cos^2\frac{\theta}{2}\sin g)\\
c_1 &= \frac{\sin\theta}{\sqrt{2}}(e^{-i\gamma}\cos g + e^{i\gamma}\sin g)\\
c_2 &= e^{-i\phi}(e^{-i\gamma}\cos^2\frac{\theta}{2}\cos g + e^{i\gamma}\sin^2\frac{\theta}{2}\sin g) \qquad (14)
\end{aligned}
$$

Two angles, θ and ϕ, define the orientation of the classic spin vector. The angle, γ, is the angle of rotation of the quadrupole moment about the spin vector. The parameter, g, defines change of the spin vector magnitude and that of the quadrupole moment.

Finally, we give GCS defined on the noncompact manifold $SU(1,1)/U(1)$, the two-sheet hyperboloid, $\mathbf{S}^{1,1}$:

$$|\zeta>=(1-|\zeta|^2)^k e^{\zeta\hat{K}^+}|0> \qquad (15)$$

where again ζ is a complex number, $\hat{K}^+ = \hat{K}^x + i\hat{K}^y$, and $|0>$ is the ground state: $\hat{K}^-|0>=0$, i.e. $|0>=|k,k>$.

3. Averaged spin operators and their products

Here we consider classical counterparts of the spin operators and their products contained in the Hamiltonians (1) and (2).

The vector

$$\vec{S}=<\Psi|\hat{\vec{S}}|\Psi> \qquad (16)$$

can be regarded as a classical spin vector, and

$$Q^{ij}=<\Psi|\hat{S}^i\hat{S}^j|\Psi> \qquad (17)$$

as a component of the quadrupole moment.

Because the spin operators at different lattice sites commute, we have for all such products

$$<\Psi|\hat{S}^i_j\hat{S}^k_{j+1}|\Psi>=<\Psi|\hat{S}^i_j|\Psi><\Psi|\hat{S}^k_{j+1}|\Psi>, \qquad (18)$$

where

$$|\Psi>=|\Psi>_j\,|\Psi>_{j+1}\,.$$

Corresponding expressions for the $\mathbf{CP}^1$ CGS are well-known [2,3]:

$$S^+=\bar{S}^-=2j\frac{\bar{\psi}}{1+|\psi|^2}, \qquad S^z=-j\frac{1-|\psi|^2}{1+|\psi|^2} \qquad (19)$$

$$
\begin{aligned}
Q^{zz} &= <\Psi|\hat{S}^z\hat{S}^z|\Psi>=\frac{j^2(1-|\psi|^2)+2j|\psi|^2}{(1+|\psi|^2)^2}=\\
&= (S^z)^2[1+\frac{2}{j}\frac{|\psi|^2}{(1-|\psi|^2)^2}]\equiv(S^z)^2+\frac{1}{2j}S^+S^-. \qquad (20)
\end{aligned}
$$

The quasiclassical limit requires

$$2j \gg \frac{S^+S^-}{(S^z)^2} \quad .$$

For the same reason for

$$Q^{+-} = S^+S^-(1 + \frac{1}{2j}\frac{j+S^z}{j-S^z}) \quad or \quad Q^{-+} = S^+S^-(1 + \frac{1}{2j}\frac{j-S^z}{j+S^z}) \tag{21}$$

the quasiclassical limits mean

$$2j \gg \frac{j+S^z}{j-S^z} \quad or \quad 2j \gg \frac{j-S^z}{j+S^z} \quad .$$

One can also see that any quadrupole moment component, Q^{ij}, is expressed, though nonlinearly, through two components of the spin vector $\vec{S}$. Higher order moments, Q^{ijk}, and so on are readily shown to be expressed, in a similar way. Components of the classical spin vector, $\vec{S}$, and of the quadrupole moment, Q^{ij}, for other sets of GCS are less known, though straightforward:

$$S^+ = \bar{S}^- = \sqrt{2}\frac{\bar{\psi}_1 + \psi_1\bar{\psi}^2}{1+|\psi_1|^2+|\psi_2|^2} \qquad (j=1), \tag{22}$$

$$S^z = \frac{|\psi_2|^2 - 1}{1+|\psi_1|^2+|\psi_2|^2} \qquad (j=1), \tag{23}$$

$$Q^{zz} = \frac{1+|\psi_2|^2}{1+|\psi_1|^2+|\psi_2|^2} \qquad (j=1) \tag{24}$$

(all other components of Q^{ij} can be expressed in terms of S^+, S^z and Q^{zz}).

$$S^+ = \bar{S}^- = \frac{\sqrt{3}\psi_2\bar{\psi}_3 + 2\psi_1\bar{\psi}_2 + \sqrt{3}\bar{\psi}_1}{1+|\psi_1|^2+|\psi_2|^2+|\psi_3|^2} \qquad (j=\frac{3}{2}), \tag{25}$$

$$S^z = \frac{3|\psi_3|^2 + |\psi_2|^2 - |\psi_1|^2 - 3}{1+|\psi_1|^2+|\psi_2|^2+|\psi_3|^2} \qquad (j=\frac{3}{2}), \tag{26}$$

$$Q^{zz} = \frac{1}{4}\frac{9+|\psi_1|^2+|\psi_2|^2+9|\psi_3|^2}{1+|\psi_1|^2+|\psi_2|^2+|\psi_3|^2} \qquad (j=\frac{3}{2}). \tag{27}$$

We note here that the averaged Casimir operators are

$$\hat{\mathbf{C}}_2 = \frac{1}{2}(Q^{+-}+Q^{-+}) + Q^{zz} = j(j+1) \quad \text{for the } \mathbf{CP}^1 \text{ GCS ,}$$

$$\hat{\mathbf{C}}_2 = j(j+1) = 2 \quad \text{for the } \mathbf{CP}^2 \text{ GCS}$$

and

$$\hat{\mathbf{C}}_2 = j(j+1) = \frac{15}{4} \quad \text{for the } \mathbf{CP}^3 \text{ GCS.}$$

More important to notice that for the $\mathbf{CP}^1$ GCS,

$$\vec{S}^2 = const = j^2. \tag{28}$$

For the $\mathbf{CP}^2$ GCS,

$$\vec{S}^2 + q^2 = 1 \qquad (j = 1), \tag{29}$$

with

$$q^2 = Q^{z+}Q^{z-} + Q^{+z}Q^{-z} + Q^{++}Q^{--} + (1 - Q^{zz})^2 \quad . \tag{30}$$

The analogous formula for the $\mathbf{CP}^3$ GCS is

$$\vec{S}^2 + q^2 + t^2 = const,$$

with more cumbersome expressions for q^2 involving Q^{ij} and t^2 through Q^{ijk}.
We emphasize that, in the $\mathbf{CP}^1$ case (j=1) by use of the angle parametrization (15), one has [5]:

$$\begin{aligned} S^+ &= e^{i\phi} \cos 2g sin\theta, \\ S^z &= \cos 2g cos\theta, \\ q^2 &= \sin^2 2g, \\ Q^{zz} &= 1 - \frac{1}{2} sin^2\theta + \frac{1}{2} sin^2\theta \cos 2\gamma \sin 2g. \end{aligned} \tag{31}$$

The identity $S^2 + q^2 = 1$ is trivially satisfied.
Finally, we give the expressions for the operators $\hat{K}^{\pm}$ and $\hat{K}^z$ (the generators of the group SU(1,1)) averaged over the corresponding $\mathbf{L}^{1,1} = SU(1,1)/U(1)$ GCS:

$$K^+ = \bar{K}^- =< |\hat{K}^+| >= 2k\frac{\zeta}{1 - |\zeta|^2} \tag{32}$$

$$K^z = k\frac{1 + |\zeta|^2}{1 - |\zeta|^2} \tag{33}$$

$$K^{zz} =< |\hat{K}^z\hat{K}^z| >= (K^z)^2(1 + \frac{2}{k}\frac{|\zeta|^2}{(1 + |\zeta|^2)^2}) = (K^z)^2 + \frac{1}{2k}K^+K^- \tag{34}$$

$$K^{+-} = K^+K^-(1 + \frac{1}{2k}|\zeta|^2), \qquad K^{-+} = K^+K^-(1 + \frac{1}{2k}|\zeta|^{-2}) \tag{35}$$

such that the averaged Casimir operator is

$$\mathbf{C}_2 = \frac{1}{2}(K^{+-} + K^{-+}) - K^{zz} = k(1 - k)$$

and the classical pseudovector $\vec{K} = (K^x, K^y, K^z)$ obeys the condition:

$$K^+K^- - (K^z)^2 = -k^2 \tag{36}$$

and lies on the two-sheet hyperboloid $\mathbf{S}^{1,1}$. (We omit the details which appear in [2]).

4. Classical lattice Hamiltonians

In this part, we derive classical lattice Hamiltonians which are just Hamiltonians (1) and (2) averaged over different GCS.

1) First we use the spin coherent states and consider model (1). As was already mentioned, the spin operators at neighboring sites commute, so the coherent state of the whole lattice is

$$|\Psi>=\prod_j |\Psi>_j \tag{37}$$

Averaging (1) with (37) and using eqs. (19) in the ferromagnetic case, one has

$$\begin{aligned} H_e &= <\Psi|\hat{H}_e|\Psi>= -J\sum_j(\vec{S}_j\vec{S}_{j+1}+\delta S_j^z S_{j+1}^z) \\ &= -j^2 J\sum_m \frac{2(\bar{\psi}_m\psi_{m+1}+c.c.)+(1+\delta)(1-|\psi_m|^2)(1-|\psi_{m+1}|^2)}{(1+|\psi_m|^2)(1+|\psi_{m+1}|^2)} \end{aligned} \tag{38}$$

the classical lattice Hamiltonian which, in the continuum limit (d=1), becomes

$$H_e = -j^2 JN + J\int_{-\infty}^{\infty}\frac{dx}{a_o}\left(\frac{a_o^2}{2}\vec{S}_x\vec{S}_x - \delta S^z S^z\right) \tag{39}$$

for the $\sigma-model$ represetation, or

$$H_e = -const + 2j^2 J a_o\int_{-\infty}^{\infty} dx\frac{|\psi_x|^2+\rho|\psi|^2}{(1+|\psi|^2)^2} \tag{40}$$

for the stereographic projection with $\delta = \frac{a_o^2}{2}\rho$. N is the total number of lattice sites.

From eq. (40), we see that classical energy above the ground state (a large negative constant) is positive and classical excitations will possess positive energy. On the contrary, in the antiferromagnetic case, direct application of the spin GCS leads to excitations with negative energy, i.e. such a system should be unstable (more correct, the quantum vacuum over which we construct the excitations is unstable). Apparently, this is the reason why the search for the vacuum in the antiferromagnetic case is such a complicated problem. To avoid this difficulty and provide excitations with positive energy, following [2], we use the following trick. Rewrite (1) via operators of the $su(1,1)$ algebra

$$\hat{K}^{\pm} = i\hat{S}^{\pm}, \qquad \hat{K}^z = \hat{S}^z. \tag{41}$$

Then we have the pseudospin representation for antiferromagnet:

$$\hat{H}_e = -J\sum_j\left[\frac{1}{2}(\hat{K}_j^+\hat{K}_{j+1}^- + h.c.) - \hat{K}_j^z\hat{K}_{j+1}^z(1+\delta)\right]. \tag{42}$$

Now, we treat this model applying the above scheme and using $\mathbf{L}^{1,1}$ GCS to obtain the classical lattice model:

$$H_e = -k^2 J \sum_j \frac{2(\bar{\zeta}_j \zeta_{j+1} + \zeta_j \bar{\zeta}_{j+1}) - (1+\delta)(1+|\zeta_j|^2)(1+|\zeta_{j+1}|^2)}{(1-|\zeta_j|^2)(1-|\zeta_{j+1}|^2)}. \tag{43}$$

The continuum limits are:

$$\begin{aligned} H_e &= -J\sum_j \vec{K}^2 + \frac{a_o}{2} J \int dx (\vec{K}_x \vec{K}_x + \rho K^z K^z) \\ &= Jk^2 N + \frac{a_o}{2} \int dx (\vec{K}_x \vec{K}_x + \rho K^z K^z) \end{aligned} \tag{44}$$

for the $\sigma - model$ representation, or

$$H_e = const + 2k^2 a_o J \int dx \frac{|\zeta_x|^2 + \rho|\zeta|^2}{(1-|\zeta|^2)^2} \tag{45}$$

for the stereographic projection.

Thus, we avoid the problem of excitations with negative energy but obtain, instead, the problem of treating noncompact groups and manifolds ($\sigma - model$ representation) or singular expressions (stereographic projection).

Models (38)-(40) and (43)-(45) can be regarded as appropriate if the anisotropy constant δ is very small: $\delta \ll 1$. Then, symmetries of the quantum Hamiltonians and of GCS manifolds should be very close: spherical for the ferromagnet (1) and pseudospherical for the antiferromagnet (1). While taking the continuum limit, we require $a_o/\lambda \ll 1$ where λ is the wavelength considered.

The same procedure can be applied to (2) to give

$$\begin{aligned} H_s &= <\Psi|\hat{H}|\Psi> = -J\sum_j (\vec{S}_j \vec{S}_{j+1} + \delta Q_j^{zz}) \\ &= -J\sum_m \{\vec{S}_m \vec{S}_{m+1} + \delta[(S_m^z)^2 + \frac{1}{2j} S_m^+ S_m^-]\}, \end{aligned} \tag{46}$$

or

$$\begin{aligned} H_s = &- j^2 J \sum_m [\frac{2(\bar{\psi}_m \psi_{m+1} + c.c.) + (1-|\psi_m|^2)(1-|\psi_{m+1}|^2)}{(1+|\psi_m|^2)(1+|\psi_{m+1}|^2)} + \\ &+ \delta \frac{(1-|\psi_m|^2)^2 + \frac{2}{j}|\psi_m|^2}{(1+|\psi_m|^2)^2}]. \end{aligned} \tag{47}$$

In the continuum limit,

$$H_s = -j^2 JN(1+\frac{\delta}{2j}) + \frac{a_o}{2} J \int dx [\vec{S}_x \vec{S}_x - \delta(1-\frac{1}{2j})(S^z)^2] \tag{48}$$

for the $\sigma - model$ representation, or

$$H_s = -j^2 JN(1+\frac{\delta}{2j}) + 2a_o j^2 J \int dx \frac{|\psi_x|^2 + \rho_1|\psi|^2}{(1+|\psi|^2)^2}, \quad \rho_1 = \rho(1-\frac{1}{2j}) \tag{49}$$

for the stereographic projection.

These formulas imply both systems are equivalent, up to renormalization of the following constants: the ground state energy level and the anisotropy rate. Hence, the classical dynamics of both systems is the same (with the exception of the $j = \frac{1}{2}$ case).

If we take the (-) sign in Hamiltonian (2), then

$$H_s = k^2 JN(1 - \frac{\delta}{2k}) + \frac{a_o}{2} J \int dx [\vec{K}_x \vec{K}_x + \delta(1 + \frac{1}{2k})(K_z)^2] \tag{50}$$

for the $\sigma - model$ representation, or

$$H_s = k^2 JN(1 - \frac{\delta}{2k}) + +2a_o k^2 J \int dx \frac{|\zeta_x|^2 + \rho_2 |\zeta|^2}{91 - |\zeta|^2)^2}, \quad \rho_2 = \rho(1 + \frac{1}{2k}) \tag{51}$$

for the stereographic projection.

Again both systems are equivalent.

Let us now proceed to other GCS.

2) Here we use $\mathbf{CP}^2$ GCS for treating model (1) with $s = 1$. Then we have the same expressions for the Hamiltonians in the $\sigma - model$ representation, viz. (38),(39) and (44). In the complex representation, we have the lattice Hamiltonian:

$$\begin{aligned} H_e = & - j^2 J \sum_m [\frac{(\bar{\psi}_{1m} + \psi_{1m}\bar{\psi}_{2m})(\psi_{1m+1} + \bar{\psi}_{1m+1}\psi_{2m+1}) + c.c.}{(1 + |\psi_{1m}|^2 + |\psi_{2m}|^2)(1 + |\psi_{1m+1}|^2 + |\psi_{2m+1}|^2)} + \\ & + \frac{(1+\delta)(1 - |\psi_{2m}|^2)(1 - |\psi_{2m+1}|^2)}{(1 + |\psi_{1m}|^2 + |\psi_{2m}|^2)(1 + |\psi_{1m+1}|^2 + |\psi_{2m+1}|^2)}]. \end{aligned} \tag{52}$$

The corresponding continuum expression is very complicated and we do not give it here for the general case. But to understand the difference between the two models, we must study their *ground states.* It is natural to suggest that, at least in the ferromagnetic case, this state will be close to the one for which

$$\psi_m = \psi_l, \quad at \quad m \neq l. \tag{53}$$

Then, Hamiltonian (38) becomes

$$H_e = -J \sum_m (\vec{S}_m^2 + \delta(S_m^z)^2) = -j^2 JN - \delta J \sum_m (S_m^z)^2 \tag{54}$$

and the energy assumes its minimum when

a) $\delta > 0$ (easy-axis model)

$$\vec{S} = (0, 0, S^z) \equiv (0, 0, j): \quad E = -j^2 JN(1 + \delta). \tag{55}$$

b) $\delta < 0$ (easy-plane model)

$$\vec{S} = (S^x, S^y, 0); \qquad E = -j^2 JN. \tag{56}$$

For the $\mathbf{CP}^2$ model, we have (due to $s^2 + q^2 = 1$ at j=1)

$$H_e = -J\sum_m(\vec{S}_m^2 + \delta(S_m^z)^2) = -JN - J\sum_m(\delta(S_m^z)^2 - q_m^2) \tag{57}$$

with q_m^2 given by (30).

The ground states again are the states with $q_m^2 = 0$. However, in looking for excitations in the frame of the $\mathbf{CP}^1$ model, the term,

$$H_{e1} = -J\sum_m \delta(S_m^z)^2, \tag{58}$$

is varied. In the $\mathbf{CP}^2$ case,

$$H_{e2} = -J\sum_m[\delta(S_m^z)^2 - q_m^2] \tag{59}$$

i.e. here an additional term, $\sum_m q_m^2$, proportional to the quadrupole moment and anisotropic in nature, appears. Moreover in the first case, classical dynamics is an orientation dynamics such that the classical spin vector being of constant value merely alters its direction (lies on the sphere $\mathbf{S}^2$). In the second case the vector can alter its value (along with quadrupole moment) as well as the direction.

The $\mathbf{CP}^3$ model, besides the quadrupole moment, q, has an octupole moment, t, and, hence, one more term:

$$H_{e3} = -J\sum_m[\delta(S_m^z)^2 - q_m^2 - t_m^2] \tag{60}$$

and so on.

Consider model (2). Here, in the $\mathbf{CP}^1$ case, we have Hamiltonian (58) with $\delta_1 = \delta(1 - \frac{1}{2j})$. But for the $\mathbf{CP}^2$ model,

$$H_s = -J\sum_j(\vec{S}_j\vec{S}_{j+1} + \delta Q_j^{zz}) \tag{61}$$

and, in the vicinity of the ground state, we have

$$H_s = -JN - J\sum_j(\delta Q_j^{zz} - q_j^2), \tag{62}$$

the Hamiltonian containing a pair product of spin operators.

Let us consider again the ground state of the easy-axis type. We have (see (22))

$$S^+ = \sqrt{2}\frac{\bar{\psi}_1 + \psi_1\bar{\psi}_2}{1 + |\psi_1|^2 + |\psi_2|^2} = 0$$

if

1. $\psi_1 = \psi_2 = 0$ and $S^z = 1, q^2 = 0$; this state is similar to the $\mathbf{CP}^1$ case.

2. $\psi_1 = 0$, ψ_2 is arbitrary:

$$S^z = \frac{|\psi_2|^2 - 1}{|\psi_2|^2 + 1} < 1, \quad q^2 = 4\frac{|\psi_2|^2}{1 + |\psi_2|^2}.$$

3. by expressing $\psi_j = |\psi_j|e^{i\phi_j}$, one has $e^{-i\phi_1} + |\psi_2|e^{i(\phi_1 - \phi_2)} = 0$ or ψ_1 is arbitrary and $\psi_2 = e^{i(2\phi_1 - \pi)}$.

It is easy to check that the first solution has the minimal energy:

$$E = -JN(1 + \delta), \quad \delta > 0 .$$

The same result can be expressed in terms of the angle parametrization (15), where

$$H_s = -\sum_i [\cos^2 2g + \delta(1 - \frac{1}{2} sin^2\theta(1 - \cos 2\gamma \sin 2g))] \tag{63}$$

and if $\delta > 0$, we have $sin\theta = 0$ or $\theta = 0$ and $g = 0$, i.e. the known $\mathbf{CP}^1$ easy-axis vacuum.

Longer calculations show that, for $\delta < 0$, we arrive at the easy-plane ground state: $S^z = 0$ but now there is a spin value reduction,

$$S^2 = 1 - \frac{1}{16}\delta^2, \qquad |\delta| < 4 , \tag{64}$$

and we have the minimal energy,

$$E = -JN(1 - \frac{1}{2}\delta + \frac{1}{16}\delta^2) \tag{65}$$

at

$$sin^2 g = \frac{1}{16}\delta^2, \qquad \theta = 0. \tag{66}$$

The same expressions in the complex parametrization are readily given by

$$|\psi_2|^2 = 1 \tag{67}$$

$$|\psi_1|^2 = 2\frac{4 - \delta}{4 + \delta} \tag{68}$$

$$\phi_2 = 2\phi_1 \tag{69}$$

For the $\mathbf{CP}^3$ case, it is easy to show that ground states for model (1) are the same in terms of the classical spin vectors and $q^2 = g = 0$. Moreover, here again $\partial_t S^2 = 0$. In this case, model (2) requires further study.

5. Classical lattice equations of motion

To obtain classical equations of motion we, following [2], consider the transition amplitude from the CS $|\psi>$ at time t to the state $|\psi_1>$ at instant t_1:

$$P(\psi_1, t_1|\psi, t) =< \psi_1|exp\{-\frac{i}{\hbar}\hat{H}(t_1 - t)\}|\psi >$$

Dividing up the interval $t_1 - t$ into n equal subintervals $\epsilon = \frac{1}{n}(t_1 - t)$ and passing to the limit $n \to \infty$, we have

$$P(\psi_1, t_1|\psi, t) = \lim_{n\to\infty} < \psi_1|(1 - \frac{i}{\hbar}\hat{H}\epsilon_1)(1 - \frac{i}{\hbar}\hat{H}\epsilon_2)\dots(1 - \frac{i}{\hbar}\hat{H}\epsilon_n)|\psi_n > \quad .$$

By using the fact that the GCS obey the relation

$$\int d\mu_i(\psi)|\psi >< \psi| = I,$$

we obtain

$$\begin{aligned} P &= \lim_{n\to\infty}\int\dots\int[\prod_{k=1}^{n-1} d\mu(\psi_k)]\prod_{k=1}^{n} < \psi_k|(1 - \frac{i}{\hbar}\hat{H}\epsilon)|\psi_{k-1} > \\ &= \lim_{n\to\infty}\int\dots\int[\prod_{k=1}^{n-1} d\mu(\psi_k)]\prod_{k=1}^{n} < \psi_k|\psi_{k-1} > (1 - \frac{i\epsilon}{\hbar}\frac{< \psi_k|\hat{H}|\psi_{k-1} >}{< \psi_k|\psi_{k-1} >}), \end{aligned}$$

with $\psi_o = \psi$ and $\psi_n = \psi_1$. In the limit $\epsilon \to 0$, we have

$$1 - \frac{i\epsilon}{\hbar}\frac{< \psi_k|\hat{H}|\psi_{k-1} >}{< \psi_k|\psi_{k-1} >} \simeq exp\{-\frac{i\epsilon}{\hbar}\frac{< \psi_k|\hat{H}|\psi_{k-1} >}{< \psi_k|\psi_{k-1} >}\} .$$

Now since $\epsilon \ll 1$, we have $\psi_{k-1} = \psi_k - \Delta\psi_k$ and then

$$\begin{aligned} < \psi_k|\psi_{k-1} > &= 1 - \partial_{\psi'} < \psi_k|\psi' > |_{\psi'=\psi_k}\Delta\psi_k - \partial_{\bar{\psi}'} < \psi_k|\psi' > |_{\psi'=\psi_k}\Delta\bar{\psi}_k \\ &= 1 + \frac{1}{2}\frac{\psi_{1k}\Delta\bar{\psi}_{1k} + \psi_{2k}\Delta\bar{\psi}_{2k} - c.c.}{1 + |\psi_1|^2 + |\psi_2|^2} + O[(\Delta\psi_i)^2], \end{aligned}$$

or

$$< \psi_k|\psi_{k-1} > \cong exp\{\frac{1}{2}\frac{\psi_{1k}\Delta\bar{\psi}_{1k} + \psi_{2k}\Delta\bar{\psi}_{2k} - c.c.}{1 + |\psi_1|^2 + |\psi_2|^2}\},$$

then

$$\begin{aligned} \prod_{k=1}^{n} < \psi_k|\psi_{k-1} > &= exp\{\frac{1}{2}\sum\frac{\psi_{1k}\Delta\bar{\psi}_{1k} + \psi_{2k}\Delta\bar{\psi}_{2k} - c.c.}{1 + |\psi_1|^2 + |\psi_2|^2}\} \\ &= exp\{\frac{1}{2}\int dt\frac{\psi_{1k}\frac{d}{dt}\bar{\psi}_{1k} + \psi_{2k}\frac{d}{dt}\bar{\psi}_{2k} - c.c.}{1 + |\psi_1|^2 + |\psi_2|^2}\}. \end{aligned}$$

Finally, combining all the expressions, we obtain

$$P(\psi_1, t_1|\psi, t) = \int D\mu(\psi)exp\{\frac{i}{\hbar}\int_t^{t'} d\tau\mathcal{L}(\psi_1, \psi_2, \bar{\psi}_1, \bar{\psi}_2)\},$$

where

$$\mathcal{L} = \frac{\frac{i\hbar}{2}}{1 + |\psi_1|^2 + |\psi_2|^2}\sum_{j=1}^{2}(\psi_j\frac{d}{dt}\bar{\psi}_j - c.c.) - \mathcal{H} \qquad (70)$$

with

$$\mathcal{H} = < \psi_k | \hat{H} | \psi_k > \tag{71}$$

at a lattice site.

The total classical lattice Lagrangian is the sum over all sites

$$L = \sum_{m=1}^{N} L_m \quad , \tag{72}$$

where $L_m = \mathcal{L}$ at site m. By varying (72) with respect to $\bar{\psi}_1$ and $\bar{\psi}_2$, one has at a site

$$\frac{i}{1+|\psi_1|^2+|\psi_2|^2}\{\dot{\psi}_1(1+|\psi_2|^2) - \psi_1\bar{\psi}_2\dot{\psi}_2\} = \frac{\partial\mathcal{H}}{\partial\bar{\psi}_1}$$
$$\frac{i}{1+|\psi_1|^2+|\psi_2|^2}\{\dot{\psi}_2(1+|\psi_1|^2) - \psi_2\bar{\psi}_1\dot{\psi}_1\} = \frac{\partial\mathcal{H}}{\partial\bar{\psi}_2}$$

or for the lattice

$$i\dot{\psi}_{1m} = (1+|\psi_{1m}|^2+|\psi_{2m}|^2)\{(1+|\psi_{1m}|^2)\frac{\partial H}{\partial\bar{\psi}_{1m}} + \psi_{1m}\bar{\psi}_{2m}\frac{\partial H}{\partial\bar{\psi}_{2m}}\}, \quad (1 \rightleftharpoons 2), \tag{73}$$

where $H = \sum_m H_m$ is the classical lattice Hamiltonian.

In the simplest case of weak exchange anisotropy, we have (the $\mathbf{CP}^1$ case)

$$\frac{1}{j^2 J}H_{em} = -2\frac{\bar{\psi}_m\psi_{m+1} + c.c.}{(1+|\psi_m|^2)(1+|\psi_{m+1}|^2)} - (1+\delta)\frac{(1-|\psi_m|^2)(1-|\psi_{m+1}|^2)}{(1+|\psi_m|^2)(1+|\psi_{m+1}|^2)} \tag{74}$$

and the equations of motion (see [2])

$$i\dot{\psi}_m = (1+|\psi_m|^2)^2\frac{\partial H}{\partial\bar{\psi}_m}$$

i.e.

$$\begin{aligned}\frac{i}{2}\dot{\psi}_m &= \frac{\psi_{m-1}}{1+|\psi_{m-1}|^2} + \frac{\psi_{m+1}}{1+|\psi_{m+1}|^2} \\ &- \psi_m^2(\frac{\bar{\psi}_{m-1}}{1+|\psi_{m-1}|^2} + \frac{\bar{\psi}_{m+1}}{1+|\psi_{m+1}|^2}) \\ &- 2(1+\delta)\psi_m\frac{1-|\psi_{m-1}|^2|\psi_{m+1}|^2}{(1+|\psi_{m-1}|^2)(1+|\psi_{m+1}|^2)},\end{aligned} \tag{75}$$

which, in the small amplitude region, is

$$\begin{aligned}\frac{i}{2}\dot{\psi}_m &= \psi_{m-1} + \psi_{m+1} - 2(1+\delta)\psi_m \\ &- (\psi_{m-1}|\psi_{m-1}|^2 + \psi_{m+1}|\psi_{m+1}|^2) \\ &- \psi_m^2(\bar{\psi}_{m-1} + \bar{\psi}_{m+1}) + \psi_m(|\psi_{m-1}|^2 + \psi_{m+1}|^2).\end{aligned} \tag{76}$$

the equation of the so-called ϕ^4-theory. It is easy to verify that eq.(76) possesses two ground states solutions:

$$\psi_m = \psi_{m\pm1}$$

then

$$-2\delta\psi_m \frac{1-|\psi_m|^4}{(1+|\psi_m|^2)^2} = 0$$

and

$$\begin{aligned} \psi_m &= 0 \quad \textit{being the easy-axis vacuum state,} && (77)\\ |\psi_m|^2 &= 1 \quad \textit{being the easy-plane vacuum state.} && (78) \end{aligned}$$

Eq. (76) only has the easy-axis vacuum.

For *spin* $= \frac{3}{2}$, one has instead of (73)

$$\begin{aligned} i\dot{\psi}_{1m} &= (1+|\psi_{1m}|^2+|\psi_{2m}|^2+|\psi_{3m}|^2)\{(1+|\psi_{1m}|^2)\frac{\partial H}{\partial\bar{\psi}_{1m}} + \\ &+\psi_{1m}\bar{\psi}_{2m}\frac{\partial H}{\partial\bar{\psi}_{2m}} + \psi_{1m}\bar{\psi}_{3m}\frac{\partial H}{\partial\bar{\psi}_{3m}}\},\ (1\rightleftharpoons 2)\ (1\rightleftharpoons 3)\ , \end{aligned} \tag{79}$$

or for arbitrary s (and $\mathbf{CP}^{2s}$ GCS):

$$i\dot{\psi}_{jm} = (1+\sum_{k=1}^{2s}|\psi_{km}|^2)\{(1+|\psi_{jm}|^2)\frac{\partial H}{\partial\bar{\psi}_{jm}} + \psi_{jm}\sum_{k=1}^{2s}\bar{\psi}_{jk}\frac{\partial H}{\partial\bar{\psi}_{jk}}\alpha_{1k}\}, \tag{80}$$

with

$$\alpha_{1k} = \begin{cases} 0 & if \quad k=1 \\ 1 & if \quad k\neq 1 \end{cases}$$

6. Classical continuum equations of motion

To obtain equations of motion in the continuum approximation, we apply the conventional procedure of expanding lattice functions, $\psi_{n\pm1}$, in a Taylor series, up to the second order derivatives (the first nonvanishing terms), assuming $\lambda a_o \ll 1$:

$$\psi_{n\pm1} = \psi(x) \pm a_o\psi'(x) + \frac{1}{2}a_o^2\psi''(x) + O(a_o^3\psi'''),$$

where $x = a_o n$.

Consider first for the sake of simplicity, the $\mathbf{S}^2$ and $\mathbf{S}^{1,1}$ cases. Then the equations of motion become

$$i\dot{\psi} + \Delta\psi - 2\frac{(\nabla\psi)^2\bar{\psi}}{1+|\psi|^2} + \delta\frac{1-|\psi|^2}{1+|\psi|^2} = 0 \tag{81}$$

for the ferromagnetic case and

$$i\dot{\zeta} + \Delta\zeta + 2\frac{(\nabla\zeta)^2\bar{\zeta}}{1-|\zeta|^2} + \delta\frac{1+|\zeta|^2}{1-|\zeta|^2} = 0 \tag{82}$$

for the antiferromagnetic case, where $\Delta = \nabla^2$ is the Laplace operator. For one-space dimensional systems, these equations, being the stereographic projections of compact and noncompact Landau-Lifshitz models defined respectively on the sphere $\mathbf{S}^2$ and the hyperboloid $\mathbf{S}^{1,1}$, are gauge equivalent to various integrable versions of the Nonlinear Schrodinger Equation (NSE) [7]. In the particular case of $\delta > 0$, (81) is equivalent to the cubic attraction type NSE and (82), to the repulsive type NSE. The latter, usually called the Ginzburg-Landau equation, describes superfluid phenomenology and gives the correct Bogolubov excitation spectrum. What is probably more intriguing is that the $\sigma - model$ version of (82) even gives a correct quasiclassical description of the Bogolubov condensate, thereby pointing out at the intimate coupling of antiferromagnetism and superfluidity. Quite a detailed discussion of this equivalence along with solutions of the equations can be found in [2] and [7].

Naturally, the question occurs as to what extent these continuum models can be related to the initial lattice models. It is easy to show that at $D \geq 2$ such continuum systems lose stationary localized solutions. In what follows, we focus on the $\mathbf{CP}^2$ model. In this concern, as we have already seen, the most interesting is model (2) with single-ion anisotropy. Equations of motion then assume the simplest form in terms of real functions $(\theta, \phi, \gamma, g)$ and $D = 1$:

$$\begin{aligned}
\dot{\phi} &= \frac{cos2g}{sin\theta}\theta_{xx} - cos2g cos\theta\phi_x^2 - 4\frac{sin2g}{sin\theta}g_x\theta_x \\
&+ \delta\frac{cos\theta}{cos2g}(4sin2g cos2\gamma - 1) \\
\dot{\theta} &= -cos2g(sin\theta\phi_{xx} + 2cos\theta\phi_x\theta_x) + 4sin2g sin\theta g_x\theta_x \\
&- \frac{\delta}{2}tan2g sin2\theta sin2\gamma \\
\dot{\gamma} &= 2sin2g(g_{xx} + 2cotan\theta g_x\theta_x) - cos2g(cotan\theta\theta_{xx} - 4g_x^2 - \theta_x^2 - \phi_x^2) - 2cos2g \\
&+ \delta[\frac{cos^2\theta}{cos2g}(1 - sin2g cos2\gamma) + \frac{1}{2}cotang sin^2\theta cos2\gamma] \\
\dot{g} &= \frac{\delta}{2}sin2\gamma sin^2\theta
\end{aligned} \tag{83}$$

In order to make the models comparable we give the equations for the exchange anisotropy model (1):

$$\begin{aligned}
\dot{\phi} &= \frac{cos2g}{sin\theta}\theta_{xx} - cos2g cos\theta\phi_x^2 - 4\frac{sin2g}{sin\theta}g_x\theta_x - 2\delta cos2g cos\theta \\
\dot{\theta} &= -cos2g(sin\theta\phi_{xx} + 2cos\theta\phi_x\theta_x) + 4sin2g sin\theta g_x\theta_x \\
\dot{\gamma} &= 2sin2g(g_{xx} + 2cotan\theta g_x\theta_x) - cos2g(cotan\theta\theta_{xx} - 4g_x^2 - \theta_x^2 - \phi_x^2) - 2cos2g \\
\dot{g} &= 0
\end{aligned} \tag{84}$$

One can readily see that system (84) reduces to the Landau-Lifshitz equations when $g = 0$. Also with the aid of systems (83) and (84), the first two items of the next section follow.

7. Conclusion

Summaring the results, we conclude the following:
1) Spherically symmetric systems and systems with exchange anisotropy, regardless of its magnitude and sign, can be treated via $\mathbf{CP}^1$ GCS because the Hamiltonian in this case does not contain pair products, and hence S^2 and q^2 are conserved separately: $\partial_t S^2 = \partial_t q^2 = 0$ (q is a cyclic coordinate). Since in both ground states (or in the domain regions) $q^2 = 0$, the same thing occurs in the transition (intermediate) region, the domain wall. From the physical point of view, these nonlinear systems behave, in a sense, as linear for they do not excite higher "harmonics", in our case, higher moments. The Landau-Lifshitz phenomenology should work well for these and if, nevertheless, they display some spin reduction, it should be attributed, in the scope of our consideration, to stochastic (chaotic) processes. This is one of the objectives of investigating the underlying lattice models.
2) Systems with single-ion anisotropy, regardless of the anisotropy coefficient and its sign, should be treated via $\mathbf{CP}^2$ GCS for $s = 1$ and higher $\mathbf{CP}^{2s}$ for higher spins. Here pair products of the spin operators at the same site are in the Hamiltonians and therefore $\partial_t S^2 \neq 0$. Ground states can also depend on q (see (66) and (68)) so the spin reduction can occur even in the ground states and, in principle, may be attributed to both mechanisms: excitations of higher moments and stochastic processes.
3) The same conclusions are valid for antiferromagnets and SU(1,1)/U(1) GCS.
4) For the systems with large enough exchange anisotropy, it is necessary to develop techniques based on the q-deformed algebras, in the sense that the algebra deformation is related to anisotropy coefficient.

Along with all the impressive results, obtained for the models considered in the scope of the continuum approximation, especially concerning with their dynamical properties, there is still, in fact, nearly nothing about the dynamics of the corresponding lattice models. Therefore, the study of the dynamical behavior of such models is of a great interest from both the theoretical point of view and physical applications. This study should consist of analytical research as well as computer modelling.

8. Acknowledgment

This work was done under auspices of the U.S. Department of Energy at the Center for Nonlinear Studies at Los Alamos National Laboratory. I wish to thank the Center for its generous hospitality. The work was completed thanks to valuable support of Gary Doolen who I am grateful to. Discussions with him and Alan Bishop on the problems discussed above are appreciated as well.

References

[1] A. Izergin and V. Korepin. Quantum Method of Inverse Problem. *Particles and Nuclei, Moscow*, **13**, (1982), pp.501-541.

THE INTERACTION OF LOCALIZED SOLUTIONS IN A SURFACE REACTION

HELMUT R. BRAND [†,*] and ROBERT J. DEISSLER [*]

[†] *Theoretische Physik III, Universität Bayreuth*
D 95440 Bayreuth, F.R. Germany

[*] *Center for Nonlinear Studies, MS-B 258, Los Alamos National Laboratory*
University of California, Los Alamos, NM 87545

ABSTRACT

We discuss similarities and differences between spatially localized solutions of the quintic complex Ginzburg-Landau equation applicable near a weakly inverted bifurcation to traveling waves and localized states observed during the catalytic oxidation of CO on Pt(110). It turns out that both, the properties of a single localized state as well as their interaction, share many common features in comparing the two systems. We point out that the experimental results rule out a solitonic mechanism and are only compatible with the picture of a dissipative driven system.

1. Introduction

In this contribution we expand on one point touched upon in the review lecture given by one us (R.J.D.) on spatially localized solutions and their interaction as they result from the study of the quintic complex Ginzburg-Landau equation applicable near a weakly inverted bifurcation.

The properties of stable pulses with fixed shape and their interactions in one spatial dimension has been discussed in detail in refs.1-4. For localized solutions that have reached their asymptotic shape four types of interaction have been seen[2,3]: interpenetration and annihilation for stabilizing cross-coupling between localized states traveling to the right and to the left, a transition to an absolutely unstable state leading to a cell filled with a pattern for destabilizing cross-coupling and a compound state for sufficiently small approach velocity of the pulses with waves traveling to the right in the left half of the state and waves traveling to the left in the right half of the compound state. In addition we have investigated the influence of nonlinear gradient terms[5,6] on the properties of the pulses and their interaction[3] and we have shown that they make the pulses asymmetric, change their speed in magnitude and even sign and can also lead to the destruction of the stable pulses for large enough values of the nonlinear gradient terms. It turns out that all the phenomena described above for localized solutions and their interaction carry over to two spatial dimensions[7]. This behavior is in strong contrast to solitonic systems[8-10], which are non-dissipative and for which solitons can only interpenetrate with size and shape unchanged after the collision in one spatial dimension, but which are typically unstable for two dimensions. Furthermore it turns out that there is a new parameter in two dimensions, the impact parameter familiar from scattering theory and atomic physics. This additional parameter leads to the possibility of passing by each other as the impact parameter is increased. We also note that we have shown recently[11] that the existence of localized solutions in these driven dissipative and dispersive systems is not restricted to nonlinearities of polynomial type, but can also be found for systems with saturating nonlinearities as they arise for example for the dye laser with a saturable absorber in nonlinear optics when the atomic variables (polarization and inversion) are eliminated adiabatically[11].

Very recently we have shown[12] that the quintic Ginzburg-Landau equation with complex coefficients does not have only localized solutions with fixed shape, but also stable breathing localized solutions for which the modulus oscillates around a positive mean value. These oscillations can be periodic, quasiperiodic, of high periodicities including period 3, period 6, period 12 and period 24, and even chaotic. This is the first time that a spatially localized chaotic solution has been found in a spatially extended system. We will describe elsewhere in detail[13] how these breathing localized states can interact. It turns out that their interactions are even richer than those for the fixed shape solutions for which one always gets the same result as a consequence of an interaction for fixed parameter values in the equation. For the breathing localized solutions the outcome of the interaction depends on the phase difference and on the distance between the two states before the collision. It even depends sensitively on the initial conditions for the chaotic localized states[13].

2. Experimental Observations and their Comparison with Solitonic Behavior

In an interesting and stimulating recent note[14], the authors describe their observation of the interaction of localized solutions occurring during the catalytic oxidation of CO on Pt(110). These initial results have been further elucidated and generalized in a number of more recent publications[15-17]. Upon colliding, one or both of the observed pulses annihilate, or the pulses interpenetrate unchanged in size and shape. The authors suggest an interpretation of the pulses and their interaction they observe in the framework of solitons. This concept has found widespread applications in nonlinear science since the pioneering work of Zabusky and Kruskal[8] (see ref.9 for an excellent review).

Here we suggest a different interpretation of the important experimental results

given in ref.14 and generalized in refs.15-17 by pointing out that the observations show
a) strong overlap with behavior found recently[1−4] for localized solutions in one and two spatial dimensions near a weakly inverted bifurcation, and
b) features incompatible with truly solitonic behavior[8−10].

All the equations (including the Korteweg - de Vries, sine-Gordon and non-linear Schrödinger equations) giving rise to truly solitonic behavior[9,10] are purely reversible in nature, that is they do not incorporate dissipative processes. All these equations can be characterized by an infinite number of conservation laws. Their pulse solutions never completely or partially annihilate and can have a large range of possible velocities. In addition there is typically a specific relation between width and height (for example narrower solitons are higher).

3. Analysis of Experimental Observations

The system encountered in the experiment is an externally driven system, which incorporates both dissipative and non-dissipative processes[14−17]. The width of the pulses shows only small variations and there is only one value for their speed within the experimental error bar. The regime over which the pulses are observed is narrow in parameter space and complete interpenetration or annihilation of one or both of the pulses is observed. In the case of interpenetration, the total intensity during interaction is less than the intensity of the sum of the individual pulses before and after the interaction. Finally the authors note that there is a dip in intensity in the propagation direction of the pulse.

Here we point out that all features of the experiments, except for the last one, closely parallel those obtained recently[1−4,7] for the one- and two-dimensional localized solutions of the envelope equations near a weakly inverted bifurcation. That the appearance of a dip is not contained in the solution of the envelope equations, however, might simply be due to the fact that one has already split off the rapidly varying shape of the critical mode to arrive at the envelope equation, which contains only the spatially slowly varying parts. This prototype envelope equation, which arises for driven systems, contains dissipative as well as non-dissipative contributions[1−4,7]:

$$A_t + vA_x = \chi A + \gamma A_{xx} - \beta|A|^2 A - \delta|A|^4 A - \zeta|B|^2 A \qquad (1)$$

$$B_t - vB_x = \chi B + \gamma B_{xx} - \beta|B|^2 B - \delta|B|^4 B - \zeta|A|^2 B \qquad (2)$$

where the two complex fields A and B correspond to waves traveling to the right and to the left. They are slowly varying complex amplitudes and the coefficients (except for the group velocity v) are in general complex, i.e. of the form $z = z_r + iz_i$. The coefficient χ may be taken as real, since the imaginary part can be transformed away with a simple transformation. We take $\chi < 0$ and $\beta_r < 0$ so that the system is subcritical and take $\delta_r > 0$ to guarantee saturation. In writing down eq.(1) we have discarded nonlinear gradient terms[5,6,3]. The properties of the localized solutions and their collision behavior can be summarized as follows. For fixed values of the external conditions, the single pulses possess – after an initial transient – unique values for speed, width, and height. They only occur over a finite range of parameter values, since their existence requires that two states of different amplitude are locally stable. After the single pulses have reached their asymptotic shape, they annihilate (Fig.1) or interpenetrate (Fig.2) completely with their shape and size unchanged depending on the relative strength of the cross-coupling between counterpropagating waves. In the case of interpenetration with stabilizing cross-coupling, the amplitude of the individual pulses during interaction is less than their amplitudes before and after the interaction (Fig.1).

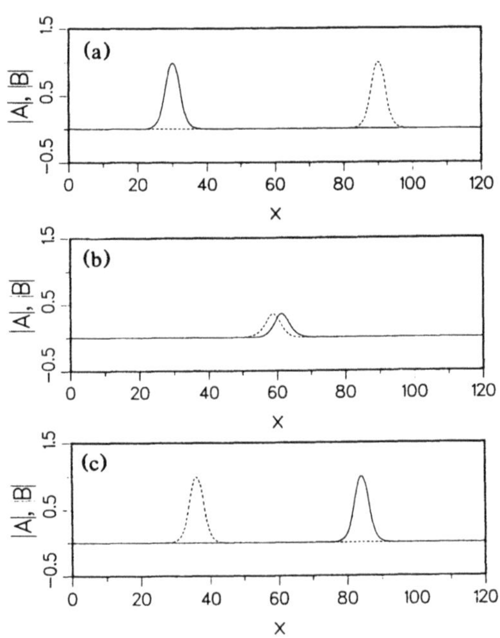

Fig. 1: Complete interpenetration of two one-particle solutions (Fig.1 of ref.2). The parameter values are: $\chi = -0.1$, $v = 1$, $\gamma_r = 1.0$, $\gamma_i = 0$, $\beta_r = -3.0$, $\beta_i = -1.0$, $\delta_r = 2.75$, $\delta_i = -1.0$, $\zeta_i = 1$ and $\zeta_r = 2.5$. (a) initial approach, (b) interaction, and (c) final state after interaction.

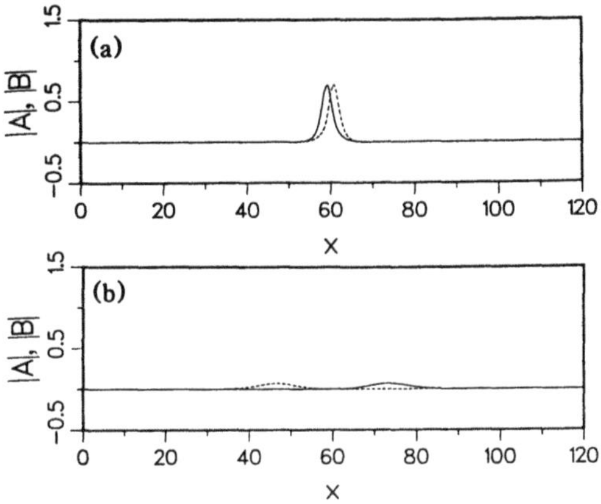

Fig. 2: Annihilation of two one-particle solutions (Fig.2 of ref.2). The initial condition is identical to Fig.1a. The parameters are as in Fig.1 except for $\zeta_r = 3$. (a) interaction and (b) decaying remnants of the two one-particle solutions.

If one or both of the colliding pulses have not yet reached their asymptotic shape, the annihilation of only one pulse in the regime where annihilation occurs is also possible (Fig.3).

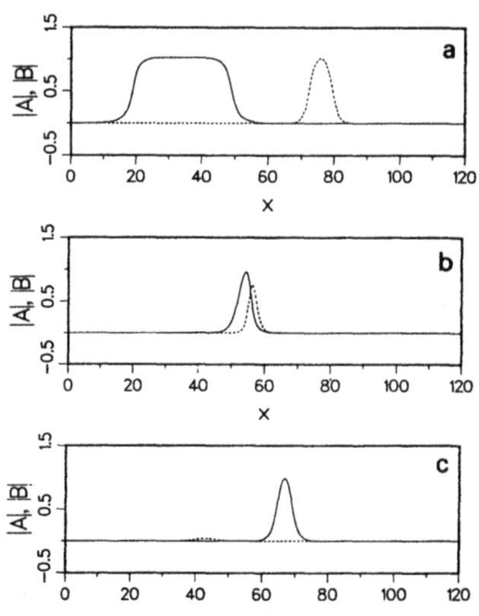

Fig. 3: Annihilation of a one-particle solution by another solution that has not yet reached its asymptotic shape (Fig.3 of ref.3). The parameters are as in Fig.2 except for $v = 2$. (a) initial approach, (b) interaction, and (c) the surviving one-particle solution propagates to the right.

To underscore these statements we present in Figs.4 and 5 results for the collision of stable two-dimensional pulses obtained from two coupled envelope equations which have the same structure as eqs.(1) and (2) except for replacing the second derivative in the x-direction by the transverse Laplace operator in the $x-y$ - plane: $\Delta_\perp = \partial_x^2 + \partial_y^2$.

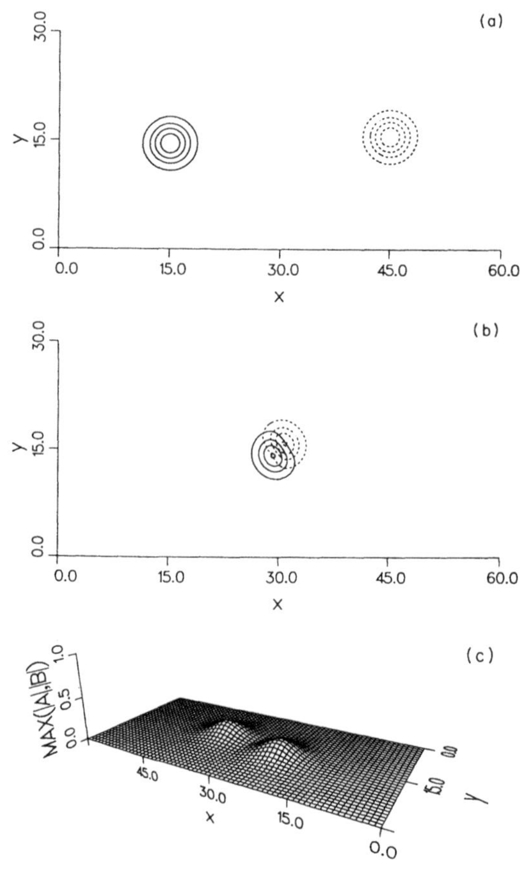

Fig. 4: Complete annihilation of two localized states of impact parameter $b = 1$ and stabilizing cross-coupling $\zeta_r = 2.0$ and $\zeta_i = 0$; other parameter values as in Fig.1 (Fig.2 of ref.7). (a) contour plot well before the interaction, (b) contour plot during the interaction, and (c) 3-D plot of the decaying remnants after the interaction.

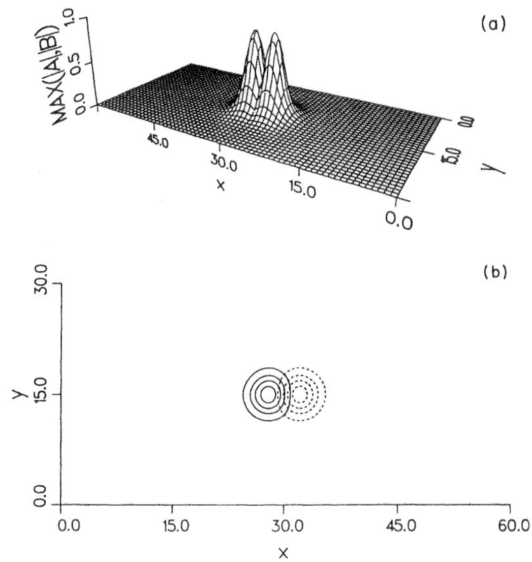

Fig. 5: Stationary compound state as a result of a head-on collision for stabilizing cross-coupling $\zeta_r = 1.0$ and sufficiently small group velocity $v = 0.1$; other parameter values as in Fig.4 (Fig.5 of ref.7). (a) 3-D plot and (b) contour plot

Fig.4 shows that even for a finite impact parameter annihilation of pulses is possible. Fig.5 demonstrates that for low enough approach velocity even the compound states mentioned above becomes possible. We would also like to stress that these two-dimensional results can be compared to the experimental results since they have been obtained for the envelope equation applicable to systems that are anisotopric in the plane such as for example electroconvection in nematic liquid crystals[18]. Similarly the system studied in ref.14 is anisotropic in the plane. By rescaling the two directions in the plane, i.e. x and y, one easily gets pulses with elliptic cross-section instead of the pulses with circular cross-section shown in Figs.4 and 5.

All these properties[1−4] are consistent with those presented for the pulses in ref.14.

4. Conclusions

In conclusion we have argued that the pulses and their interaction observed recently[14−17] near a surface reaction share many features with the localized solutions and their interaction studied recently[1−4] for the prototype envelope equation which arises near a weakly inverted bifurcation to traveling waves. It will be most interesting to check the amount of hysteresis encountered experimentally and to provide a stronger link between the reaction-diffusion equations applicable to this reaction and the reduced description in terms of an envelope equation.

In addition it would be very important to check whether it is possible to observe in this autocatalytic chemical surface reaction breathing localized states and their interaction.

5. Acknowledgements

The work done at the Center for Nonlinear Studies, Los Alamos National Laboratory has been performed under the auspices of the United States Department of Energy. We also thank the Deutsche Forschungsgemeinschaft for partial support of this work through the Graduiertenkolleg 'Nichtlineare Spektroskopie und Dynamik'.

6. References

1. O. Thual and S. Fauve, *J. Phys. (Paris)* **49** (1988) 1829.
2. H.R. Brand and R.J. Deissler, *Phys. Rev. Lett.* **63** (1989) 2801.
3. R.J. Deissler and H.R. Brand, *Phys. Lett.* **A146** (1990) 252.
4. V. Hakim, P. Jakobsen, and Y. Pomeau, *Europhys. Lett.* **11** (1990) 19.
5. H.R. Brand, P.S. Lomdahl, and A.C. Newell, *Phys. Lett.* **A118** (1986) 67.
6. H.R. Brand, P.S. Lomdahl, and A.C. Newell, *Physica* **D23** (1986) 345.
7. R.J. Deissler and H.R. Brand, *Phys. Rev.* **A44** (1991) R3411.
8. N.J. Zabusky and M.D. Kruskal, *Phys. Rev. Lett.* **15** (1965) 240.
9. A.C. Newell, *Solitons in Mathematics and Physics* (Society for Industrial and Applied Mathematics, Philadelphia, 1985).
10. E. Infeld and G. Rowlands, *Nonlinear Waves, Solitons and Chaos* (Cambridge University Press, Cambridge, 1990).
11. H.R. Brand and R.J. Deissler, *Physica* **A204** (1994) 87.
12. R.J. Deissler and H.R. Brand, *Phys.Rev.Lett.* **72** (1994) 478.
13. R.J. Deissler and H.R. Brand, submitted for publication.
14. H.H. Rotermund, S. Jakubith, A. von Oertzen, and G. Ertl, *Phys.Rev.Lett.* **66** (1991) 3083.
15. H.H. Rotermund, *Physica Scripta* **T49** (1993) 549.
16. M.D. Graham, I.G. Kevrikidis, K. Asakura, J. Lauterbach, K. Krischer, H.-H. Rotermund, and G. Ertl, *Science* **264** (1994) 80.
17. M.D. Graham, M. Bär, I.G. Kevrikidis, K. Asakura, J. Lauterbach, H.-H. Rotermund, and G. Ertl, *Phys.Rev.E*, submitted for publication.
18. I. Rehberg, S. Rasenat, M. de la Torre, W. Schöpf, F. Hörner, G. Ahlers and H.R. Brand, *Phys. Rev. Lett.* **67** (1991) 596.

Nonlinear Amplification of Solitons in High Dispersion Fiber Transmission Systems

I. Gabitov*, D. D. Holm**, B. Luce***, and Arnold Mattheus****

** L.D. Landau Institute for Theoretical Physics, Kosigina 2, 117940, Moscow, Russia. E-mail: gabitov@sun.rhrk.uni-kl.de*

***Theoretical Division, Los Alamos National Laboratory, Los Alamos, NM 87545, E-mail: dholm@lanl.gov*

****Center for Nonlinear Studies, Los Alamos National Laboratory, Los Alamos, NM 87545, E-mail: luceb@goshawk.lanl.gov*

*****FTZ Telekom, Am Kavalleriesand 3, D-64295, Darmstadt Germany, E-mail:fz241d@vmxa.fz.telekom.de*

Abstract

We study the use of nonlinear amplifying loop mirrors (NALM's) as elements of high dispersion fiber transmission systems to reshape soliton pulses in order to overcome broadening. We approach this problem as a mapping problem of input pulse to output pulse, for a segment of transmission fiber followed by a combination of linear and nonlinear amplification. For a wide range of amplifier spacings, we find that an optimal soliton input pulse can be found which is well reproduced at output. These optimal input pulses produce only about three percent continuous radiation in the output pulse.

1 Introduction

The fundamental elements of optical soliton transmission systems are segments of optical fiber, followed by amplifiers to compensate fiber losses. In order to function properly, each element must be able to reproduce the input pulses at output, without creating excessive continuous radiation. We may consider the process of propagation followed by amplification as a mapping of input pulse to output pulse. The problem of optical soliton transmission is then to find stationary points of the mapping.

In practice, this problem falls into two categories: long distance soliton transmission based on dispersion shifted fiber with periodic amplification [1, 2], and the improvement of existing networks based on standard monomode fibers.

Long distance transmission has been studied greatly in recent years [2]. The most important characteristic of these systems is that if the dispersion length z_d, the distance for which a small amplitude linear pulse will approximately double its pulsewidth, is much greater than the amplifier spacing z_a, the system supports what is known as center guiding soliton transmission, in which the average envelope of the pulses are described by the usual nonlinear Schrödinger equation [3].

According to recent investigations [4], the capacity of existing networks might be improved by operating in the soliton regime, at a wavelength of $1.55\mu m$, for which commercially available Erbium doped fiber amplifiers exist. At this wavelength, standard monomode fiber is highly dispersive. In this case, at the desired data transmission rates, the dispersion length will be of the order of the amplifier spacing (resonance case), and center guiding soliton transmission will not be achieved. The solution to this problem proposed in [4] is to use an "over-soliton", that is, an input pulse of the form $\eta sech(\beta T)$, where $\beta > \eta$, so that nonlinear compression during the initial stage of propagation will compensate broadening later on. A drawback of this method is that the initial nonlinear compression is accompanied by substantial production of continuous radiation, leading to nonlocal interaction of bit pulses.

In this paper we study a suggestion made in [5] to use a nonlinear amplifying loop mirror (NALM) which is specially configured to provide radiationless nonlinear amplification of solitons. It was shown in [5] that a properly configured NALM will, with high accuracy, cube the amplitude of the incident pulse, leaving the phase unchanged. This effect is due to incomplete cancellation when counter-propagating pulses in the loop are recombined, due to asymmetrical placement of an amplifier within the loop. Recently, in [6], it was shown that in the center guiding soliton case, the NALM is effective in reducing the signal to noise ratio in long distance transmission systems.

We find that the solution of the mapping problem for the NALM followed by an additional linear amplifier, for a wide range of amplifier spacings and fiber losses, can be approximated very well with soliton input pulses of the form $\eta sech(\eta T)$.

The evolution of optical pulses is described by [2]

$$\frac{\partial q}{\partial Z} = \frac{i}{2}\frac{\partial^2 q}{\partial T^2} + i|q|^2 q - \gamma q. \tag{1}$$

These nondimensional variables are related to the electric field amplitude E and fiber parameters via

$$Z = z/z_a, \quad T = t/\sqrt{-z_a k''}, \quad \gamma = \Gamma z_a, \quad q = \sqrt{\frac{\pi z_a n_2}{\lambda_0}}E. \tag{2}$$

Here, k'' is the group velocity dispersion, n_2 is the Kerr coefficient, and Γ is an attenuation coefficient related to the fiber loss L_{dB} in decibels per kilometer by $\Gamma = -ln(10^{-.1L_{dB}})/2[km]$. A typical value of L_{dB} for silica fiber at $\lambda_0 = 1.55\mu m$ is $L_{dB} = 0.2dB/km$.

The group velocity dispersion k'' is related to the group delay parameter D of the fiber by $k'' = -\lambda_0^2 D/2\pi c$. Values of D for high dispersion standard monomode fiber are about 17 ps/nm/km.

The nondimensional pulsewidth T_s (FWHM) is given by $T_s = 1.76/\eta$. The ratio of the dispersion length to the amplifier spacing for an optimal pulse parameterized by η, in the nondimensional units defined by equations (2), is conveniently given by

$$Z_d = \frac{z_d}{z_a} = \frac{1}{\eta^2}. \tag{3}$$

2 Numerical Results

We seek a solution to the mapping problem with the NALM by numerically determining the gain parameters and parameters of soliton pulses that are restored after propagation under equation (1) over a unit nondimensional distance and amplification by the fiber mirror and the additional linear amplifier. We lump the linear gain of the NALM and the linear amplifier into the parameter A, so that the total amplification process is represented by the mapping

$$q(T) \to A|q(T)|^2 q(T). \tag{4}$$

For a fixed value of Γ, the nondimensional attenuation coefficient γ is linear in the amplifier spacing, and we are interested only in values of γ between 0 and 1. Values of γ larger than unity too greatly decrease the pulse over the interval $\Delta Z = 1$. We look for optimal input pulses of the form

$$q_{in} = \eta(\gamma) sech(\eta(\gamma) T), \tag{5}$$

which we find to be sufficient for our purpose. This choice corresponds to the fundamental soliton solution of equation (1) with $\gamma = 0$.

Our procedure consisted of fixing the value of γ, and measuring the percentage of dispersive energy in the output pulse (defined below) as a function of η, where A is chosen for each value of η such that the output pulse has the same amplitude as the input pulse. Our task then reduces to finding the optimal choice of η and A for each γ.

The optimal choice is defined to be the one that produces the least dispersive wave energy in the output pulse. The dispersive wave energy E_{disp} is given by the difference of total output and soliton energies, $E_{disp} = E_{tot} - E_{sol}$, where

$$E_{tot} = \int_{-\infty}^{+\infty} |q_{out}|^2 dT, \tag{6}$$

and E_{sol} is numerically determined from the spectrum of the linear eigenvalue problem associated to the NLS equation

$$\frac{d\psi_1}{dT} + i\xi\psi_1 = q_{out}\psi_2 \tag{7}$$

$$\frac{d\psi_2}{dT} - i\xi\psi_2 = -q^*_{out}\psi_1. \tag{8}$$

As in [5], we measure both the percentage of dispersive energy and the percentage of dispersive components in NLS Hamiltonian. The NLS Hamiltonian is

$$H = \frac{1}{2}\int_{-\infty}^{+\infty}\left(|q|^4 - |\frac{dq}{dT}|^2\right)dT, \tag{9}$$

so that equation (1), with $\gamma = 0$, is expressible as $\frac{\partial q}{\partial Z} = i\delta H/\delta q$.

These percentages are defined via the ratios $\Delta E = E_{disp}/E_{tot}$ and $\Delta H = H_{disp}/H_{tot}$, respectively, where $H_{disp} = H_{tot} - H_{sol}$. Note that ΔH may be greater than one in magnitude because H_{tot} may be negative. ΔH may also be negative because it is possible for $|H_{tot}| < |H_{disp}|$ to occur. Expressions for E_{sol} and H_{disp} in terms of the eigenvalues of system (8) are given in [5].

Our primary result is that for any choice of γ in the interval [.3, 1] there is a single optimal pulse, corresponding to a particular value of A, with a minimum dispersive energy output of about two or three percent, with a minimum of dispersive components of the Hamiltonian in the output pulse of between 0.3 percent and about 6 percent. This optimal choice results from the fact that values of A less than the optimal values yield output pulses broader than the input pulse, and larger values yield narrower outputs. The reasons for these behaviors can be understood from soliton perturbation theory, and will be reported elsewhere [7]. There is significant degradation of the output pulse for values of γ outside of the interval [0.3, 1].

The optimal values of A, and the associated values of η are plotted in Figure 1(a). The percentages ΔE and ΔH are given in Figure 1(b). From Figure 1(b) it can be seen that there is a particularly optimal choice of γ around 0.45. All of these optimal values were found by minimizing only ΔE, but these values were found to correspond well with the minima of ΔH, as well as to an additional least squares measure (not shown) of the difference between the input and output pulses.

Note that the trends displayed in Figure 1 indicate that the NALM is only useful for reshaping pulses if the product of the attenuation and the amplifier spacing (whose product is γ) is of an intermediate value. This is because pulse broadening, indirectly due to attenuation, is essential for the existence of an approximate fixed point for soliton input pulses in this approach to the optical soliton transmission system.

We calculate from equation (3), and from the optimal values of η displayed in Figure 1, that the ratio of the dispersion length to the amplifier spacing for the optimal pulses ranges from about 1/16 for $\gamma = 0.3$, to about 1/4 for $\gamma = 1$. This suggests that greater amplifier spacing can be achieved, as compared to networks operating in the linear regime or with center guiding soliton transmission, while maintaining the data transmission rate, or higher data transmission rates can be achieved, while maintaining the same amplifier spacing.

It is easy to see that the approximate fixed points we find for the mapping problem are dynamically unstable due to the cubing process: If the input amplitude is greater or lesser than optimal for a given choice of A, then subsequent amplifications will further increase or decrease amplitude, respectively. Figure 2 shows the amplitude input/output map for soliton input pulses for $\gamma = .4$ and $A = 1.2$. The fact that the slope of this map is greater than one at the fixed point (where the map crosses the identity line) means that the fixed point must be unstable. This instability should not be a problem if the number of amplifiers is not too great.

3 Acknowledgements

We would like to thank Y. Kodama for discussions which stimulated this work, and V. Makhankov and S. Burtsev for useful suggestions. I. Gabitov would like to thank the Center for Nonlinear Studies and the T-7 Theoretical Division at Los Alamos National Laboratory for supporting him during this work.

References

[1] L.F. Mollenauer, M.J. Neubelt, S.G. Evangelides, J.P. Gordon, J.R. Simpson, and L.G. Cohen. Experimental study of soliton transmission over more than 10,000 km in dispersion-shifted fiber. *Optics Letters*, 15(21), November 1 1990.

[2] Y. Kodama and A. Hasegawa. Theoretical foundation of optical-soliton concept in fibers. *Progress in Optics*, XXX:205–259, 1992.

[3] Y. Kodama, A. Maruta, and A. Hasegawa. Long distance communications with solitons. *Journal of the European Optical Society*, 1994. to appear.

[4] Gabitov et. al. Soliton propagation on standard monomode fibers with lumped amplifiers: Basic features and performance analysis. *Proceedings of LEOS Summer Topical Meeting on Impact of Fiber Nonlinearities on Lightwave Systems*, July 26 1993. Santa Barbara, California, USA.

[5] M. Matsumoto, Y. Kodama, and A. Hasegawa. Adiabatic amplification of solitons by means of nonlinear amplifying loop mirrors. 1994. to appear.

[6] M. Matsumoto, H Ikeda, and A. Hasegawa. Suppression of noise accumulation in bandwidth-limited soliton transmission by means of nonlinear loop mirros. *Optics Letters*, 19(3):183–185, February 1994.

[7] I. Gabitov, D. Holm, B. Luce, and A. Mattheus. Theoretical basis for optimal pulses in optical fiber transmissian systems utilizing nonlinear amplification. In preparation.

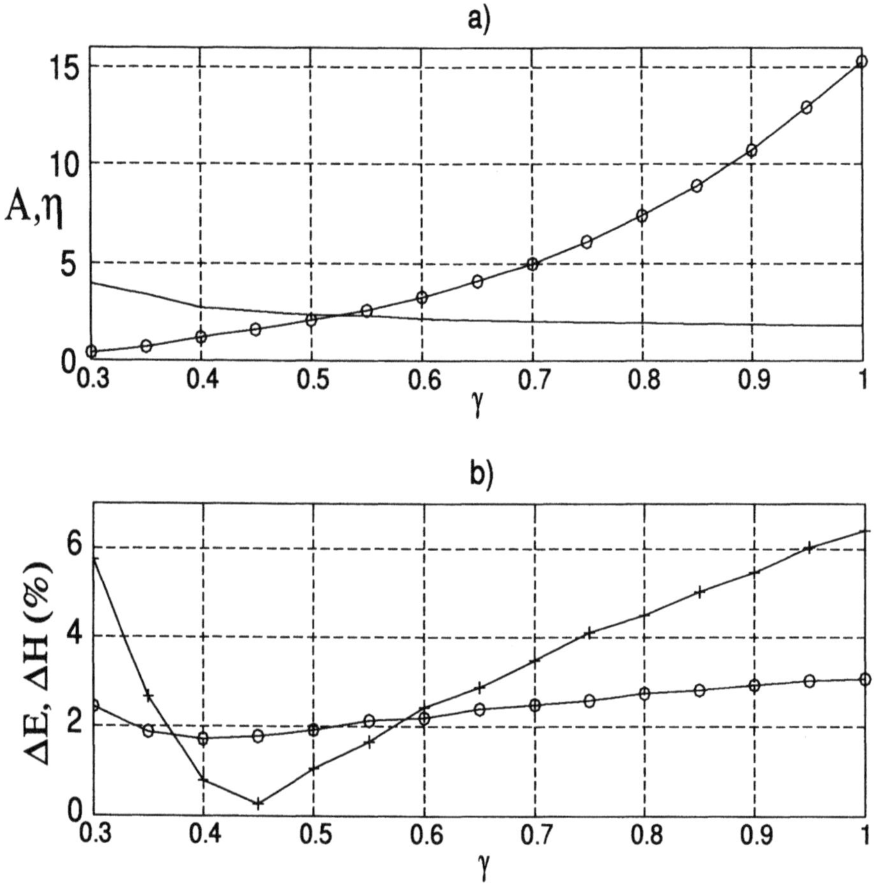

Figure 1: a) Optimum values of A(-o-) and η(-) vs. γ. b) Percentages of dispersive energy ΔE(-o-) and dispersive contribution to the Hamiltonian ΔH(-+-).

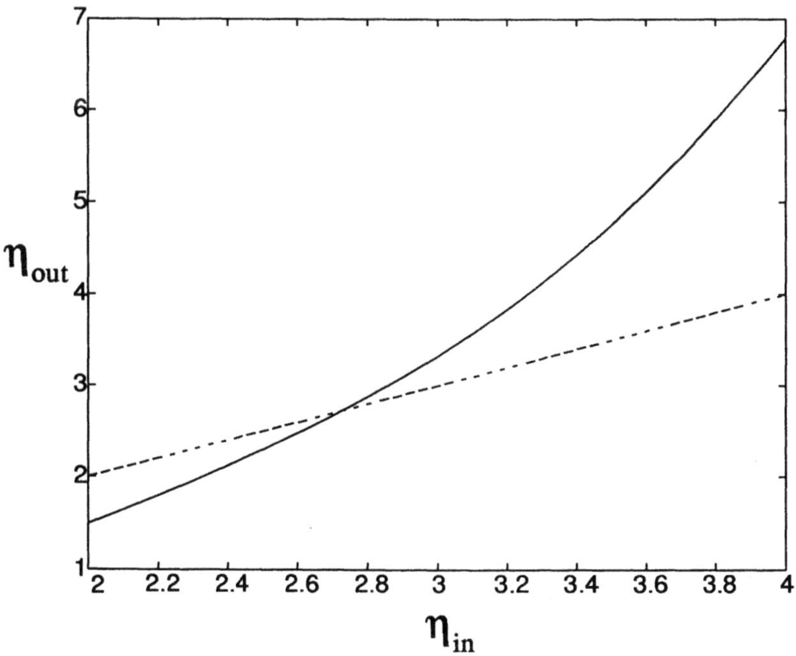

Figure 2: Peak amplitude input/output map for $\gamma = .4$, $A = 1.2$. The dotted line represents the identity line.

Morphology of Laplacian growth processes and statistics of equivalent many-body systems

Raphael Blumenfeld

Center for Nonlinear studies and Theoretical Division, MS B258

Los Alamos National Laboratory, Los Alamos, NM 87545, USA

Abstract

I propose a theory for the nonlinear evolution of two dimensional interfaces in Laplacian fields. The growing region is conformally mapped onto the unit disk, generating an equivalent many-body system whose dynamics and statistics are studied. The process is shown to be Hamiltonian, with the Hamiltonian being the imaginary part of the complex electrostatic potential. Surface effects are introduced through the Hamiltonian as an external field. An extension to a continuous density of particles is presented. The results are used to study the morphology of the interface using statistical mechanics for the many-body system. The distribution of the curvature and the moments of the growth probability along the interface are calculated exactly from the distribution of the particles. In the dilute limit, the distribution of the curvature is shown to develop algebraic tails, which may, for the first time, explain the origin of fractality in diffusion controlled processes.

LA-UR-94-xxxx

1. Introduction

Diffusion controlled growth processes that result in fractal and self-similar patterns abound in nature but the morphologies of these interfaces are not well understood theoretically. Well known examples are solidification of supercooled liquid, diffusion-limited aggregation, electrodeposition and growth of bacterial colonies.[1]Currently, there is no theory that starts from the local equations of motion (EOM) and derives the asymptotic patterns that such growths evolve into, or even whether they should become fractal or not. Traditionally, renormalization group approaches are employed,[2]which recently managed to yield quite close values for the fractal dimension of several growths.[3],[4]

It was proposed[5]that two-dimensional free interfaces namely, without surface tension) that grow in Laplacian fields can be conformally mapped onto the unit circle and be studied via the evolution of the map. This idea has been used repeatedly since the forties. Richardson showed that the process supports a set of constants,[6]Shraiman and Bensimon[7]showed explicitly how to convert the problem into a many-body system and pointed out that the free interface evolution is ill-posed in the Cauchy sense. Namely, the formalism breaks down after a finite time due to unstable growth of perturbations on ever short lengthscales along the interface, which develop into cusp singularities along the physical interface. Efforts to avoid this catastrophe employed surface tension to cut off the short lengthscales in a renormalizable manner,[8]but these met with another difficulty: adding small capilarity effects turns out to be a singular perturbation to the EOM of the system. This means that solutions derived by this approach are strongly sensitive to initial conditions,[9]and therefore so are the morphologies of these interfaces. Yet, observations show that the final morpholgical proprties are robust to the details of the growth and to initial conditions. Therefore the singular pertubation solutions seem to miss some of the basic physics. Alternatively, it has been recently found that tip-splitting reduces high surface curvature and it was proposed that cusps are naturally prevented by implementing this into the EOM of the equivalent many-body system.[10] An issue that attracted little attention in this context is whether the process is Hamiltonian or Lagrangian. Although the problem is known to support a set of conserved quantities,[6],[10],[11] it is not clear how these help in finding a Hamiltonian structure or a Lyapunov function.

In this presentation I first introduce the EOM of the interface and show that there is an underlying Hamiltonian structure. I show that the Hamiltonian is directly related to the physical two dimensional potential. I introduce the EOM of the particles of the equivalent many-body system, which are the the zeros and poles of the analytic (actually, meromorphic) map. An extension of the formalism to a continuous density of two species of particles is derived, the continuous EOM is explicitly derived, as well as the evolution equation for this density. Having a Hamiltonian allows a natural introduction of surface effects in the form of a field that repels particles from the interface. This overcomes the ill-posedness of the problem, extending its validity to time$\to\infty$. Thes Hamiltonian structure also allows a statistical mechanical analysis of the many body system, and in particular the distribution of the particles can be investigated. Finally, I show how to extract information on the morphology of the interface from this distribution. As an example I calculate the moments of the growth probability distribution along the interface and the distribution of the curvature.

2. Formulating the problem and the EOM

Consider a two-dimensional simply connected curve (the interface), $\gamma(s)$, that is parametrized by $0 \leq s < 2\pi$, and which encloses a region Γ. Outside Γ there is a Laplacian field

$$\nabla^2 \Phi = 0 \tag{2.1}$$

and γ is fixed at a given value of Φ (the concentration for diffusion controlled processes or the electrostatic potential in electrodeposition). A higher value of Φ is assigned to a circular boundary whose radius is very large compared to the linear size of Γ. The surface is constituted to advance at a rate that is proportional to the local gradient of the field normal to the surface

$$v_n = -\nabla\Phi \cdot \hat{\mathbf{n}} \ . \tag{2.2}$$

When this rate is sufficiently slow the field can be considered to be static. Denoting the physical (complex) plane by $\zeta = x + iy$, we conformally map γ at each instant of time, t, onto the unit circle via

$$\zeta = F(z,t) \quad ; \quad \gamma(s,t) = \lim_{z\to e^{is}} F(z,t) \ , \tag{2.3}$$

where z is a mathematical complex plane. The field gradient along the curve is

$$-\nabla\Phi(\zeta) = -\left(\frac{\partial\Phi(\zeta)}{\partial\zeta}\right)^* = \frac{-i}{(zF')^*} \ , \tag{2.4}$$

where * stands for complex conjugate and the prime indicates differentiation with respect to z. Since $z \to e^{is}$ one can write down the EOM for γ:[7]

$$\partial_t\gamma(s,t) = -i\partial_s\gamma(s,t)\left[|\partial_s\gamma(s,t)|^{-2} + ig(s)\right] \ . \tag{2.5}$$

The first term within the square brackets on the r.h.s. of (2.5) represents the normal growth, (2.2), but the second term is introduced by hand (although uniquely determined) and represents 'sliding' of a point along the curve. The entire square brackets on the r.h.s. of (2.5) is the limit of an analytic function $G(z,t)$ (see below) and the EOM for F becomes

$$\dot{F} = zF'G \ . \tag{2.6}$$

It is convenient to write the logarithimc derivative of this equation with respect to z:

$$\frac{d}{dt}\ln F' = \frac{1}{F'}\frac{d}{dz}(zF'G) \ . \tag{2.7}$$

Generally, the map must satisfy the following constraints:[10]i) The boundary conditions far away should remain unchanged under the map, enforcing $\lim_{z\to\infty} F \sim z$; ii) The map may have no branch cuts. These conditions are satisfied by the general form

$$F' = A(t) \prod_{n=1}^{N} \frac{z - Z_n(t)}{z - P_n(t)} , \tag{2.8}$$

where the time dependence appears in the scaling factor, A, and the locations of the zeros and poles (the particles). It is possible to show that with this form and with a proper choice of the number N, one can describe *any* initial simply connected cuspless curve, indicating that this form is *general* and does not represent merely a special class of maps.[(12)] Integrating (2.8) we have

$$\zeta = F(z,t) = A(t) \left[1 + \sum_{n=1}^{N} R_n \ln(z - P_n) \right] , \tag{2.9}$$

and

$$G = G_0 + \sum_{n=1}^{N} \frac{Q_n}{z - Z_n} . \tag{2.10}$$

The quantities R_n are the residues of the product (without the prefactor A) in (2.8) at P_n, and

$$Q_n = 2 \prod_{m=1}^{N} \frac{(1/Z_n - P_m^*)(Z_n - P_m)}{(1/Z_n - Z_m^*)(Z_n - Z_{m'})}$$

$$G_0 = \sum_{m=1}^{N} \frac{Q_m}{2Z_m} + \prod_{m=1}^{N} \frac{P_m}{Z_m} .$$

By contour-integrating around the location of the poles in either (2.6) or (2.7), and by contour integrating around the locations of the zeros in (2.7), one finds their EOM:

$$\begin{aligned} -A^2(t)\dot{Z}_n &= Z_n \Big\{ G_0 + \sum_{m'} \frac{Q_n + Q_{m'}}{Z_n - Z_{m'}} \Big\} + Q_n \Big\{ 1 - \sum_{m} \frac{Z_n}{Z_n - P_m} \Big\} \equiv f_n^{(Z)}(\{Z\};\{P\}) \\ -A^2(t)\dot{P}_n &= P_n \Big\{ G_0 + \sum_{m} \frac{Q_m}{P_n - Z_m} \Big\} \equiv f_n^{(P)}(\{Z\};\{P\}) . \end{aligned} \tag{2.11}$$

From (2.6), and taking into account that there are no branch cuts, one can show that the quantities $A(t)R_n(t)$ are constants of the motion that are independent of each other and which are determined only by initial conditions. These constants are directly related to those found by Richardson[6] and later authors.[10],[11] The time evolution of $A(t)$ is found from (2.6) as follows

$$\dot{A}(t) = A(t)G_0 \ . \tag{2.12}$$

3. A continuous distribution of zeros and poles

In the forgoing formulation the EOM do not allow production of new zeros and poles. Such production is desirable because it would better reflect the physical growth where tip-splitting and side-branching occur. A mechanism that incorporates such processes has been recently introduced and studied by Blumenfeld and Ball[10] (see also below).

The approach that I propose here starts from a *continuous* density of particles, $\mathcal{N}(Z,P)$, whose continuous arguments correspond to the locations in the two-dimensional mathematical plane of the zero- and pole-species. This approach obviates the above problem because the number of particles is infinite to begin with. Clearly, the evolution of the interface is unaffected by the new description and can be used again to derive the evolution of the density $\mathcal{N}$. First, write down the map and its derivative in the continuous form:

$$\begin{aligned} F' &= A(t)\exp\left\{\int \mathcal{N}(Z',P')\left[\ln(z-Z') - \ln(z-P')\right] dZ'dP'\right\} \\ F &= A(t)\left\{z + \int \mathcal{N}(Z',P')R(P')\ln(z-P')dZ'dP'\right\} \ , \end{aligned} \tag{3.1}$$

where

$$R(P') = \exp\left\{\int \mathcal{N}(Z',P')\left[\ln(P'-Z') - [1-\delta(P'-P')]\ln(P'-P')\right] dZ'dP'\right\}$$

corresponds to the residue of F'. The quantities Q_n, G_0 and G can be similarly generalised

$$\begin{aligned} Q_n \to Q(Z) &= 2\exp\left\{\int \mathcal{N}(Z',P')[1-\delta(Z-Z')]\ln\frac{(1-ZP'^*)(Z-P')}{(1-ZZ'^*)(Z-Z')}dZ'dP'\right\} \\ G_0 &= \int \mathcal{N}(Z',P')\frac{Q(Z')}{2Z'}dZ'dP' - \exp\left\{\int \mathcal{N}(Z',P')\left[\ln P' - \ln Z'\right]dZ'dP'\right\} \\ G(z) &= G_0 + \int \mathcal{N}(Z',P')\frac{Q(Z')}{z-Z'}dZ'dP' \ . \end{aligned} \tag{3.2}$$

Using these expressions we obtain for the local velocities

$$\begin{aligned}
-A^2(t)\dot{Z} &= Z\left\{G_0 + \int (1-\delta(Z-Z'))\,\mathcal{N}_+(Z')\frac{Q(Z)+Q(Z')}{Z-Z'}dZ'\right\} \\
&\quad + Q(Z)\left\{1 - \int \mathcal{N}_-(P')\frac{Z}{Z-P'}dP'\right\} \\
-A^2(t)\dot{P} &= P\left\{G_0 + \int \mathcal{N}_+(Z')\frac{Q(Z')}{P-Z'}dZ'\right\} ,
\end{aligned} \tag{3.3}$$

where

$$\mathcal{N}_+(Z') \equiv \int \mathcal{N}(Z',P')dP' \quad \text{and} \quad \mathcal{N}_-(Z') \equiv \int \mathcal{N}(Z',P')dZ'$$

are, respectively, the distributions of the zero-species and the pole-species. The evolution equation for the density then reads

$$\frac{\partial \mathcal{N}}{\partial t} + \dot{Z}\cdot\nabla_Z\mathcal{N} + \dot{P}\cdot\nabla_P\mathcal{N} = 0 , \tag{3.4}$$

where ∇_R is the gradient in the two-dimensional plane R_1-R_2, and where R stands for either Z or P. These equations are nonlocal as can be seen from (3.3). This master equation can be addressed either in the continuous or discrete form.

4. The Hamiltonian structure

The fact that the process is Hamiltonian is suggested already by EOM (2.5), where only the normal rate is considered:

$$\partial_t\gamma(s,t) = -i\partial_s\gamma(s,t)\,|\partial_s\gamma(s,t)|^{-2} = -i\frac{\delta s}{\delta\gamma^*} . \tag{4.1}$$

Expression (4.1) is equivalent to Hamilton's equations, suggesting that the (rescaled) length of the actual growth, s, plays the role of a Lyapunov functional, while γ is a complex field. This observation is very appealing and it is tempting to ask whether the EOM for the map, F, where a tangential propagation is included, does also enjoy such a structure. It is not difficult to show[13] that this is indeed so: Simply multiply the complex conjugate of (2.5) by γ_s and rewrite its imaginary part in the form

$$\mathrm{Im}\{F_t^* F_s\} = 1 \ .$$

Writing $F = F_1 + iF_2$, with F_1 and F_2 real functions, this equation implies that the Jacobian of the transformation from the coordinates $t - s$ to the coordinates $F_1 - F_2$ is unity,

$$\frac{\partial(F_1, F_2)}{\partial(t, s)} = 1 \ . \tag{4.2}$$

With this it is straightforward to show that

$$\dot{F}_1 = \frac{\partial s}{\partial F_2} \qquad \text{and} \qquad \dot{F}_2 = -\frac{\partial s}{\partial F_1} \ , \tag{4.3}$$

or

$$\dot{F} = \frac{\partial s}{\partial F^*} \ , \tag{4.4}$$

and the map indeed follows Hamiltonian dynamics as z goes to the unit circle. The notion that s is some sort of conserved Hamiltonian may look strange but I argue here that it is related to a real energy by being the true physical potential field (in electrostatic terms). The way to see this goes through generalising the above argument by taking $z \to e^{is}$. Eq. (4.4) then generalises to

$$\dot{F} = -i\frac{\partial \mathrm{Im}\{\ln z\}}{\partial F^*} = -i\frac{\partial \mathrm{IM}\{\Phi\}}{\partial F^*} \ , \tag{4.5}$$

where Φ is *exactly the complex potential* in the physical plane and therefore $\mathrm{Im}\{\Phi\}$ denotes the stream lines in the two dimensional potential.

An alternative way to obtain a Hamiltonian structure is by defining a complex function $\Psi \equiv F' + iz$ and manipulating Eq. (2.7),[14],[15] which gives

$$\frac{\partial \Psi}{\partial t} = -i\frac{\partial H_0}{\partial \Psi^*} \ , \tag{4.6}$$

where $H_0 \equiv zF'G$ is a complex Hamiltonian.

The foregoing shows that the p.d.e. that describes the curve's evolution follows indeed Hamiltonian dynamics. Since the many-body formulation is an equivalent, one-to-one, description of the curve's dynamics it follows that the system of the particles, the poles and zeros, must also support a Hamiltonian structure. Finding the explicit Hamiltonian for the two-dimensional many-body system proves very difficult and this problem is currently unsolved. But the very realisation that such a Hamiltonian exists allows access to a whole new bag of tools.

5. Surface effects

It has been mentioned in the introduction that without surface tension (or capilary forces) cusps form along the interface, which correspond to a real phenomenon rather than to a mathematical artefact.[16] The stability analysis that shows this is well known and will not be repeated here. Real growth processes, however, do not really admit cusps either due to atomistic cutoffs, or because the system must pay surface energy as the curvature increases. Since the foregoing formalism is continuous we are outside the microscopic regime, and I focus on the macroscopic surface effects. Consider a local protrusion at s that grows into a potential cusp - surface effects act to eliminate small length scales that appear with sharpening and therefore to effectively introduce a small, but macroscopic, length scale. In the many-body system, the sharpening corresponds to a zero approaching the unit circle at a point s. Thus the appearance of a small cutoff corresponds to a minimal distance in the (possibly rescaled) mathematical plane below which the zero cannot approach the unit circle† It follows that, in the many-body system, surface energy effects must correspond

† In the present formalism this distance needs to be renormalized with time simply because the region grows, while the map is always to the unit circle. This point can be taken into account by realising that the map is rescaled with time by the factor $A(t)$ in Eq. (2.8).

to an *effective repulsive potential* between the particles and the interface defined by the unit circle. It should be stressed that only the existence of a Hamiltonian makes it possible to use the term 'repulsive potential' with a proper physical meaning.

There are various ways to employ this idea: One is by simply introducing a surface potential term in the many-body Hamiltonian. This direct approach is currently difficult since the explicit form of this Hamiltonian is unknown as yet. Nevertheless, one can insert such a term in the Hamiltonian of the p.d.e., $H = H_0 + V$ in (4.6), derive the EOM for the particles with the modified Hamiltonian and analyse the effect. The functional form of the surface potential term is determines to a large extent the asymptotic morphology. For example, the stronger the repulsion, the smoother the resulting interfaces. Another approach is to make use of the fact that the boundary induces a field that acts on the particles. This concept can be extended by letting the field effect *creation and annihilation* of zeros and poles (in pairs because of the constraints on the map). This extension allows to introduce tip-splitting and side-branching.[10]This treatment is downstream of the main thrust of this presentation, but I will just comment that: i) In many processes a number of zeros and poles that increases with time has the same advantage as in the continuous formulation; ii) The form (2.8) naturally accommodates creation and annihilation of zeros and poles: Once Z_j and P_k occupy the same position in the plane the term $(z-Z_j)/(z-P_k)$ cancels out and these two particles disappear. Similarly, creation of a pair can be regarded as follows: One can consider each point in the z-plane as occupied by both a zero and a pole, contributiong a factor of 1 to the product in (2.8). Fluctuations in the field can then separate the pair slightly so that the degeneracy is lifted and they appear as separate entities. It is straightforward to verify from the EOM that once such a pair is created the two newborn particles immediately drift apart and maintain their identities.

It is convenient to let the field (or the repulsive potential) depend on the curvature.

For example, a possible (complex) potential term would be

$$V = \sigma \lim_{z \to e^{is}} \ln \left[K(\{Z\},\{P\}) \right] , \tag{5.1}$$

where K is the (complex) curvature in terms of the locations of the zeros and the poles[10]

$$K(z,\{Z\},\{P\}) = \lim_{z \to e^{is}} |F'|^{-1} \Big\{ 1 + \sum_{n=1}^{N} \{ \frac{Z_n}{z - Z_n} - \frac{P_n}{z - P_n} \} \Big\} , \tag{5.2}$$

whose real part is the conventional physical curvature along the interface. This particular surface potential is easy to handle in term of zeros and poles because it contributes a *constant* repelling term to the EOM of the particles, (2.11), as they come sufficiently close to the unit circle. In the absence of tip-splitting phenomena, which is not accounted for by such a form, this leads to a considerable simplification of the dynamics. However, it was recently shown that a better choice would be by a term that diverges in the EOM as a particle approaches the unit circle.[15]

6. Noise and statistical analysis

In the above discussion the evolution has been taken to be deterministic. However, the evolution under these EOM is chaotic in the sense that if one starts from very close initial conditions, one ends up very quickly with different deterministic structures.[17]But having a Hamiltonian underlying the dynamics one expects to be able to employ statistical mechanics and in particular to construct a partition function using Gibbs' measure

$$\mathcal{Z} = \int e^{-\beta(H_0 + V)} \mathcal{DR} , \tag{6.1}$$

where $\mathcal{DR} \equiv \prod_{n=1}^{N} d^2 Z_n d^2 P_n$ is a an infenitesimal volume in phase space, and where the parameter β reflects the 'noise' in the system. Averages of quantities such as the moments of the curvature and moments of the field gradient, $|\nabla\Phi|$, can be found by calculating expectation values with this partition fuction

$$\mathrm{E}\{X\} = \frac{1}{\mathcal{Z}} \int X \; e^{-\beta(H_0 + V)} \mathcal{DR} . \tag{6.2}$$

Since, at present, the form of H_0 is unknown, this approach is not easy to implement.

Alternatively, one can construct the master equation for the evolution of the distribution of the particles, $\mathcal{N}(\{Z\},\{P\})$, using Liouville's theorem

$$\frac{\partial \mathcal{N}}{\partial t} + \sum_{n=1}^{N} f_n^{(Z)} \frac{\partial \mathcal{N}}{\partial Z_n} + f_n^{(P)} \frac{\partial \mathcal{N}}{\partial P_n} = 0 \, . \tag{6.3}$$

Note that this equation is exactly the discrete analogue of equation (3.4). Anticipating a steady state after rescaling by $A(t)$ we can discard the explicit time derivative in the new coordinates (which probably correspond to a similarity solution). This leaves us with a Vlasov-type equation, whose solution should yield all there is to know about the interface's statistics. As usual in statistical mechanics, a generally exact solution is a hopeless task, which is why one needs to solve it in some limit. Once this is achieved, this distribution can be directly converted into information about the asymptotic morphology. For example, the distribution of the curvature, $\mathcal{P}_K$, is directly derivable from $\mathcal{N}$ by

$$\mathcal{P}_K = \int \mathcal{N}(\{Z\},\{P\})\, \delta \left\{ K_1 - \left[1 + \mathrm{Re} \sum_{n=1}^{N} \left(\frac{1}{1 - Z_n e^{-is}} - \frac{1}{1 - P_n e^{-is}} \right) \right] \right\} \mathcal{DR} \, , \tag{6.4}$$

where K_1, is the real part of K and corresponds to the measurable curvature along the interface. Similarly, any morphology-related quantity that can be written in terms of the locations of the particles is attainable through this method. In particular, one can use this formalism to calculate the values of the moments of the growth probability distribution along the interface. This distribution occupies a central place in most of studies on diffusion-limited-aggregation and similar phenomena. Its moments immediately yield the so-called multifractal spectrum. This calculation has been recently carried out by Blumenfeld and Ball[10] and is only briefly outlined here. The growth probability at a point s along the interface is

$$p(s) = C_0 |\nabla \Phi(s)| = \lim_{z \to e^{is}} C_0 \, |F'(z)| \quad ; \quad C_0 = 1 / \oint_{\text{along the interface}} |\nabla \Phi(l)| dl \, .$$

The moments of this distribution are exactly

$$M_q = \oint_{\zeta \in \text{interface}} |\nabla \Phi|^q \, dl = \frac{1}{2\pi i} \oint_{|z|=1} |F'(z)|^{1-q} \frac{dz}{z} \, . \tag{6.5}$$

Substituting from Eq. (2.8) we have

$$M_q = \frac{A(t)^{1-q}}{2\pi i} \oint \prod_{n=1}^{N} \left(\frac{z - Z_n}{z - P_n} \right)^{\frac{1-q}{2}} \prod_{n=1}^{N} \left(\frac{1 - zZ_n^*}{1 - zP_n^*} \right)^{\frac{1-q}{2}} \frac{dz}{z} \,. \tag{6.6}$$

The second product in the integrand is analytic within the unit disk while the first contains N poles of order $(1-q)/2$. A simple pole also exists at the origin. Thus it is straightforward to evaluate this integral for odd values of q:

$$M_{q>1} = A(t)^{-2\nu} \left\{ \left(\prod_{n=1}^{N} \frac{P_n}{Z_n} \right)^{\nu} + \frac{1}{\nu!} \sum_{n=1}^{N} \frac{d^{\nu-1}}{dz^{\nu-1}} \left[\frac{(z-P_n)^{\nu}}{zJ(z)^{\nu}} \prod_{k \neq n} \left(\frac{z-P_k}{z-Z_k} \right)^{\nu} \right]_{z=Z_n} \right\} , \tag{6.7}$$

where $\nu \equiv (q-1)/2$. Other values of q can be obtained either by taking into account the branch-cuts that are involved in the calculations, or by simply continuing the results obtained for odd integer values. Thus the third moment is exactly

$$M_3 = A(t)^{-2} \left[\prod_{n=1}^{N} \frac{P_n}{Z_n} + \sum_{n=1}^{N} \frac{Q_n}{2Z_n} \right] \,. \tag{6.8}$$

This moment is directly related to the fractal dimension[(18),(19)] by

$$D_f = \frac{\ln M_3}{\ln R_g} \,,$$

where R_g is the size (or the radius of gyration) of the growth. In a similarl way one finds for $q < 1$

$$M_{q<1} = A(t)^{2\tilde{\nu}} \left\{ \left(\prod_{n=1}^{N} \frac{Z_n}{P_n} \right)^{\tilde{\nu}} + \frac{1}{\tilde{\nu}!} \sum_{n=1}^{N} \frac{d^{\tilde{\nu}-1}}{dz^{\tilde{\nu}-1}} \left[\frac{(z-Z_n)^{\tilde{\nu}} J(z)^{\tilde{\nu}}}{z} \prod_{k \neq n} \left(\frac{z-Z_k}{z-P_k} \right)^{\tilde{\nu}} \right]_{z=P_n} \right\} , \tag{6.9}$$

where $\tilde{\nu} \equiv -\nu$. To gain some insight on some features of the morphology, I now proceed to discuss an the distribution of the curvatures using a dilute limit approximation for the distribution of the particles. Although the approximation is crude, the result is interesting and quite robust to the specifics of the solution for $\mathcal{N}$.

7. The dilute boundary layer approximation

Inspecting the dependence of the curvature (see Eq. (5.2)) on the locations of the zeros and the poles we can see that if a zero is very close to the unit circle, $Z_n = (1-\rho)e^{is_n}$ with $\rho \ll 1$, then the (real) curvature at S_n, K_1, is dominated by this zero and it diverges with ρ approximately as

$$K_1(S_n) \approx \frac{C}{\rho^2} , \tag{7.1}$$

where, to a good accuracy, C is independent of ρ. The same consideration regarding a pole, $P_n = (1-\eta)e^{is_n}$, with $\eta \ll 1$, shows that the local curvature is *regular* in η. Thus for the purpose of finding the distribution of the curvature these are the zeros with $\rho \ll 1$ that dominate and *not all the particles*. Therefore, let us consider only the zeros that lie within the ring $\rho_c < |z| < 1$, where $1-\rho_c \ll 1$. To make progress we assume that the typical distance between neighbouring particles within the ring are larger than $1-\rho_c$. This approximation corresponds to probing only the distribution of exposed external parts of the growth and not the 'fjords'. We can now calculate the one-point distribution of K_1 in front of the dominant zeros, $\mathcal{P}_K$, in terms of the distribution of ρ. Once $\mathcal{P}_K$ is given it is straightforward to calculate its moments. Although this is a crude approximation and the calculation is limited, it yields a result that is worth mentioning.

Since the local curvature is dominated by the location of the closest zero its distribution must be directly related to the distribution of the zeros, $\mathcal{P}_n(\rho)$, through

$$\mathcal{P}_K = Const.\ K_1^{-3/2}\mathcal{P}_n\left((C/K_1)^{1/2}\right) . \tag{7.2}$$

At the moment we have no solution for $\mathcal{P}_n$, but we can nevertheless observe that even if this distribution is well behaved (i.e., has exponential tails) the distribution of K_1 entertains *algebraically decreasing tails*. This may hint at the origin of fractality in the system, namely, that it may generically originate from the occurence of these long tails. If indeed $\mathcal{P}_n$ were well-behaved then the above calculation would imply that the moments of the curvature distribution can be written as

$$\mu_q \equiv \int_0^{K_{max}} K^q \mathcal{P}_K dK \sim K_{max}^{(2q-1)/2} \sim a^{(1-2q)/2} , \tag{7.3}$$

where a is some small cutoff lengthscale. But since the entire aggregate grows this smaller cutoff also reduces with time by a factor of $1/A^2(t)$ and therefore

$$\mu_q \sim A^2(t)^{(2q-1)/2} . \tag{7.4}$$

It is quite plausible though that $\mathcal{P}_n$ also supports algebraic tails, which can directly affect the behaviour of M_q. Then a multifractal, rather than fractal, behaviour of these moments should appear. Such a behaviour, however, can only be predicted on the basis of a solution to the master equation (6.3) for $\mathcal{N}$.

By exploiting the dilute boundary limit one can also evaluate the moments of the growth probability, M_q, discussed above. This calculation is currently under way and will be reported elsewhere.

8. Discussion and conclusions

To summarise I showed that the p.d.e. that governs the growth of an interface in a Laplacian field is Hamiltonian and pointed out the relation between this Hamiltonian and the two-dimensional complex potential field. I discussed the implications of the Hamiltonian on converting the problem into an equivalent many-body system, whose dynamics reflects in a one-to-one manner the evolution of the curve. The many-body formalism has been extended to a continuous distribution of zeros and poles thus making way for introduction of tip-splitting, side-branching and other phenomena that involve creation of particles. Surface energy was proposed to give rise to a repulsive potential term in the Hamiltonian of the many-body system, which naturally prevents cusp singularities from forming along the physical interface and regularises the growth against the inherent ill-posedness of the problem. This method of regularisation avoids the main problem of the traditional singular perturbation approach, i.e., the sensitivity to initial conditions. Next I discussed the distribution of the zeros and poles and related it to the morphology of the physical interface. The construction of the partition function and the solution of the relevant master equation are difficult, but this is no different than for any master equation in statistical mechanics, where one needs to resort to some physically valid approximations and special limits to obtain exact solutions. I related the distribution of the curvatures to that of the particles and presented an exact calculation of the moments of the growth probability distribution along the interface. I used an approximation which, however crude, showed that the distribution of the curvature in the exposed regions of the growth develops an *algebraic tail.* Based on this, it is conjectured here that the appearance of such a tail points to the onset of fractality in this, and related, growth processes. This idea should hopefully give insight into understanding how the scale invariance and fractality become so prominent in such systems. This first-principles approach is promising for formulating a theory to describe and predict the morphology of Laplacian growth, and other diffusion-controlled, processes.

REFERENCES

1. For review see, e.g., P. Pelce, *Dynamics of Curved Fronts* (Academic Press, San Diego, 1988); D. A. Kessler, J. Koplik and H. Levine, *Adv. Phys.* **37** (1988) 255; P. Meakin, in *Phase Transitions and Critical Phenomena* Vol. 12 Eds. C. Domb and J. L. Lebowitz (Academic Press, 1988); T. Vicsek, *Fractal Growth Phenomena* (World Scientific, Singapore, 1989).
2. H. Gould, F. Family and H. E. Stanley, *Phys. Rev. Lett.* **50** (1983) 686; T. Nagatani, *Phys. Rev.* **A 36** (1987) 5812; J. Phys. **A 20** (1987) L381; X. R. Wang, Y. Shapir and M. Rubinstein, *Phys. Rev.* **A 39** (1989) 5974; J. Phys. **A 22** (1989) L507; P. Barker and R. C. Ball, *Phys. Rev.* **A 42** (1990) 6289.
3. L. Piteronero, A. Erzan and C. Evertsz, *Phys. Rev. Lett.* **61** (1988) 861, *Physica* **A151** (1988) 207.

4. T. C. Halsey and M. Leibig, *Phys. Rev.* **A 46** (1992) 7793; T. C. Halsey, *Phys. Rev. Lett.* **72** (1994) 1228.
5. L. A. Galin, *Dokl. Akad. Nauk USSR* **47** (1945) 246; P. Ya. Polubarinova-Kochina, *Dokl. Akad. Nauk USSR* **47** (1945) 254; *Prikl. Math. Mech.* **9** (1945) 79.
6. S. Richardson, *J. Fluid Mech.* **56** (1972) 609.
7. B. Shraiman and D. Bensimon, *Phys. Rev.* **A 30** (1984) 2840.
8. D. Bensimon, L. P. Kadanoff, S. Liang, B. I. Shraiman and C. Tang, *Rev. Mod. Phys.* **58** (1986) 977; W-s Dai, L. P. Kadanoff and S. Zhou, *Phys. Rev.* **A 43** (1991) 6672.
9. S. Tanveer, *Philosophical Transactions of The Royal Society (London)* **A 343** (1993) 155 and references therein.
10. R. Blumenfeld and R. C. Ball, *Phys. Rev. E*, submitted.
11. M. B. Mineev, *Physica* **D 43** (1990) 288.
12. N. Robidoux and R. Blumenfeld, in preparation.
13. M. Mineev-Weinstein, private communication.
14. R. Blumenfeld, in *Fluctuations and Order: The New Synthesis*, ed. M.M.Milonas (Springer-Verlag, 1994).
15. R. Blumenfeld, *Phys. Rev.* **E**, To be published.
16. W. W. Mullins and R. F. Sekerka, *J. Appl. Phys.* **34** (1963) 323.
17. S. D. Howison, *J. Fluid Mech.* **167** (1986) 439.
18. T. C. Halsey, *Phys. Rev. Lett.* **59** (1987) 2067.
19. R. C. Ball and R. Blumenfeld, *Phys. Rev.* **A 44** (1991) R828.

LA-UR-94-3641

Symplectic Computation of Lyapunov Exponents

Salman Habib[†] and Robert D. Ryne[*]

[†] *T-6, Theoretical Astrophysics*
and
T-8, Elementary Particles and Field Theory
Theoretical Division
Los Alamos National Laboratory
Los Alamos, NM 87545

[*] *AOT-1, Accelerator Physics and Special Projects*
Accelerator Operations and Technology Division
Los Alamos National Laboratory
Los Alamos, NM 87545

Abstract

A recently developed method for the calculation of Lyapunov exponents of dynamical systems is described. The method is applicable whenever the linearized dynamics is Hamiltonian. By utilizing the exponential representation of symplectic matrices, this approach avoids the renormalization and reorthogonalization procedures necessary in usual techniques. It is also easily extendible to damped systems. The method is illustrated by considering two examples of physical interest: a model system that describes the beam halo in charged particle beams and the driven van der Pol oscillator.

e-mail:
habib@predator.lanl.gov
ryne@lanl.gov

1 Introduction

Chaotic dynamical systems exhibit exponential divergence of initially nearby trajectories. This divergence is quantified by the Lyapunov exponents of the system which are obtained from linearizing the dynamics around a fiducial trajectory [1]. Over the past two decades or so there has been "intense activity" [2] directed toward the computation of these exponents resulting in several different numerical approaches [1][3][4]. The two obvious difficulties associated with the computation of Lyapunov exponents are: (1) exponential growth of the separation vector (between the fiducial and nearby trajectories) and (2) the exponential collapse of initially orthogonal separation vectors onto the direction of maximal growth. Most conventional methods overcome these hurdles by intermittent numerical rescaling and reorthogonalization (through, *e.g.*, the Gramm-Schmidt procedure [3]). Many chaotic systems are Hamiltonian or they can be transformed into a Hamiltonian system by suitable manipulations. However, none of the above general methods are designed to take advantage of this fact.

As is well known, the dynamics of classical Hamiltonian systems has an underlying symplectic structure [5]. In recent years symplectic methods have been applied with great success to classical dynamical problems. The field of accelerator dynamics has been revolutionized by the introduction of nonlinear symplectic maps as represented by Lie transformations [6][7]. Very long time integration of charged particle and planetary systems has been aided by the development of high order symplectic integration algorithms [8]. Recently a symplectic map-based approach for the calculation of Lyapunov exponents has been developed [9]. As shown below, this approach obviates *analytically* the need for rescaling and reorthogonalization in the numerical computation of the exponents.

2 The Method

Consider a 2-m dimensional continuous-time dynamical system governed by the equations

$$\frac{d\mathbf{z}}{dt} = \mathbf{F}(\mathbf{z}, t), \tag{1}$$

where $\mathbf{z} = (z_1, z_2, \cdots, z_{2m})$ and similarly for $\mathbf{F}$. Let $\mathbf{z}_0$ denote some given fiducial trajectory. Define deviations from this trajectory by letting $\mathbf{Z} = \mathbf{z} - \mathbf{z}_0$, and linearize the above equations. The new set of equations for the deviation variables is

$$\frac{d\mathbf{Z}}{dt} = \mathbf{DF}(\mathbf{z}_0, t) \cdot \mathbf{Z}. \tag{2}$$

The approach described below can be used whenever this linearized set of equations is derivable from a Hamiltonian. From now on suppose that this is the case, and that one can write $\mathbf{Z} = (q_1, q_2, \cdots, q_m, p_1, p_2, \cdots, p_m)$, where q_i and p_i denote canonically conjugate coordinates and momenta, respectively. It follows that

$$\frac{d\mathbf{Z}}{dt} = -\{H, \mathbf{Z}\}, \tag{3}$$

where $\{,\}$ denotes the Poisson bracket, and where H is a homogeneous quadratic polynomial in the q_i and p_i. A system such as this is governed by a symplectic matrix M that maps the initial variables $\mathbf{Z}^{in}$ into time-evolved variables $\mathbf{Z}(t)$,

$$\mathbf{Z}(t) = M(t)\mathbf{Z}^{in}. \tag{4}$$

Let Λ be given by

$$\Lambda = \lim_{t\to\infty} \left(M\tilde{M}\right)^{1/2t}, \tag{5}$$

where $\tilde{M}$ denotes the matrix transpose of M. The Lyapunov exponents then equal the logarithm of the eigenvalues of Λ [1].

It is easy to show that M satisfies the equation of motion (See, for example, Ref. [11])

$$\frac{dM}{dt} = JSM, \tag{6}$$

where S denotes the symmetric matrix given by

$$H(\mathbf{Z}, t) = \frac{1}{2}\sum_{i,j=1}^{2m} S_{ij} Z_i Z_j, \tag{7}$$

and where

$$J = \begin{pmatrix} 0 & \mathbf{1} \\ -\mathbf{1} & 0 \end{pmatrix}. \tag{8}$$

Here $\mathbf{1}$ denotes the $m \times m$ identity matrix. It follows that the evolution of $M\tilde{M}$ is governed by the equation

$$\frac{d}{dt}M\tilde{M} = JSM\tilde{M} - M\tilde{M}SJ. \tag{9}$$

Standard methods for obtaining the Lyapunov exponents deal with M, which is not real symmetric (hence the need for reorthogonalization) and which has exponentially growing elements. To avoid these difficulties one exploits the fact that M is symplectic by making use of the exponential representation of symplectic matrices [6]: Any symplectic matrix M can be written in the form

$$M = e^{JS_a} e^{JS_c} \tag{10}$$

where S_a is a symmetric matrix that anticommutes with J and S_c is another symmetric matrix that commutes with J. It is important to note that the second matrix on the right hand side of (10) is in fact unitary, so that

$$M\tilde{M} = e^{2JS_a}. \tag{11}$$

Note that this matrix has fewer degrees of freedom than M and its eigenvectors are orthogonal. Rather than attempting to directly integrate Eqn. (9) which would still have a large numbers problem, focus attention instead on the exponent JS_a in Eqn. (10). It is clear that now there is no large numbers problem since JS_a already appears as an exponent.

To proceed further, one obvious approach is to use an explicit representation of $\exp(JS_a)$. Such a representation is well known for $Sp(2)$ and has recently been found for $Sp(4)$ (see the Appendix). Generalizations to $Sp(2m)$ are in progress [10]. To illustrate the method it is convenient to restrict attention to systems with a two-dimensional phase space. When driven, these represent the simplest continuous-time systems that can exhibit chaos. The most general two-dimensional symplectic matrix can be written in the form

$$\begin{aligned} M &= e^{JS_a}e^{JS_c} \\ &= e^{\mu(B_2 \cos a + B_3 \sin a)}e^{bB_1}, \end{aligned} \tag{12}$$

where a, b and μ are real coefficients and where the B_i are basis elements of the Lie algebra $sp(2)$ [6]:

$$B_1 = \begin{pmatrix} 0 & 1 \\ -1 & 0 \end{pmatrix}, \quad B_2 = \begin{pmatrix} 0 & 1 \\ 1 & 0 \end{pmatrix}, \quad B_3 = \begin{pmatrix} 1 & 0 \\ 0 & -1 \end{pmatrix}. \tag{13}$$

It follows that

$$M\tilde{M} = e^{2\mu(B_2 \cos a + B_3 \sin a)}. \tag{14}$$

Thus, one obtains,

$$\Lambda = \lim_{t\to\infty} e^{(\mu/t)(B_2 \cos a + B_3 \sin a)}. \tag{15}$$

Finally, it is easily shown that the eigenvalues of this matrix are $e^{\pm\mu/t}$. The Lyapunov exponents are then equal to $\pm\mu/t$ in the limit $t \to \infty$. With this convenient choice of variables, the explicit representation of $M\tilde{M}$ is given by

$$M\tilde{M} = \begin{pmatrix} \cosh 2\mu + \sin a \sinh 2\mu & \cos a \sinh 2\mu \\ \cos a \sinh 2\mu & \cosh 2\mu - \sin a \sinh 2\mu \end{pmatrix}. \tag{16}$$

The unknown quantities a and μ can grow in time at most as $O(t)$. Differential equations for these quantities can be obtained by returning to Eqn. (9), the dynamical equation for $M\tilde{M}$.

For simplicity, assume that H contains no term proportional to qp, so that the matrix S in (7) is of the form

$$S = \begin{pmatrix} s_{11} & 0 \\ 0 & s_{22} \end{pmatrix}. \qquad (17)$$

After some manipulation, Eqns. (7)–(9) lead to the following:

$$\begin{aligned} \frac{d\mu}{dt} &= \frac{1}{2}(s_{22} - s_{11}) \cos a, \\ \frac{da}{dt} &= s_{11} + s_{22} - (s_{22} - s_{11}) \sin a \coth \mu. \end{aligned} \qquad (18)$$

From the initial condition $M(0) = I$, if one chooses $\mu(0) = 0$, then $\cos^2 a(0) = 1$, *i.e.*, $a(0) = 0$ or π. These differential equations form the basis of the method for calculating the Lyapunov exponents of Hamiltonian systems: They are stepped forward in time numerically till some desired convergence for the exponents, $\pm\mu/t$, is achieved. It will be shown later how to apply the method to certain non-Hamiltonian systems.

3 Applications: Two Examples

As a first concrete example, consider the newly developed "core-halo" model which describes beam halo in mismatched charged particle beams [12]. The transverse equation of motion for a halo particle in this model, assuming constant external focusing, is

$$\ddot{x} + x - (1 - \eta^2) f(x, r(t)) = 0 \qquad (19)$$

where x is the position variable for a halo particle, $f(x, r(t))$ is the force due to the space charge of the beam core, and $r(t)$ is the time dependent *rms* radius of the core. The core radius is assumed to follow the envelope equation

$$\ddot{r} + r - \frac{1 - \eta^2}{r} - \frac{\eta^2}{r^3} = 0. \qquad (20)$$

Here units have been chosen so that the time independent solution of (20) (*i.e.*, a matched beam) is given by $r = 1$. In these units $\eta = 0$ corresponds to the space charge dominated regime, while $\eta = 1$ corresponds to the emittance dominated regime. Now assume

$$f(x, r) = \frac{x}{x^2 + r^2} \qquad (21)$$

which has the correct asymptotic behavior: the force is linear when $x \ll r$, and it is inversely proportional to x when $x \gg r$. The Eqns. (19) and (20) describe a driven nonlinear system with a mixed phase space as demonstrated by the stroboscopic

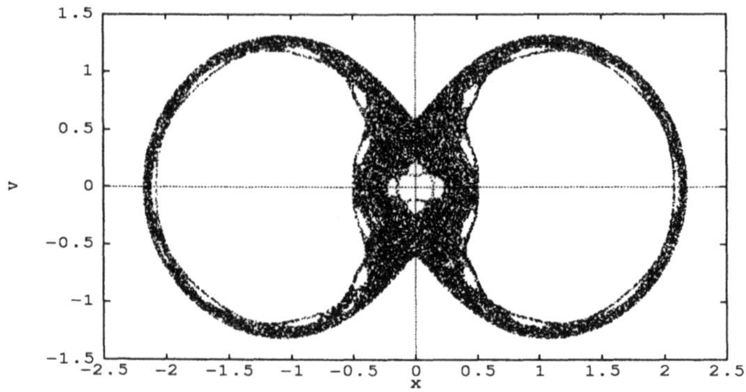

Figure 1: Stroboscopic plot of the chaotic sea in the core-halo model. Snapshots were taken at successive beam minima for 32 test particles. Parameter values were $r(0) = 0.6$, $\dot{r}(0) = 0$, and $\eta = 0.2$.

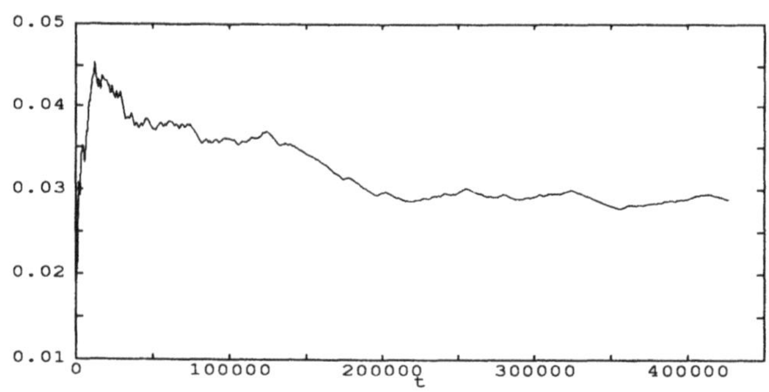

Figure 2: Positive Lyapunov exponent for the core-halo model in a typical run. Parameters are the same as in Fig. 1. The simulation was run for 10^5 periods of the driving force using 100 integration steps per period for Eqns. (18) with a third-order Runge-Kutta algorithm. The envelope equation and the fiducial trajectory were integrated with a fourth-order symplectic algorithm using 200 steps per period.

plot shown in Fig. 1. The presence of a chaotic band is important because particles initially in the core can leak through the broken separatrix and be carried to large amplitudes. The presence of such large amplitude particles can cause unacceptably high radioactivation levels in high intensity linacs planned for future accelerator-driven technologies [13]. Leakage through the separatrix can be enhanced through particle collisions and recent work has shown that this rate is controlled by the Lyapunov exponent [14]. The Lyapunov exponent for this system may be computed by integrating (18) with

$$\begin{aligned} s_{11} &= 1 - \left(1-\eta^2\right)\left(\frac{1}{x_0^2+r^2} - \frac{2x_0^2}{\left(x_0^2+r^2\right)^2}\right) \\ s_{22} &= 1 \end{aligned} \tag{22}$$

where x_0 denotes the fiducial trajectory. Fig. 2 displays the result for the Lyapunov exponent against time. The slow convergence of the exponent to its asymptotic value is typical of Hamiltonian systems.

So far only explicitly Hamiltonian systems were considered. However, the only real requirement for using the method is that the linearized deviation equations in *some* variables be Hamiltonian. This allows for the inclusion of damped systems in the scheme. As an example, consider the following general driven nonlinear oscillator

$$\ddot{x} + \lambda\left(1 - \epsilon x^2\right)\dot{x} + V'(x) = a\cos(\omega t). \tag{23}$$

By appropriate choices of λ, ϵ, and $V(x)$, this reduces to an assortment of well-known equations including van der Pol ($\lambda < 0$, $\epsilon = 1$, $V(x) = (1/2)x^2$), Duffing ($\lambda > 0$, $\epsilon = 0$, $V(x) = \alpha x^2 + \beta x^4$), and the damped driven pendulum ($\lambda > 0$, $\epsilon = 0$, $V(x) = 1 - \cos(x)$). In terms of the deviation variable δ, the linearization of (23) yields

$$\ddot{\delta} + \lambda\left(1 - \epsilon x_0^2\right)\dot{\delta} + \left(V''(x_0) - 2\epsilon\lambda x_0\dot{x}_0\right)\delta = 0 \tag{24}$$

where x_0 represents the fiducial trajectory. Introducing the new variable Δ defined through

$$\Delta = \delta e^{-g(t)} \tag{25}$$

where

$$\dot{g} = -\frac{1}{2}\lambda\left(1 - \epsilon x_0^2\right), \tag{26}$$

Eqn. (24) reduces to that describing an undamped oscillator with time dependent frequency,

$$\ddot{\Delta} + \left(V''(x_0) - \epsilon\lambda x_0\dot{x}_0 - \frac{1}{4}\lambda^2\left(1 - \epsilon x_0^2\right)^2\right)\Delta = 0. \tag{27}$$

It is now straightforward to proceed in the usual way: for linear damping ($\epsilon = 0$) the Lyapunov exponents $\chi_\pm$ of this system are given by

$$\chi_\pm = -\frac{1}{2}\lambda \pm \lim_{t\to\infty} \frac{1}{t}\mu_0(t) \tag{28}$$

where μ_0 follows from solving (18) for the system defined by (27). When $\epsilon \neq 0$,

$$\chi_\pm = \lim_{t\to\infty} \frac{1}{t}\left(g(t) \pm \mu_0(t)\right) \tag{29}$$

modulo terms that are exponentially suppressed at late times.

Figs. 3(a) and 3(b) show the Lyapunov exponents of the van der Pol system. For the chosen set of parameters these results are in agreement with those of Ref. [4]. In contrast with the results shown in Fig. 2, the convergence of the exponents is much faster, as is typical of non-weakly damped systems.

4 Conclusion

To summarize, a method for computing Lyapunov exponents that exploits the underlying symplectic structure of Hamiltonian dynamics has been reviewed. Just as symplectic integrators are not a panacea for all time integration problems, this method does not have universal applicability and advantages. However, when applicable, the method has certain advantages over standard techniques, most importantly the lack of systematic errors associated with intermittent reorthogonalization and rescaling.

5 Acknowledgements

The authors acknowledge useful discussions with Alex Dragt, Michael Mattis, and Michael Nieto. This work was supported by the U. S. Department of Energy at Los Alamos National Laboratory and by the Air Force Office of Scientific Research. The authors also acknowledge use of the facilities provided by the Advanced Computing Laboratory at Los Alamos National Laboratory.

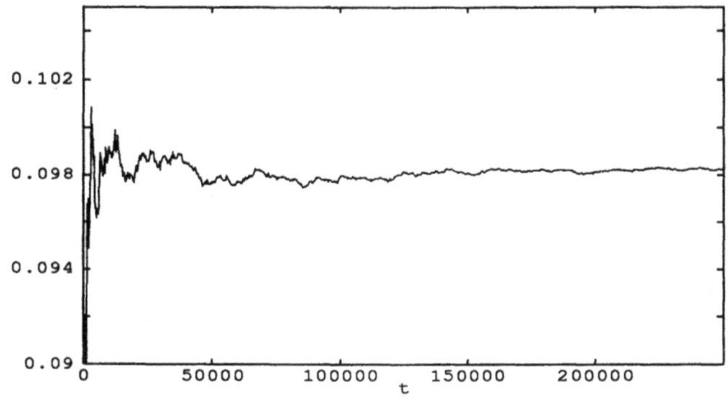

Figure 3: (a)Positive Lyapunov exponent for the van der Pol oscillator with parameters $\lambda = -5$, $a = 5$, and $\omega = 2.466$ (taken from Ref. [4]). The simulation was run for 10^5 periods of the driving force using 100 million time steps for Eqns. (18) and 200 million time steps for the fiducial trajectory. The integrators were third-order Runge-Kutta and fourth-order symplectic, respectively.

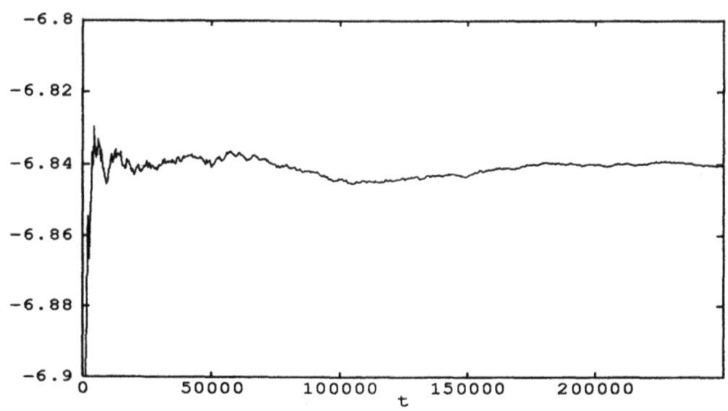

Figure 3: (b)Negative Lyapunov exponent for the van der Pol oscillator with the same parameters as in Fig. 3(a).

References

[1] J.-P. Eckmann and D. Ruelle, *Rev. Mod. Phys.* **57**, 617 (1985) and references therein.

[2] J. Guckenheimer and P. Holmes, *Nonlinear Oscillations, Dynamical Systems, and Bifurcations of Vector Fields* (Springer-Verlag, New York, 1986).

[3] A. Wolf, J. B. Swift, H. L. Swinney, and J. A. Vastano, *Physica* **16D**, 285 (1985).

[4] K. Geist, U. Parlitz, and W. Lauterborn, *Prog. Theo. Phys.* **83**, 875 (1990).

[5] V. I. Arnold, *Mathematical Methods of Classical Mechanics* (Springer, New York, 1978).

[6] A. J. Dragt, in *Lectures on Nonlinear Orbit Dynamics*, American Institute of Physics Conference Proceedings No. 87, edited by R. A. Carrigan, F. R. Huson, and M. Month (AIP, New York, 1982).

[7] E. Forest, *J. Math. Phys.* **31**, 1133 (1990).

[8] E. Forest and R. D. Ruth, *Physica* **43D**, 105 (1990); H. Yoshida, *Phys. Lett. A* **150**, 262 (1990); E. Forest, J. Bengtsson, and M. F. Reusch, *Phys. Lett. A* **158**, 99 (1991); J. Wisdom and M. Holman, *Astron. J.* **102**, 1528 (1991).

[9] S. Habib and R. D. Ryne, *Phys. Rev. Lett.* (in press).

[10] S. Habib and R. D. Ryne, (in preparation).

[11] A. J. Dragt and E. Forest, *J. Math. Phys.* **24**, 2734 (1983).

[12] See, *e.g.*, R. L. Gluckstern, *Phys. Rev. Lett.* (in press); T. P. Wangler, LANL preprint, LA-UR-94-1135 (1994); J.-M. Lagniel, *Nucl. Inst. Meth. A* (in press).

[13] R. A. Jameson, G. P. Lawrence, and S. O. Schriber, *Accelerator-driven Transmutation Technology for Energy Production and Nuclear Waste Treatment*, in Proceedings of the European Particle Accelerator Conference, Berlin, 1992.

[14] S. Habib, H. E. Kandrup, and E. Mahon, Los Alamos/Florida preprint, LA-UR-94-2124 (1994).

A Appendix: Summary of Sp(4) Results

There exists a particularly convenient way of organizing the basis elements of $sp(4)$ in terms of a triplet of matrices (the F_i, G_i, and the B_i) each triplet consisting in turn of three matrices. The B_i belong to the unitary sector which is irrelevant to the computation of the Lyapunov exponents. The most general (4×4) symplectic matrix then turns out to be of the form $\exp(\mathbf{a} \cdot \mathbf{F} + \mathbf{b} \cdot \mathbf{G})$ times a unitary matrix, where

$$\begin{aligned} \mathbf{a} &= \{a_1, a_2, a_3\}, \quad \mathbf{b} = \{b_1, b_2, b_3\} \\ \mathbf{F} &= \{F_1, F_2, F_3\}, \quad \mathbf{G} = \{G_1, G_2, G_3\}. \end{aligned} \tag{30}$$

Here, the a_i and the b_i are scalars, and the explicit forms of the matrices F_i and G_i are given below.

It turns out to be possible to resum the formal exponential above and to obtain the following result [10]:

$$\begin{aligned} M &\equiv \exp(\mathbf{a} \cdot \mathbf{F} + \mathbf{b} \cdot \mathbf{G}) \\ &= \frac{1}{2}(\cosh[u_x] + \cosh[u_y])\mathbf{I} \\ &\quad + \frac{1}{2}(\mathrm{sch}[u_x] + \mathrm{sch}[u_y])\,(\mathbf{a} \cdot \mathbf{F} + \mathbf{b} \cdot \mathbf{G}) \\ &\quad + \frac{4}{(u_x^2 - u_y^2)}(\cosh[u_x] - \cosh[u_y])\,(\mathbf{a} \times \mathbf{b}) \cdot \mathbf{K} \\ &\quad + \frac{4}{(u_x^2 - u_y^2)}(\mathrm{sch}[u_x] - \mathrm{sch}[u_y])\,([\mathbf{b} \times (\mathbf{a} \times \mathbf{b})] \cdot \mathbf{F} + [\mathbf{a} \times (\mathbf{b} \times \mathbf{a})] \cdot \mathbf{G}) \end{aligned} \tag{31}$$

where "$\cdot$" represents the usual vector dot product, "$\times$" represents the vector cross product, $\mathbf{I}$ is the (4×4) identity matrix and

$$\begin{aligned} \mathrm{sch}[x] &\equiv \sinh[x]/x, \\ u_x &= \sqrt{2s^2 + 4v}, \\ u_y &= \sqrt{2s^2 - 4v}, \\ s^2 &= \mathbf{a} \cdot \mathbf{a} + \mathbf{b} \cdot \mathbf{b}, \\ v^2 &= (\mathbf{a} \times \mathbf{b}) \cdot (\mathbf{a} \times \mathbf{b}). \end{aligned} \tag{32}$$

A new triplet of matrices, the K_i appears. This is because the F_i and the G_i do not form a closed subalgebra. The K_i are idempotent and unitary. Note that all the F_i, G_i and K_i are traceless. Thus, quite trivially, $Tr(M) = 2(\cosh[u_x] + \cosh[u_y])$ and the eigenvalues of M are $\exp[u_x], \exp[-u_x], \exp[u_y], \exp[-u_y]$.

Finally, the explicit forms of the above matrices are given by:

$$K_1 = \begin{pmatrix} -1 & 0 & 0 & 0 \\ 0 & 1 & 0 & 0 \\ 0 & 0 & -1 & 0 \\ 0 & 0 & 0 & 1 \end{pmatrix}, K_2 = \begin{pmatrix} 0 & -1 & 0 & 0 \\ -1 & 0 & 0 & 0 \\ 0 & 0 & 0 & -1 \\ 0 & 0 & -1 & 0 \end{pmatrix}, K_3 = \begin{pmatrix} 0 & 0 & 0 & -1 \\ 0 & 0 & 1 & 0 \\ 0 & 1 & 0 & 0 \\ -1 & 0 & 0 & 0 \end{pmatrix} \tag{33}$$

$$F_1 = \begin{pmatrix} 0 & -1 & 0 & -1 \\ -1 & 0 & -1 & 0 \\ 0 & -1 & 0 & 1 \\ -1 & 0 & 1 & 0 \end{pmatrix}, F_2 = \begin{pmatrix} 1 & 0 & 1 & 0 \\ 0 & -1 & 0 & -1 \\ 1 & 0 & -1 & 0 \\ 0 & -1 & 0 & 1 \end{pmatrix}, F_3 = \begin{pmatrix} 1 & 0 & -1 & 0 \\ 0 & 1 & 0 & -1 \\ -1 & 0 & -1 & 0 \\ 0 & -1 & 0 & -1 \end{pmatrix} \tag{34}$$

$$G_1 = \begin{pmatrix} 0 & -1 & 0 & 1 \\ -1 & 0 & 1 & 0 \\ 0 & 1 & 0 & 1 \\ 1 & 0 & 1 & 0 \end{pmatrix}, G_2 = \begin{pmatrix} 1 & 0 & -1 & 0 \\ 0 & -1 & 0 & 1 \\ -1 & 0 & -1 & 0 \\ 0 & 1 & 0 & 1 \end{pmatrix}, G_3 = \begin{pmatrix} -1 & 0 & -1 & 0 \\ 0 & -1 & 0 & -1 \\ -1 & 0 & 1 & 0 \\ 0 & -1 & 0 & 1 \end{pmatrix} \tag{35}$$

Closed–Form Expressions for the Noncompact Part of $Sp(2n)$

Ivan Gjaja

Theoretical Division and Center for Nonlinear Studies, Los Alamos National Laboratory,

Los Alamos, NM 87545

The Iwasawa decomposition of the symplectic group is used to write down closed–form expressions for the noncompact part of the group. The results make possible the computation of Lyapunov exponents using the method described in [1] in systems with more than two degrees of freedom.

In these Proceedings [1] Habib and Ryne propose a method for computing Lyapunov exponents in cases where the linearized dynamics is Hamiltonian. The method requires a closed form expression for the noncompact part of the symplectic group, denoted in [1] by $Sp(2n)$, in the defining representation. (Other common notations for the symplectic group are $Sp(n, \mathbb{R})$ and $Sp(2n, \mathbb{R})$.) The authors provide such an expression for $Sp(2)$ and $Sp(4)$ and state that they are working on generalizations to $Sp(2n)$. In this note I point out that suitable closed form expressions for the noncompact part of $Sp(2n)$, $n \geq 1$, can be found rather easily using the Iwasawa decomposition of this group [2].

Denoting the set of matrices of the form JS_c and JS_a (Eq. (10) of [1]) by K and P, I first demonstrate that they represent the Cartan decomposition of the symplectic Lie algebra $sp(2n)$. In the calculations that follow use is made of the Killing form for $sp(2n)$ which is given by $(X_1, X_2) = (2n+2)\text{Tr}(X_1X_2)$ for any $X_{1,2} \in sp(2n)$ [3]. Then, since any element of K can be written in the form $JS+SJ$ where S is a symmetric $2n \times 2n$ matrix and J is defined in [1], and any element of P in the form $JS - SJ$ [4], it can be verified that $[K, K] \subset K$, $[K, P] \subset P$, $[P, P] \subset K$, $(X, X) < 0$ for $X \neq 0$, $X \in K$, and $(Y, Y) > 0$ for $Y \neq 0$, $Y \in P$. Thus $sp(2n) = K + P$ is indeed the Cartan decomposition of $sp(2n)$. K is the maximal compact subalgebra, which in this case is the unitary Lie algebra $u(n)$. To compute the closed form expression for the noncompact factor of $Sp(2n)$ displayed in Eq. (11) of [1], I begin by writing down the Iwasawa decomposition of $sp(2n)$, $sp(2n) = u(n) + H_P + N$. H_P is an abelian and N a nilpotent Lie algebra for which the generators can be taken as

$$h_i = e_{ii} - e_{n+i,n+i}, \quad 1 \leq i \leq n \quad \text{for } H_P; \tag{1a}$$

$$m_{ij} = e_{i,n+j} + e_{j,n+i}, \quad 1 \leq i,j \leq n \tag{1b}$$

and

$$n_{ij} = e_{ij} - e_{n+j,n+i}, \quad 1 \leq i,j \leq n, \quad i < j \quad \text{for } N. \tag{1c}$$

Here e_{ij} is the usual Weyl basis for $2n \times 2n$ matrices, $(e_{ij})_{rs} = \delta_{ir}\delta_{js}$. (The method for computing the Iwasawa decomposition for any semisimple real Lie algebra is given in [2].)

An arbitrary element of $Sp(2n)$ can now be written as $e^X e^Z e^W$, where $X \in u(n)$, $Z \in H_P$, $W \in N$ [3]. Since H_P is abelian, the exponential series in e^Z is easily evaluated, whereas the nilpotency of N converts the infinite series in e^W into a polynomial of order $2n-1$. The factor $e^Z e^W$ is thus

$$e^Z e^W = \text{diag}(\lambda_1, \ldots, \lambda_n, \lambda_1^{-1}, \ldots, \lambda_n^{-1})\Big(I + \sum_{s=1}^{2n-1} \frac{1}{s!}\big(\sum_{i,j} a_{ij} m_{ij} + \sum_{i<j} b_{ij} n_{ij}\big)^s\Big). \tag{2}$$

The parameters a_{ij} and b_{ij} are in $\mathbb{R}$ and $\lambda_1, \ldots \lambda_n^{-1} > 0$. The quantity of interest in [1], $M\tilde{M}$, takes the form

$$M\tilde{M} = \widetilde{\left(e^Z e^W\right)} e^Z e^W. \tag{3}$$

As an example of the use of Eq. (2) I write down the matrix display for $Sp(4)$:

$$e^Z e^W = \text{diag}(\lambda_1, \lambda_2, \lambda_1^{-1}, \lambda_2^{-1}) \begin{pmatrix} 1 & b_{12} & 2a_{11} - \frac{1}{3}a_{22}b_{12}^2 & a_{12} + a_{22}b_{12} \\ 0 & 1 & a_{12} - a_{22}b_{12} & 2a_{22} \\ 0 & 0 & 1 & 0 \\ 0 & 0 & -b_{12} & 1 \end{pmatrix}. \tag{4}$$

REFERENCES

[1] S. Habib and R. D. Ryne, these Proceedings and Phys. Rev. Lett. (in press).

[2] A. O. Barut and R. Rączka, *Theory of Group Representations and Applications* (World Scientific, Singapore, 1986) pp. 41 and 104.

[3] S. Helgason, *Differential Geometry, Lie Groups, and Symmetric Spaces* (Academic Press, San Diego, 1978) pp. 189 and 431.

[4] A. J. Dragt, in *Physics of High Energy Accelerators*, AIP Conf. Proc. No. 84, edited by R. A. Carrigan, F. R. Huson, and M. Month (AIP, New York, 1982) p. 147.

LA-UR-94-3816

Φ^4 Kinks: Statistical Mechanics[1]

Salman Habib

T-6, Theoretical Astrophysics
and
T-8, Elementary Particles and Field Theory
Los Alamos National Laboratory
Los Alamos, NM 87545

Abstract

Some recent investigations of the thermal equilibrium properties of kinks in a 1+1-dimensional, classical Φ^4 field theory are reviewed. The distribution function, kink density, correlation function, and certain thermodynamic quantities were studied both theoretically and via large scale simulations. A simple double Gaussian variational approach within the transfer operator formalism was shown to give good results in the intermediate temperature range where the dilute gas theory is known to fail.

e-mail:
habib@predator.lanl.gov

---

[1]This paper is based on a presentation by the author at NEEDS '94, Los Alamos, NM (September 1994)

1 Introduction

The statistical mechanics of nonlinear coherent structures such as solitons and solitary waves has been a subject of study for quite some time [1]. More recent interest has been fueled by applications not only in condensed matter physics [2][3], but also by potential applications in particle physics (sphalerons) [4] and cosmology (domain walls, baryogenesis) [5].

A particularly simple model that displays behavior representative of a large class of soliton-bearing systems is the "double-well" Φ^4 scalar field theory in 1+1-dimensions. This theory admits exact solitary wave solutions ("kinks") and our purpose here is to describe the equilibrium statistical mechanics of these objects as studied recently in Refs. [6][7].

The Lagrangian for this theory is

$$L = \frac{1}{2}(\partial_t\Phi)^2 - \frac{1}{2}(\partial_x\Phi)^2 + \frac{1}{2}m^2\Phi^2 - \frac{1}{4}\Lambda\Phi^4. \tag{1}$$

The equations of motion that follow from (1) admit static kink solutions (centered at $x = x_0$)

$$\Phi_k = \frac{m}{\Lambda}\tanh\left(\frac{m}{\sqrt{2}}(x - x_0)\right). \tag{2}$$

The total energy of an isolated kink is $E_k = \sqrt{8/9}m^3/\Lambda$ and the negative of a kink, also a solution, is called an antikink.

In numerical work it is customary to use the dimensionless form of this theory, given by the transformations: $\Phi \to a\phi$, $x \to x/m$, and $t \to t/m$, where $a^2 = m^2/\lambda$. The equation of motion then becomes

$$\partial^2_{tt}\phi = \partial^2_{xx}\phi - \phi\left(\phi^2 - 1\right). \tag{3}$$

The statistical mechanics of kinks has been studied by two approaches. In the first, and phenomenological, approach one assumes that the kinks and the fluctuations ("phonons") about the asymptotic field minima may be treated as weakly interacting elementary excitations. Provided the kink density is low, the canonical partition function can then be found by standard methods [1][8][9]. Alternatively, as shown by Krumhansl and Schrieffer (KS) [8], building on earlier work of Sears, Scalapino, and Ferrell [10], it is possible to calculate the partition function, in principle exactly, by exploiting a transfer operator technique. KS showed that in the low temperature ("dilute gas") limit the partition function naturally factorizes into two contributions both having counterparts in the phenomenological theory; a tunneling term which they were able to identify with the kink contribution, and the remainder which they identified as linearized phonons. The ideas of KS were further refined and extended to a wider class of systems by Currie *et al* [9]. In particular, interactions of kinks with linearized phonons were considered, leading to substantial corrections to the results of KS.

A key result of these efforts is the prediction that the spatial density of kinks

$$n_k \propto \sqrt{E_k\beta}\exp(-E_k\beta), \tag{4}$$

where E_k is the kink energy ($E_k = \sqrt{8/9}$ for the dimensionless form of the theory) and β, the inverse temperature (for the dimensionless case, $\beta \to \bar{\beta} \equiv \beta/(a^3\sqrt{\lambda})$). A related prediction is that at low temperatures the field correlation length λ defined by

$$\langle\phi(0)\phi(x)\rangle \sim \exp\left(-x/\lambda\right) \tag{5}$$

has an exponential temperature dependence [9],

$$\lambda = \frac{1}{4}\sqrt{\frac{\pi}{3}}\frac{1}{\sqrt{E_k\beta}}\exp(E_k\beta). \tag{6}$$

Computer simulations to verify these results date back to the work of Koehler *et al* [11] who found only a qualitative agreement. More recent investigations [12][13][14] led to more detailed comparisons; however, significant discrepancies were reported. This led to theoretical speculation [15][16] regarding possible corrections to the dilute gas theory of kinks. It was speculated that these discrepancies could be due to finite size effects and phonon dressing of the bare kink energy (breather contributions to the free energy could also be significant [17]). However, these simulations were not carried out at low enough temperatures and the authors interpreted their results in terms of WKB (dilute gas) formulas that, it turns out, were simply not valid over the range of temperatures studied [6][7]. In Refs. [6][7], by going to low enough temperatures it was shown that the standard dilute gas results were indeed valid. Furthermore, an earlier claim of substantial phonon dressing even at these temperatures [11] was shown not to be correct.

The equilibrium statistical mechanics of kinks in the Φ^4 model (1) was investigated by implementing a Langevin code on a massively parallel computer [6]. The key idea behind Langevin simulations is to supplement the equation of motion (3) with noise and viscosity terms obeying an appropriate fluctuation-dissipation theorem so that the system is driven to thermal equilibrium at the desired temperature [18]. To understand the numerical results in the high and intermediate temperature region not susceptible to a dilute gas analysis, a double Gaussian wave function approximation for the quantum mechanical problem which results from applying the transfer operator method was used. Not only is this method accurate but it also suggests a natural decomposition of nonlinear phonon and kink degrees of freedom in the intermediate temperature regime [7].

The main results of these investigations are: (1) the dilute gas predictions for the kink density and the correlation length are in good agreement with the simulations below a certain (theoretically estimable) temperature, (2) above this temperature the double Gaussian results for the kink number and the correlation length agree with the simulations, (3) kinks are found to "disappear" into the thermal phonon background above a characteristic temperature, in good agreement with theoretical prediction, (4) the double Gaussian approximation accurately describes the classical single point field distribution function at high and intermediate temperatures where the dilute gas (WKB) approximation breaks down, and (5) the internal energy and the specific heat calculated in the double Gaussian approximation show an interesting energy sharing process between kinks and nonlinear phonons in an intermediate-temperature range below the characteristic temperature at which kinks appear: a peak in the specific heat in this temperature range is shown to be due essentially to kinks.

2 Numerical Results

The Langevin equation for the dimensionless theory is

$$\partial^2_{tt}\phi = \partial^2_{xx}\phi - \eta\partial_t\phi - \phi(1-\phi^2) + F(x,t). \tag{7}$$

To guarantee an approach to equilibrium, the Gaussian, white noise F and the viscosity η are related via the fluctuation-dissipation theorem:

$$\langle F(x,t)F(x',t')\rangle = 2\eta\bar{\beta}^{-1}\delta(x-x')\delta(t-t'). \tag{8}$$

Numerical simulations were performed on lattices with 16384 sites and the Langevin equation was solved numerically using standard methods [6]. The system size was 1 to 2 orders of magnitude larger than that in most previous simulations. Large system sizes are necessary to get acceptable statistics at low temperatures and to avoid finite size effects. Systems were evolved from a random initial condition to equilibrium.

To compute the kink number an operational way to identify kinks is needed even though the exact kink solution is available theoretically. Several possible definitions were examined, all of which relied on a knowledge of the kink size. From the classical solution for a kink centered at x_0, $\phi = \tanh((x-x_0)/\sqrt{2})$, the kink scale L_k was approximately 8 lattice units in the simulations. Raw kink configurations are shown in Fig. 1. At low temperatures ($\bar{\beta} > 5$), kinks may be identified easily, however at higher temperatures this is clearly not the case.

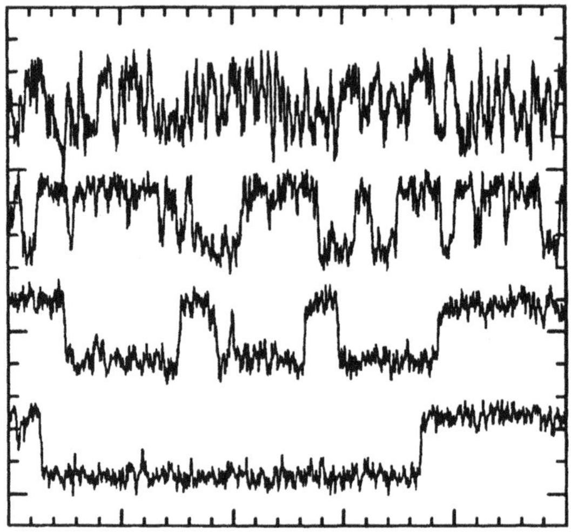

Figure 1: Sample field configurations, from top to bottom, at $\bar{\beta} = 2, 4, 5.5, 8$. Only a 1000 lattice unit sample of the total lattice size of 16384 is shown.

The simplest thing to do is to count the number of zero-crossings of the field, since one may argue that these are the "tunneling events" which correspond to kinks. However, at higher temperatures there are zero-crossings due to thermal noise (phonons), and counting all zero-crossings would lead to a gross overestimation of the number of kinks. At high temperatures it is not possible to distinguish unambiguously between kinks and nonlinear phonons (the overcounting problem occurs even at intermediate temperatures where the kinks are more or less distinct). A possible solution is to use a smoothed field by either averaging or block-spinning the actual field configuration over a length of the order of the kink scale. The latter approach was taken in previous simulations [12][13][14]. This solution is not without flaws either, as rapid fluctuations can still appear as kinks. In Refs. [6][7] the following method was used: at a particular time find all zero-crossings. To test the legitimacy of a given zero-crossing check for zero-crossings one kink scale (8 lattice units) to its right and to its left. If no zero-crossings are found, count it as a kink; otherwise not.

The number of kinks is plotted against $\bar{\beta}$ in Fig. 2. Above $\bar{\beta} \sim 6$, the two methods for counting kinks agree. Moreover, in this (low temperature) range, the dilute gas expression for the kink number (4) is in excellent agreement with the data. At elevated temperatures, there is a clear disagreement between the two methods of counting kinks. In this temperature regime the number of kinks computed with the

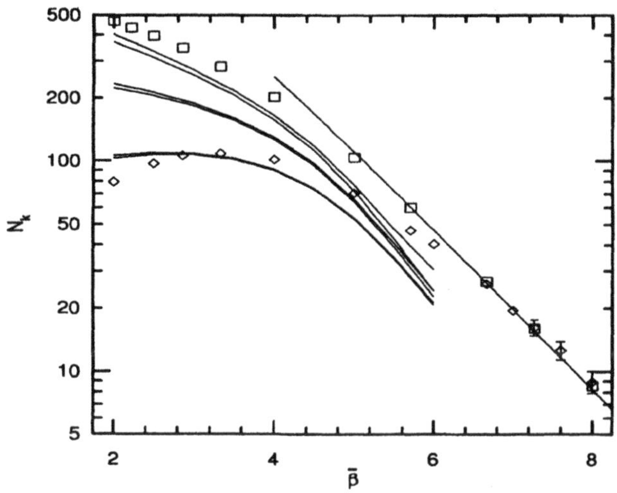

Figure 2: Total number of kinks as a function of $\bar{\beta}$. Squares denote counts with a smoothed field (averaging length of 8 lattice units) definition of kinks, diamonds for the zero-crossing counting method discussed in Sec. 2, and the solid beginning at $\bar{\beta} = 4$ is a fit to the WKB prediction (4). Also shown are three predictions from the double Gaussian theory (for details, see Ref. [7]).

smoothing method depends strongly on the smoothing scale. For $\bar{\beta} < 6$, the number of kinks cannot be extracted with any confidence from the smoothing method.

The correlation length λ is plotted against $\bar{\beta}$ in Fig. 3. For $\bar{\beta} > 6$, the WKB prediction (5) holds, whereas for $\bar{\beta} < 5$, where the wave function overlaps are not negligible, the double Gaussian approximation is valid. (Fortunately, there are no ambiguities here with regard to measurements at higher temperatures.)

3 Transfer Operator and Double Gaussian Approximation

The canonical partition function for the Lagrangian (1) is given by the functional integral

$$Z = \int D\phi D\pi \ \exp(-\beta H(\phi, \pi)) \tag{9}$$

where π is the canonical momentum of the field and H, the field Hamiltonian. The transfer operator technique [10] reduces the calculation of the partition function in the thermodynamic limit to simply finding the ground state energy of the double well

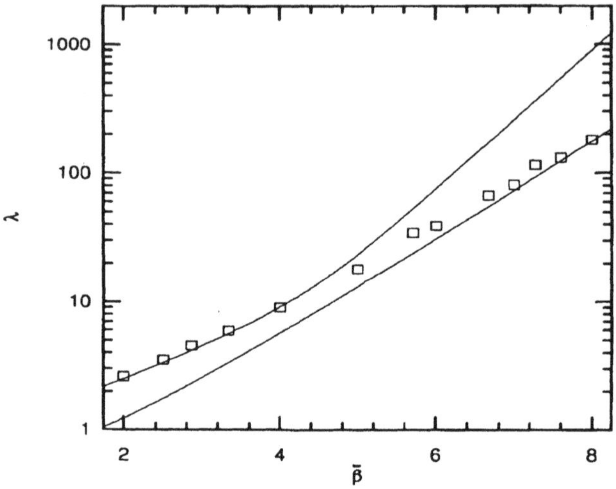

Figure 3: The correlation length λ as a function of $\bar{\beta}$. The two solid lines are the theoretical predictions (top, double Gaussian and bottom, WKB). The crossover of the ranges of validity of the two theories occurs at $\bar{\beta} \sim 5$.

quantum Hamiltonian (here written for the dimensionless case),

$$H_Q = \frac{1}{2}p^2 - \frac{1}{2\bar{\beta}^2}\bar{\phi}^2 + \frac{1}{4\bar{\beta}^4}\bar{\phi}^4 \tag{10}$$

where $\bar{\phi} = \bar{\beta}\phi$. At low temperatures the two wells are widely separated and the ground state energy is given by the oscillator ground state energy for one of the wells minus the tunnel-splitting term, usually calculated by WKB methods. The dilute gas or WKB approximation for the kink number is valid when the tunnel-splitting is small enough such that only the first two energy eigenstates are necessary to estimate the ground state energy of the Hamiltonian (10)[19]. At higher temperatures where kinks still exist, higher energy states cannot be ignored. Since kinks are associated with tunneling, one expects them to vanish when the ground state energy is higher than the classical barrier height: this intuition is confirmed by the simulations [6][7].

One can compare the simulations of the kink system with numerical solutions for the energy eigenvalues of the Hamiltonian H_Q. Instead, a different approach may be taken by implementing a double Gaussian variational method (see Ref. [7] for details) which is an order of magnitude more accurate than the simple Gaussian approximation [20] for this problem and correctly accounts for the reduction of energy due to overlap terms in the wave functions of the two wells, at least for moderate to large overlaps. Three qualitatively different regimes exist: (1) all the energy eigenvalues lie above the classical barrier, (2) the ground state energy lies below the classical barrier height, and (3) the energy difference between the ground and first excited state becomes negligible in comparison with the energy difference between the ground and the second excited state (this occurs for $\bar{\beta} > 6$). The simulations [6] confirm the theoretical expectations that there are no kinks in region (1), that there are kinks, but that the dilute gas approximation is invalid in region (2), and finally, that the dilute gas approximation is accurate in region (3) (a regime unexplored in detail by previous simulations).

The classical single point field distribution function $P[\phi]$ was measured directly in the simulations. For the analogous quantum mechanical problem this is just the square of the ground state wave function Ψ_0. Results from the simulations and the double Gaussian approximation are compared in Fig. 4. The presence of kinks implies a double peak in $P[\phi]$ [21](the converse is false) while a single peak at the origin means that kinks and thermal phonons can no longer be distinguished. From the simulations such a transition occurs at $\bar{\beta} \simeq 1.7$, in agreement with the theoretical calculation of when ${\Psi_0}^2$ goes over from a double to single peaked distribution. As expected, this is also the temperature ($\bar{\beta} = 1.734$) where the ground state energy crosses the classical barrier height (a discussion of various methods to determine the characteristic temperature is given in Ref. [22]). The double peaks in the distribution function move inward from the classical minimum as the temperature increases (this is clearly seen in Fig. 4). Physically this can be understood as nonlinear phonon corrections due to the fact that near each minimum, the potential is not symmetric under reflection around the minimum. The double Gaussian approximation underestimates the tunneling or overlap contribution at low temperatures ($\sim \exp(-\sqrt{2}\bar{\beta})$ for double Gaussian versus $\sim \exp(-\sqrt{8/9}\bar{\beta})$ for WKB). This can be seen already at $\bar{\beta} = 4$ in Fig. 4.

The equilibrium kink number has traditionally been calculated only in the phenomenological approach. In Ref. [7] an effective kink number was defined starting from the transfer operator formalism. In the WKB limit the formula agrees with the standard result up to a prefactor (which is actually not known explicitly in the standard calculation). At intermediate temperatures the double Gaussian approximation can be used to estimate the kink number expected in the simulations. The results are consistent with the simulations but show a dependence on an averaging length which again points to the problem of defining precisely what is a kink at elevated temperatures (Fig. 2).

The correlation length as computed in the double Gaussian approximation is in excellent agreement with the simulations for $\bar{\beta} < 5$. There is a smooth crossover in the numerical results to the exponential dependence on $\bar{\beta}$ predicted by the dilute gas theory from $\bar{\beta} \sim 6$ onwards. Thus the two theories agree with the simulations in the

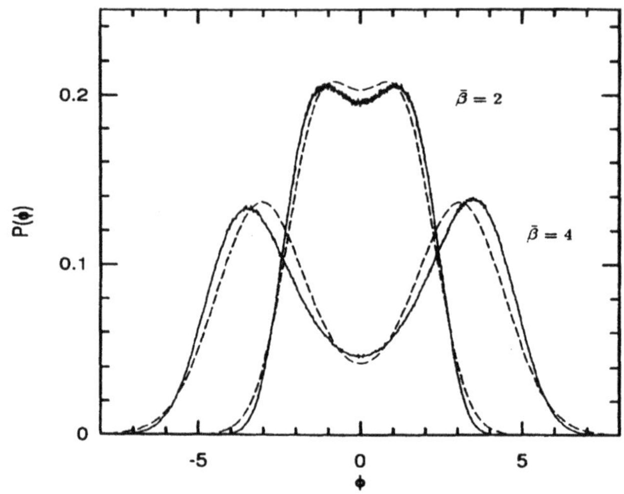

Figure 4: The classical distribution function $P[\phi]$ from the simulation compared with the distribution Ψ^2 from the double Gaussian approximation (dashed lines). The underestimation of the wave function overlap in this approximation is already visible at $\bar{\beta} = 4$.

appropriate temperature ranges.

Finally, in the double Gaussian approximation one can compute the internal energy and the specific heat [7]. The Schottky anomaly in the specific heat is clearly seen at $\bar{\beta} \simeq 5.4$. Disentangling the overlap contribution from the diagonal contribution allows one to associate the peak in C_v with the overlap contribution, *i.e.*, with the kinks (Fig. 5). A similar computation for the internal energy shows that in this temperature range the kink contribution dominates over both linear and nonlinear phonons (Fig. 6). Thus the kinks are indeed relevant to the thermodynamics of the system at least over some finite range of temperatures (at low temperatures the thermodynamics is dominated by linear phonons).

4 Conclusions

As a consequence of the above results, we conclude that the dilute gas/WKB approximation is excellent for $\bar{\beta} > 6$ with no further phonon dressing of the bare kink energy beyond that already included in (4) and (6) at these low temperatures. At higher temperatures, the WKB analysis fails, though theoretical progress is possible with the double Gaussian technique. Using this technique we can analytically calculate the temperature range where kinks dominate the thermodynamics.

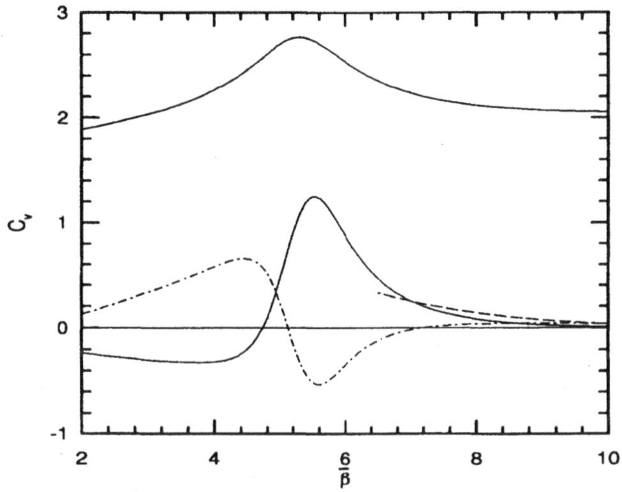

Figure 5: The specific heat C_v from the double Gaussian approximation (top curve). The peak in C_v is at $\bar{\beta} \simeq 5.4$. The contribution of the linear phonons is a constant (here, this value = 2). Individual contributions from kinks (solid line) and nonlinear phonons (dot-dashed line) are shown below. The WKB calculation for the kink contribution is shown from $\bar{\beta} = 6.5$ onwards.

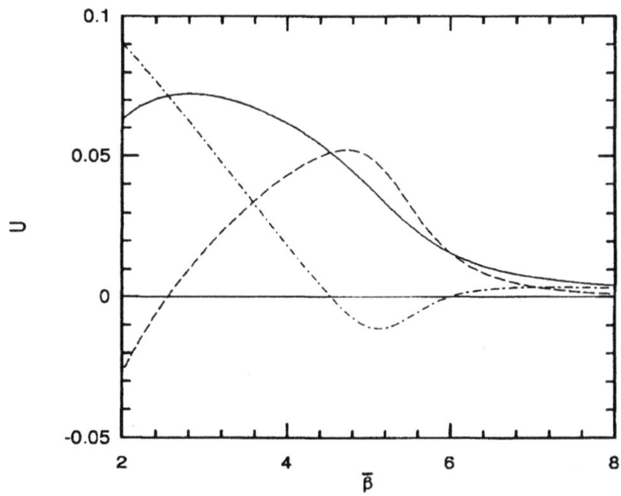

Figure 6: Internal energy against $\bar{\beta}$ using the double Gaussian approximation. Kink (dashed line) and nonlinear phonon (dot-dashed line) contributions are shown separately (modulo an irrelevant constant energy shift). Note that the kink contribution is dominant in the region where the specific heat has a maximum.

5 Acknowledgements

The work reported here was performed in collaboration with Frank Alexander and Alex Kovner. M. Alford, S. Chen, F. Cooper, G. D. Doolen, H. Feldman, R. Gupta, R. Mainieri, M. Mattis, E. Mottola, W. H. Zurek, and, especially, A. R. Bishop are thanked for helpful discussions and encouragement. This research was supported by the U. S. Department of Energy at Los Alamos National Laboratory and by the Air Force Office of Scientific Research. Numerical simulations were performed on the CM-200 at the Advanced Computing Laboratory at Los Alamos National Laboratory and the CM-2 at the Northeast Parallel Architecture Center at Syracuse University.

References

[1] See, e.g., A. Seeger and P. Schiller, in *Physical Acoustics, Vol. III* edited by W. P. Mason (Academic, New York, 1966).

[2] *Solitons and Condensed Matter Physics*, Proceedings of the Symposium on Nonlinear Structure and Dynamics in Condensed Matter, edited by A. R. Bishop and T. Schneider (Springer-Verlag, New York, 1978).

[3] A. R. Bishop, J. A. Krumhansl, and S. E. Trullinger, *Physica D* **1**, 1 (1980).

[4] V. A. Kuzmin, V. A. Rubakov, and M. E. Shaposhnikov, *Phys. Lett.* **155B**, 36 (1985).

[5] See, e.g., *The Early Universe* by E. W. Kolb and M. S. Turner (Addison-Wesley, New York, 1990).

[6] F. J. Alexander and S. Habib, *Phys. Rev. Lett.* **71**, 955 (1993).

[7] F. J. Alexander, S. Habib, and A. Kovner, *Phys. Rev. E* **48**, 4284 (1993).

[8] J. A. Krumhansl and J. R. Schrieffer, *Phys. Rev. B* **11**, 3535 (1975).

[9] J. F. Currie, J. A. Krumhansl, A. R. Bishop, and S. E. Trullinger, *Phys. Rev. B* **22**, 477 (1980).

[10] D. J. Scalapino, M. Sears, and R. A. Ferrell, *Phys. Rev. B* **6**, 3409 (1972).

[11] T. R. Koehler, A. R. Bishop, J. A. Krumhansl, and J. R. Schrieffer, *Solid State Commun.* **17**, 1515 (1975).

[12] D. Yu. Grigoriev and V. A. Rubakov, *Nucl. Phys. B* **299**, 67 (1988).

[13] A. I. Bochkarev and Ph. de Forcrand, *Phys. Rev. Lett.* **63**, 2337 (1989).

[14] M. Alford, H. Feldman, and M. Gleiser, *Phys. Rev. Lett.* **68**, 1645 (1992).

[15] F. Marchesoni, *Phys. Rev. Lett.* **64**, 2212 (1990).

[16] A. I. Bochkarev and Ph. de Forcrand, *Phys. Rev. Lett.* **64**, 2213 (1990).

[17] E. Stoll, T. Schneider, and A. R. Bishop, *Phys. Rev. Lett.* **42**, 937 (1979).

[18] c.f., T. Schneider and E. Stoll, in Ref. [2].

[19] A. R. Bishop and J. A. Krumhansl, *Phys. Rev. B* **12**, 2824 (1975).

[20] See, e.g., P. M. Stevenson, *Phys. Rev. D* **30**, 1712 (1984).

[21] A. D. Bruce, *Adv. Phys.* **29**, 111 (1980).

[22] A. R. Bishop, in *Lattice Dynamics*, Proceedings of the International Conference on Lattice Dynamics, Paris, 1977, edited by M. Balkanski (Flammarion Sciences, Paris, 1978).

LA-UR-94-4318

Statistical Mechanics of Nonlinear Coherent Structures: Kinks in the Φ^6 Model [1]

Salman Habib and Avadh Saxena

Theoretical Division
Los Alamos National Laboratory
Los Alamos, NM 87545

Abstract

We study the thermodynamics of kinks in the Φ^6 model using a Langevin code implemented on a massively parallel computer. This code can be used to study first order dynamical phase transitions which exhibit multiple length and time scales. The classical statistical mechanics of a 1 + 1-dimensional field theory reduces to a time-independent quantum problem in one dimension via the transfer integral method. Exact solutions of the Schrödinger equation exist for the Φ^6 potential (unlike the case for Φ^4) and can be used to check results from the simulations. The Φ^6 model is also much richer than the Φ^4 model in terms of the variety of coherent structures and possible phases accompanying a phase transition. Specifically, we have calculated (in a one dimensional model) such quantities as the probability density function (PDF) and field-field correlation functions. These quantities help us understand the contribution to the specific heat from coherent structures such as domain walls (kinks) and other transformation structures as opposed to the contribution from lattice vibrations. We have calibrated our results against known exact solutions for limiting cases with very high accuracy. Having understood this problem, we are now extending our Langevin code to higher dimensions.

---

[1] To appear in Proceedings of NEEDS'94, Los Alamos, September (1994)

1 Introduction

First order phase transitions are ubiquitous in nature, ranging from melting to structural transformations in crystals. Due to the discontinuity in the order parameter at a first order transition, a simple symmetric double well potential is incapable of describing such a transition and an asymmetric double well or a symmetric triple well potential (as, *e.g.*, a Φ^6 model) is needed. A symmetric double well potential (Φ^4 model) is usually employed to describe continuous (second order) transitions. The Φ^4 model and its attendant kink structure has been extensively studied in the literature using techniques such as the path integral formalism [1], Langevin dynamics [2], *etc.* The Φ^6 model, appropriate for first order phase transitions, is much richer in terms of its kink structure ("textures" in the materials context) [3][4][5][6]. However, unlike the Φ^4 model, many aspects of the Φ^6 model, specifically the variation of the probability density function (PDF) of the order parameter [7], correlation functions and structure factor within the context of Langevin dynamics, have not been studied (except in a specific case [8]).

Here we adopt a novel viewpoint to study the thermodynamics of a system based almost entirely on its PDF, *i.e.*, if the PDF is known then the ground state eigenfunction and eigenvalue are known, and therefore thermodynamic quantities, *e.g.*, the specific heat, can be determined. (Correlation functions can also be determined but not directly from the PDF.) It is clear that the PDF must be determined to a very high accuracy. Our Langevin code, implemented on a massively parallel computer, computes the PDF to the required precision (in an appropriate temperature range). This can be easily checked for the Φ^6 model because certain exact solutions of the Schrödinger equation with a Φ^6 potential exist [3][9][10]. Indeed, we find spectacular agreement between the computed and exact results for the Φ^6 model. A direct calibration of the Langevin code in this manner against the Φ^4 model is not possible because the Schrödinger equation with a Φ^4 potential has no exact solution. In addition, the field configuration for a kink can be calibrated against the exact solution for the Φ^6 kink. In this case too we find very good agreement between the Langevin code and the exact solution.

2 The Φ^6 Model

The Landau free energy density for the Φ^6 model in one dimension is given by

$$\bar{F}_L(\bar{\Phi}) = \frac{1}{2}A\bar{\Phi}^2 + \frac{1}{4}B\bar{\Phi}^4 + \frac{1}{6}\bar{\Phi}^6 , \tag{1}$$

where to describe a first order (or discontinuous) phase transition it is necessary that $B < 0$ and $C > 0$. In many physical cases (*e.g.*, soft phonon mode driven structural transformations) A is chosen to be temperature dependent [4][5]. This model also exhibits a second order (or continuous) transition for $B > 0$, $C > 0$ at $A = 0$. To include domain walls (of nonvanishing width and energy) between various phases the Landau free energy is supplemented by adding the square of the field gradient (the Ginzburg term)

$$\bar{F}_{GL}(\Phi, \bar{\Phi}') = F_L(\bar{\Phi}) + \frac{1}{2} d\bar{\Phi}'^2 \,, \tag{2}$$

where $\bar{\Phi}' = \partial\bar{\Phi}/\partial\bar{x}$ and $d > 0$. The coefficient d also pertains to a physical quantity such as the soft shear modulus of a crystal [4][5]. For convenience we use dimensionless variables [5] for the Ginzburg-Landau energy F_{GL}, the field Φ, and the spatial variable x according to

$$\bar{F}_{GL} = \lambda F_{GL} \,, \qquad \lambda = -\frac{9B^3}{16C^2} \,, \tag{3}$$

$$\bar{\Phi} = \gamma\Phi \,, \qquad \gamma = \left(-\frac{3B}{2C}\right)^{1/2} \,, \tag{4}$$

$$\bar{x} = \nu x \,, \qquad \nu = \left(\frac{4dC}{3B^2}\right)^{1/2} \,. \tag{5}$$

Thus, in dimensionless form we get

$$F_{GL} = \frac{1}{4}\tau\Phi^2 - \Phi^4 + \Phi^6 + \Phi'^2 \,, \tag{6}$$

where the prime denotes differentiation with respect to x, and

$$\tau = \frac{16AC}{3B^2} \,, \tag{7}$$

can be interpreted as a dimensionless temperature. The Φ^6 potential (1) has three minima that occur at

$$\bar{\Phi} = 0 \,, \qquad \bar{\Phi}_{min} = \pm\bar{\Phi}_0(1+z)^{1/2} \,, \tag{8}$$

$$\bar{\Phi}_0 = \frac{1}{\sqrt{3}}\gamma \,, \qquad z = \left(1 - \frac{3}{4}\tau\right)^{1/2} \,, \tag{9}$$

and two maxima that occur at

$$\bar{\Phi}_{max} = \pm\bar{\Phi}_0(1-z)^{1/2} \,. \tag{10}$$

The value of the potential (Landau free energy) at the extrema is given by

$$\bar{F}_{\pm} = \bar{F}_0(1 \mp z)^2(1 \pm 2z) \,, \tag{11}$$

where $\bar{F}_+$ represents a free energy maximum, $\bar{F}_-$, a free energy minimum, and $\bar{F}_0 = (2/27)\lambda$. For the dimensionless free energy F_{GL} and field Φ one just sets $\gamma = \lambda = \nu = 1$ in Eqs. (8-11).

Physically the central minimum represents the high temperature (parent) phase while the two side minima represent the two variants of a fully developed low temperature (product) phase. Similarly, the two maxima refer to the two variants of a partially developed product phase. For $\tau < 0$ there are only two minima and the free energy behaves as an effective Φ^4 model. For $0 < \tau < 4/3$ there are three minima. At $\tau = 1$ there are three degenerate minima which correspond to the first order transition point, *i.e.*, the parent and product phases coexist in equilibrium. For $\tau > 4/3$ there is a single well corresponding to the fact that only the parent phase is stable. At $\tau = 4/3$ there is only one minimum at $\Phi = 0$ and two points of inflection at $\pm(-B/2C)^{1/2}$.

3 Exact Kink and Domain Wall Solutions

The equilibrium field configuration is determined by minimizing the total free energy. The Euler-Lagrange equations after two integrations lead to

$$x(\Phi) = \frac{1}{2}\int \frac{d\phi}{\sqrt{\phi\left(\phi^3 - \phi^2 + \frac{\tau}{4}\phi - F_0\right)}}\,, \tag{12}$$

with $\phi = \Phi^2$ and the boundary conditions $F_0 = \lim_{x\to\pm\infty} F_L(x)$, $(F_L(x) > F_0)$, $\lim_{x\to\pm\infty} \Phi' = 0$. The following four kink and domain wall solutions exist.

1. For $\tau < 1$ and $F_0 < 0$ a kink solution between the two product variants is given by

$$\Phi(x) = \frac{\Phi_{min}\alpha \sinh \beta x}{\left(\beta^2 + \alpha^2 \sinh^2 \beta x\right)^{1/2}}\,, \tag{13}$$

where

$$\alpha = \Phi_{min}\left|2\Phi_{min}^2 - 1\right|^{1/2}, \qquad \beta = \Phi_{min}\left(3\Phi_{min}^2 - 1\right)^{1/2}. \tag{14}$$

This kink (Fig. 1) corresponds to the Φ^4 kink for $\tau < 0$.

2. For $1 < \tau < 4/3$ and $F_0 > 0$ a pulse soliton between the parent and either product variant (with the parent phase in the middle) is given by

$$\Phi(x) = \frac{\Phi_{min}\alpha}{\left(\Phi_{min}^4 - \beta^2 \tanh^2 \beta x\right)^{1/2}}\,. \tag{15}$$

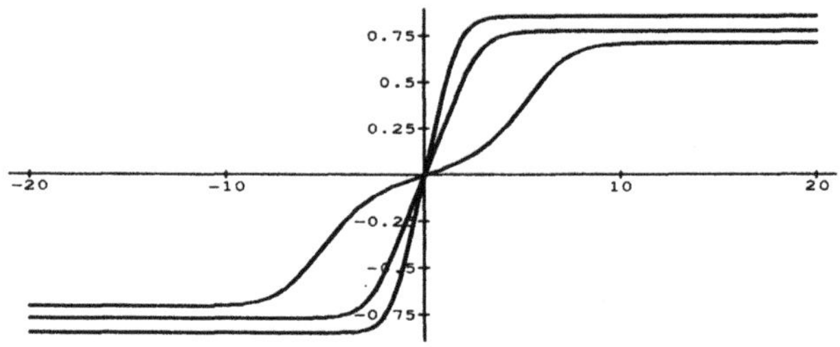

Figure 1: Kink solutions of type **1** for $\tau = .99,\ .5,\ -.5$ (asymptotically, the bottom, middle, and top curves). Note the "sticky" behavior of the solution near $\Phi = 0$ for $\tau = .99$.

This pulse solution is depicted in Fig. 2. The total free energy for solutions **1** and **2** is given by

$$F_{tot} = \frac{1}{2}\left(\beta - (\tau - 1)\ln\left(\frac{\beta + \Phi_{min}^2}{\alpha}\right)\right) . \tag{16}$$

3. For $0 < \tau < 1$ and $F_0 = 0$ a pulse soliton between either product variant and the parent phase (with either product variant in the middle) is given by

$$\Phi(x) = \frac{\Phi_2}{\left(1 + \left(1 - \frac{\Phi_2^2}{\Phi_3^2}\right)\sinh^2\sqrt{\tau}x/2\right)^{1/2}} , \tag{17}$$

where $\Phi_{2,3}^2 = (1/2)(1 \mp (1-\tau)^{1/2})$. This solution is shown in Fig. 3. The total free energy for this pulse solution is given by

$$F_{tot} = \frac{1}{4}\left(1 - (1 - \tau)\ln\left(\frac{\sqrt{2}(1 + \sqrt{\tau})}{\sqrt{1-\tau}}\right)\right) . \tag{18}$$

4. For $\tau = 1$ and $F_0 = 0$ a "half" kink between either product variant and the parent phase (at the transition point) is given by (Fig. 4)

$$\Phi(x) = \frac{1}{\sqrt{2}}\left(1 + e^{-x}\right)^{-1/2} , \tag{19}$$

with total energy $F_{tot} = 1/8$. This is the limiting case of **3** (matching with half a pulse of type **3**). Note that the above four solutions are known in the literature in a different form (and materials context) [3][4][6][11][12].

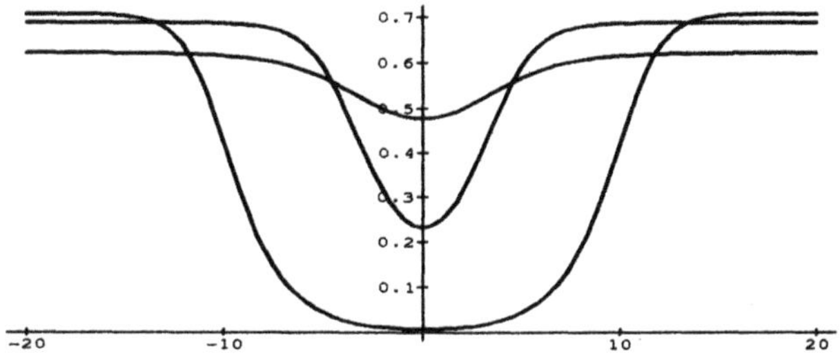

Figure 2: Pulse solutions of type **2** for $\tau = 1.0001,\ 1.1,\ 1.3$. (asymptotically, the bottom, middle, and top curves)

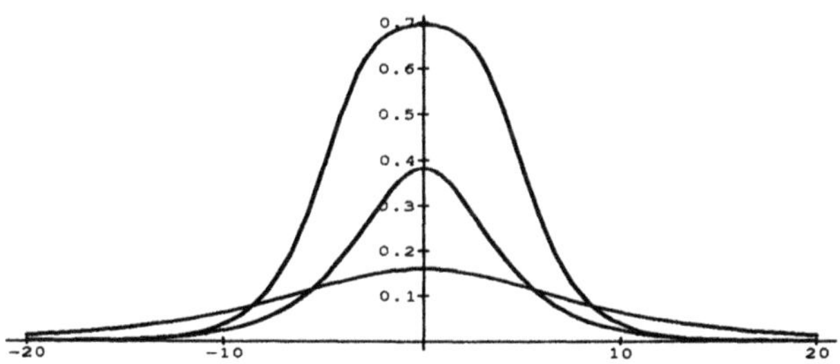

Figure 3: Pulse solutions of type **3** for $\tau = .1,\ .5,\ .999$ (the top, middle, and bottom peaks respectively).

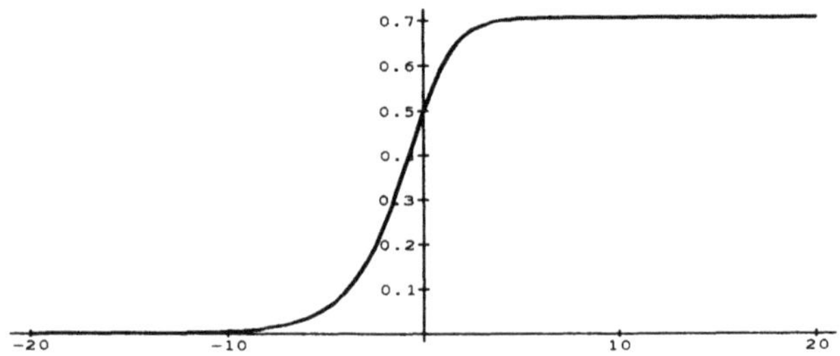

Figure 4: Half kink solution of type 4.

4 Langevin Dynamics on a Parallel Machine

Solution of field theoretic Langevin equations is particularly convenient on a massively parallel computer because little inter-processor communication is involved and because the large memory available enables the use of large lattices. For the results reported here, lattice sizes ranged from several thousands to hundreds of thousands of lattice points. The system size was kept much larger than the field correlation length (at least a factor of ten, but much larger typically). Periodic boundary conditions were used for convenience.

The Langevin equation for the Φ^6 model is

$$\partial_{tt}^2\Phi = \partial_{xx}^2\Phi - \eta\partial_t\Phi + \Phi(1-\Phi^2) + \Phi^5 + \hat{F}(x,t) \ , \tag{20}$$

where the viscosity η and the Gaussian white noise $\hat{F}$ are related by the fluctuation-dissipation theorem:

$$\left\langle \hat{F}(x,t)\hat{F}(x',t') \right\rangle = 2\eta\beta^{-1}\delta(x-x')\delta(t-t') \ . \tag{21}$$

The lattice versions of the above continuous equations were then solved using standard techniques [13]. Random initial conditions were driven to equilibrium and the results sampled in time thereafter to yield time averaged PDF's, *etc.* The use of the Langevin technique for obtaining thermodynamic quantities is straightforward and remarkably accurate. Moreover, structures such as the various kink solutions can be clearly identified and information about real time dynamical quantities such as the temporal correlation functions is also available.

5 Results

5.1 Probability Distribution Function

The exact solution for the PDF is given by (the ground state wave function [9][10] squared for the quantum mechanical problem is the PDF)

$$\Psi_0^2 = N \exp\left\{-\beta\left[\frac{1}{2}\left(\frac{dC}{3}\right)^{1/2}\Phi^4 + \frac{B}{4}\left(\frac{3d}{C}\right)^{1/2}\Phi^2\right]\right\} \tag{22}$$

where N is a normalization constant. The corresponding ground state energy [9][10] is

$$E_0 = \frac{B}{8\beta}\left(\frac{3}{dC}\right)^{1/2}. \tag{23}$$

In Fig. 5 we compare the probability density (PDF) function computed from a Langevin simulation (diamonds) with the exact solution (dashed line). The choice of parameters is $A = B = d = 1$, $C = 0.17$, and $1/\beta = 0.144146$. As is apparent, the agreement is excellent, and is indicative of the very high accuracy of the Langevin simulations. The results of our computations are, for practical purposes, numerically exact. Note that such a comparison is not directly possible in the case of the Φ^4 model [2] since there are no exact solutions in that case.

We have studied the variation of the PDF with temperature for several temperatures ranging from well below the "transition" temperature to well above it. In general, the PDF exhibits a three peak structure. However, at a certain temperature the three peaks have the same height. Above this temperature the PDF is characterized by a dominant central peak whereas below this temperature there are two dominant side peaks. In higher dimensions, this is characteristic of a first order phase transition. Since the PDF contains all essential thermodynamic information it is very important to be able to compute it accurately: our approach provides a simple and accurate method for computing thermodynamic quantities such as the specific heat, internal energy, *etc.* A detailed description is now in preparation [14].

5.2 Field Configurations

A low temperature field configuration is shown below in Fig. 6. The kinks are few in number but well defined. At higher temperatures the number of kinks increases

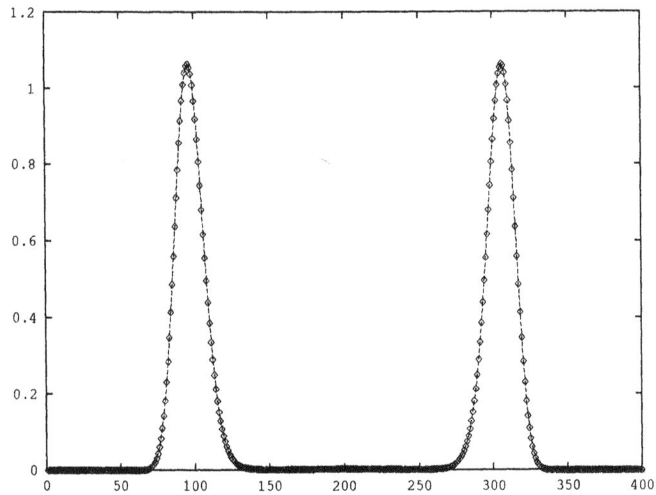

Figure 5: Comparison of the exact (dashed line) and numerical (diamonds) PDF. The agreement is excellent (parameter values are given in the text).

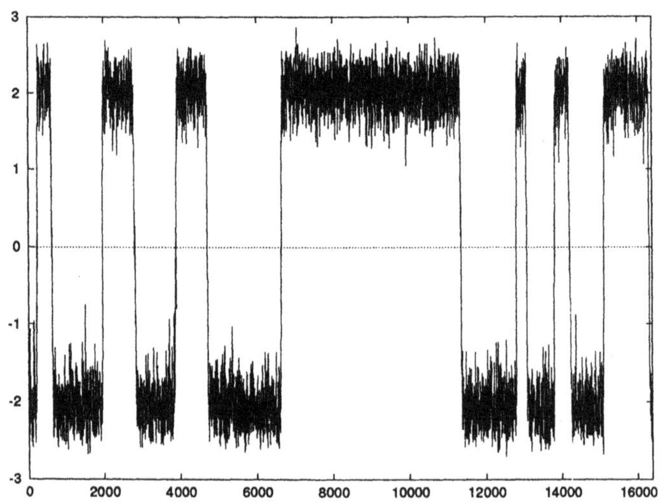

Figure 6: A sample field configuration at $1/\beta = .2$ with all other parameters the same as in Fig. 5.

but their shape is smeared by thermal noise. At still higher temperatures it becomes impossible to distinguish these nonlinear structures from nonlinear phonons. At modest temperatures, the shape of the kink computed from Langevin simulations can be compared with the exact solutions described in Sec. 3 (for example, the kinks in Fig. 6 correspond to the kinks of Fig. 1 with $\tau \simeq .91$). Here we simply report that the results from the simulations match well with theoretical expectations. Details will be given elsewhere [14]. To summarize, we have two different checks, namely the PDF and kink shape, which both show very good accuracy and provide a high level of confidence in our simulations.

5.3 Correlation Functions

The absolute value of the location of the side minima is the order parameter for the first order phase transition. Interestingly, the topological charge associated with a kink interpolating between these two minima at the phase transition point (T_c, three degenerate minima) turns out to be precisely equal to the order parameter [3]. In addition to the value of the order parameter, its spatial correlations as well as correlations of its intensity are often of interest in studying a phase transition [1][2]. The correlation functions are particularly interesting because they describe the behavior of systems which are nearly ordered but do not undergo sharp phase transitions at any finite temperature. Following the path integral (transfer operator) procedure [1][3], the correlation functions are expressed in terms of the eigenvalues and eigenstates of the transfer operator as follows:

$$C_1(x) = \sum_n |\langle \Psi_n | \Phi | \Psi_0 \rangle|^2 \exp\left[-\beta \frac{x}{a}(\epsilon_n - \epsilon_0)\right] , \tag{24}$$

$$C_2(x) = \sum_n \left|\langle \Psi_n | \delta |\Phi|^2 | \Psi_0 \rangle\right|^2 \exp\left[-\beta \frac{x}{a}(\epsilon_n - \epsilon_0)\right] , \tag{25}$$

where $\beta = 1/k_B T$, $\delta |\Phi|^2 = |\Phi(x)|^2 - \left\langle |\Phi(x)|^2 \right\rangle$, and a is the lattice constant.

For $x \gg \xi$, the lowest excited state coupled by the matrix element determines the behavior of the correlation functions. In other words, the eigenvalues set inverse correlation lengths. For $T \simeq T_c$, near degeneracy in eigenstates is reflected in an increased range of correlation (tunneling) [1][2]. At large distances $C_1(x)$ and $C_2(x)$ are dominated by the state with smallest eigenvalue for which the corresponding matrix elements are nonvanishing (excluding the $n = 0$ state). The correlation lengths for C_1 and C_2 are, respectively,

$$\frac{1}{\xi_1} \simeq \frac{\beta}{a}(\epsilon_1 - \epsilon_0) , \tag{26}$$

$$\frac{1}{\xi_2} \simeq \frac{\beta}{a}(\epsilon_2 - \epsilon_0) . \tag{27}$$

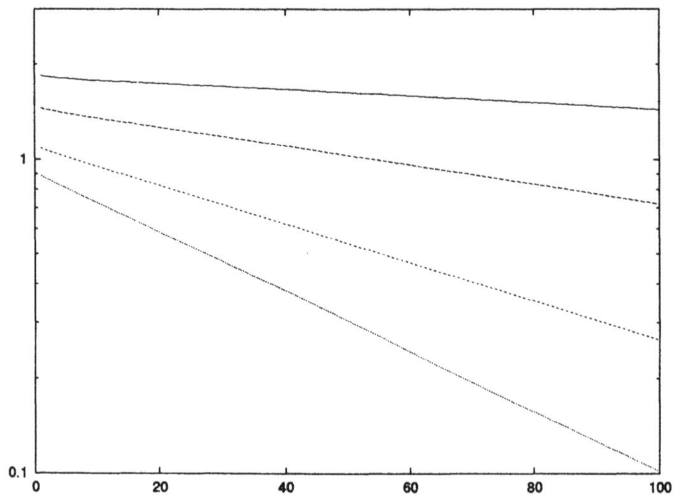

Figure 7: The (unnormalised) correlation function C_1 at four different temperatures plotted on a logarithmic scale. The parameters are the same as in Fig. 5, with only the temperature being varied: from top to bottom $1/\beta = .3,\ .35,\ .4,\ .45$.

ξ_1 is proportional to the average separation between neighboring kinks, which is the distance over which the field remains correlated. ξ_1 grows exponentially with decreasing temperature, and $\xi_1 \to \infty$ as $T \to 0$, when no kinks remain in the system. The energy density correlations are usually short range. The static structure factor (or the equal time correlation function) is the Fourier transform of C_1, and is given by

$$S(q) = \frac{1}{2\pi} \int dx e^{iqx} \langle \Phi(0)\Phi(x) \rangle . \tag{28}$$

The field-field (or order-parameter-order-parameter) correlation function C_1 for four temperatures is shown in Fig. 7. The exponential decay is apparent. The correlation length is given directly by the slope of the correlation function plotted on a logarithmic scale, while the average domain size in the system is obtained from the first zero crossing. The corresponding structure factors (fast Fourier transform of C_1) are also easy to compute but we do not display them here. More details on the correlation functions, their exact and semi-exact calculation, and comparison with numerical results will be given elsewhere [14].

6 Conclusion

In conclusion we restate some key points. First, the Φ^6 theory has sufficient structure to describe first order phase transitions especially relevant in the materials context (shape memory alloys) [4][5]. A consequence of this complexity is the appearance of several coherent nonlinear structures. The thermodynamics of the theory can be profitably studied via both the transfer operator method and Langevin simulations. The remarkable occurrence of some exact solutions in the analytic transfer operator approach for the Φ^6 theory allows [3] for a strong check on the simulations. The very accurate determination of the PDF via our simulations implies that this maybe a convenient window for a study of the thermodynamics of such systems. Finally, extension of the Langevin method to higher dimensions, and other classes of quasi-exactly solvable potentials, is simple and we expect to present our results for both the two and three dimensional cases soon.

7 Acknowledgment

We thank G. R. Barsch for fruitful discussions. This work was supported by the U.S. Department of Energy at Los Alamos National Laboratory. Numerical simulations were performed on the CM-5 at the Advanced Computing Laboratory, Los Alamos National Laboratory.

References

[1] J. A. Krumhansl and J. R. Schrieffer, *Phys. Rev. B* **11**, 3535 (1975).

[2] See, *e.g.*, F. J. Alexander and S. Habib, *Phys. Rev. Lett.* **71**, 955 (1993); F. J. Alexander, S. Habib, and A. Kovner, *Phys. Rev. E* **48**, 4284 (1993).

[3] S. N. Behera and A. Khare, *Pramana* **15**, 245 (1980).

[4] F. Falk *Z. Phys. B* **51**, 177 (1983).

[5] G. R. Barsch and J. A. Krumhansl, *Metall. Trans.* **19A**, 761 (1988).

[6] V. G. Makhankov, *Soliton Phenomenology* (Kluwer Academic Publishers, Boston, 1990), Ch. VII.

[7] A. D. Bruce, *J. Phys. C.* **14**, 3667 (1981).

[8] J. R. Morris and R. J. Gooding, *Phys. Rev. Lett.* **65**, 1769 (1990).

[9] V. Singh, S. N. Biswas, and K. Datta, *Phys. Rev. D* **18**, 1901 (1978); V. Singh *et al*, *Lett. Math. Phys.* **4**, 131 (1980).

[10] G. P. Flessas, *Phys. Lett. A* **72**, 289 (1979); *ibid* **81**, 17 (1981).

[11] A. Fousková and J. Fousek, *Phys. Stat. Sol. A* **32**, 213 (197).

[12] J. Lajzerowicz, *Ferroelectrics* **35**, 219 (1981).

[13] A. Griner, W. Strittmatter, and J. Honerkamp, *J. Stat. Phys.* **51**, 95 (1988).

[14] S. Habib and A. Saxena, (in preparation).

NUMERICALLY INDUCED SOLITON PERTURBATIONS

RUSSELL L. HERMAN
Mathematical Sciences Department, University of North Carolina Wilmington, NC 28403-3297, USA

ABSTRACT

When using finite difference schemes to simulate soliton propagation, the soliton shape or speed can be affected by the local truncation error of the scheme. In this paper we study the effects of the lowest order terms of such truncation errors for several finite difference schemes for the KdV equation. These errors are treated as perturbations of the KdV equation, using standard techniques of soliton perturbation theory. The theoretical predictions match the results of numerical simulations.

1. Introduction

The rediscovery and naming of the soliton was the result of the numerical studies of Zabusky and Kruskal[14]. They had found that an initial profile evolved into a train of solitary waves. During collisions the solitary waves interacted nonlinearly and emerged with their identities preserved. Solitons are also known for their robustness in the presence of small disturbances, which has allowed for their appearance in many phyiscal contexts, such as the study of water waves[9], ion-acoustic waves in plasma[13], and the propagation in optical fibers[3]. Two protypical evolution equations for the propagation of such nonlinear waves are the Korteweg-deVries (KdV) equation,

$$u_t + 6uu_x + u_{xxx} = 0, \tag{1}$$

and the nonlinear Schrödinger (NLS) equation,

$$iq_t = q_{xx} + 2|q|^2 q. \tag{2}$$

In order to study the properties of solitons in physical systems, one often resorts to numerical simulations. The simplest such simulations are based on finite difference schemes for the nonlinear evolution equations. Zabusky and Kruskal used such a scheme in their original studies of the KdV equation [14]. In comparing the numerical results from the Zabusky-Kruskal scheme with the known form of the KdV soliton we have found that all of the features of the soliton are accurate with the exception of the soliton speed. The computer generated soliton lags behind the analytic soliton[5]. This fact motivated us to examine other such schemes.

In this paper we will use perturbation theory to analytically study the effects of the truncation error of several finite difference schemes. We will present several finite difference schemes for the KdV and NLS equations and compute their truncation errors. We will then use the results of soliton perturbation theory to predict the changes to the soliton's shape and speed.

2. Truncation Error

We begin by studying the numerical scheme used by Zabusky and Kruskal[12,14] for studying the KdV equation (1). Writing the numerical solution as $\tilde{u}$ and the actual solution as u, we approximate the terms in equation (1) by

$$\tilde{u}_t = \frac{u(t+\Delta t) - u(t-\Delta t)}{2\Delta t}$$

$$\tilde{u} = \frac{u(x+\Delta x) + u(x) + u(x-\Delta x)}{3}$$

$$\tilde{u}_x = \frac{u(x+\Delta x) - u(x-\Delta x)}{2\Delta x}$$

$$\tilde{u}_{xxx} = \frac{u(x+2\Delta x) - 2u(x+\Delta x) + 2u(x-\Delta x) - u(x-2\Delta x)}{2\Delta x^3}. \tag{3}$$

In order to extract the local truncation errors, we employ Taylor expansions for the various terms in (3), assuming Δx to be small. This gives to leading order

$$\tilde{u}_t \approx u_t + \frac{1}{6}\Delta t^2 u_{ttt} \quad \tilde{u}_x \approx u_x + \frac{1}{6}\Delta x^2 u_{xxx}$$

$$\tilde{u} \approx u + \frac{1}{3}\Delta x^2 u_{xx} \qquad \tilde{u}_{xxx} \approx u_{xxx} + \frac{1}{4}\Delta x^2 u_{xxxxx}. \tag{4}$$

Considering that the finite difference scheme actually finds the numerical solution $\tilde{u}$ of the KdV equation, we substitute the results of (3) to obtain a perturbed KdV equation for u :

$$0 = \tilde{u}_t + 6\tilde{u}\tilde{u}_x + \tilde{u}_{xxx} \simeq u_t + 6uu_x + u_{xxx} + E(u), \tag{5}$$

where the local truncation error is given by

$$E(u) = \frac{1}{6}\Delta t^2 u_{ttt} + \Delta x^2\{2u_x u_{xx} + uu_{xxx} + \frac{1}{4}u_{xxxxx}\}. \tag{6}$$

It is well known that the Zabusky-Kruskal scheme is conditionally stable[12]. The linear stability condition is given by

$$\Delta t = (\Delta x)^3/4. \tag{7}$$

If we define

$$\epsilon = \Delta x^2 \tag{8}$$

then we see that

$$\Delta t^2 \simeq \frac{\epsilon^3}{4}. \tag{9}$$

Therefore, the leading order perturbation term in equations (5)-(6) results from the Δx^2 terms. This is summarized in the perturbed KdV equation

$$u_t + 6uu_x + u_{xxx} = -\epsilon\{2u_x u_{xx} + uu_{xxx} + \frac{1}{4}u_{xxxxx}\}. \tag{10}$$

3. Soliton Perturbation Theory

The next step in the analysis is the solution of the perturbed KdV equation (11) for the case of a single soliton. There have been numerous studies of this equation and we will just present the results. The interested reader is referred to the literature[4,6,7,8,10], and more specifically to our recent paper on this particular numerical scheme[5]. Assuming the form of the perturbed soliton to leading order as

$$u_0 = 2\eta^2 \text{sech}^2\phi, \qquad \phi = \eta(x - \frac{1}{\epsilon}x_0 - x_1), \tag{11}$$

where η, x_0, and x_1 are time dependent, we obtain

$$\eta_t = \frac{1}{4\eta}\int_{-\infty}^{\infty} d\phi E[u_0]\text{sech}^2\phi \tag{12}$$

$$x_{0t} = 4\eta^2 \tag{13}$$

$$x_{1t} = \frac{1}{4\eta^3}\int_{-\infty}^{\infty} d\phi E[u_0][\phi\text{sech}^2\phi + \tanh^2\phi + \tanh\phi]. \tag{14}$$

Equation (12) gives the change in the amplitude and width of the soliton. Equations (13) and (14) give the leading order and first order changes in the velocity, respectively. Finally, further analysis gives the further asymptotic behavior of the first order solution as

$$u_1 \sim \frac{1}{4\eta^3}\int_{-\infty}^{\infty} d\phi E[u_0]\tanh^2\phi, \tag{15}$$

where $u \approx u_0 + \epsilon u_1$. This term accounts for the well known shelf that develops behind the perturbed KdV soliton in the damped cases[4].

Similar perturbation analyses are possible for the perturbed NLS equation. Again, we will only list the pertinent results from soliton perturbation theory[4,7,8,10]. Assuming a leading order solution of the form

$$q_0 = -2i\eta e^{i(\beta - \xi\theta/\eta)} \mathrm{sech}\theta, \tag{16}$$

where

$$\theta = 2\eta(x - \frac{1}{\epsilon}x_0 - x_1) \qquad \beta = \frac{1}{\epsilon}(\beta_0 + \epsilon\beta_1) \tag{17}$$

we find that

$$\begin{aligned} x_{0t} &= 4\xi, \qquad \beta_{0t} = -4(\xi^2 + \eta^2). \\ \eta_t &= \frac{1}{2}\int_{-\infty}^{\infty} d\theta Re(\mathrm{E}[q_0]e^{i\delta})v \\ \xi_t &= \frac{1}{2}\int_{-\infty}^{\infty} d\theta Im(\mathrm{E}[q_0]e^{i\delta})v_\theta \\ x_{1t} &= \frac{1}{4\eta^2}\int_{-\infty}^{\infty} d\theta Re(\mathrm{E}[q_0]e^{i\delta})\theta v \\ \beta_{1t} &= \frac{1}{2\eta}\int_{-\infty}^{\infty} d\theta Im(\mathrm{E}[q_0]e^{i\delta})(\theta v_\theta + v). \end{aligned} \tag{18}$$

Here

$$v \equiv \mathrm{sech}\theta, \qquad \delta \equiv \xi\theta/\eta - \beta. \tag{19}$$

4. KdV Schemes

4.1 Finite Difference Schemes

In order to study the effects of other finite difference schemes, we have turned to the papers by Taha and Ablowitz[11,12]. In these papers the authors have investigated the merits of several such schemes. We have analyzed some of these schemes. In the rest of this paper we will present these schemes and the results of our analysis.

We begin by defining $u_n^m = u(n\Delta x, m\Delta t)$ as the value of the numerical solution at the specified mesh point $(n\Delta x, m\Delta t)$. In this notation the Zabusky-Kruskal scheme can be written as

$$\begin{aligned} u_n^{m+1} &= u_n^{m-1} - 2\frac{\Delta t}{\Delta x}\left(u_{n+1}^m + u_n^m + u_{n-1}^m\right)\left(u_{n+1}^m - u_{n-1}^m\right) \\ &- \frac{\Delta t}{\Delta x^3}\left(u_{n+2}^m - 2u_{n+1}^m + 2u_{n-1}^m - u_{n-2}^m\right). \end{aligned} \tag{20}$$

The corresponding perturbed KdV equation is

$$u_t + 6uu_x + u_{xxx} = -\frac{1}{6}\Delta t^2 u_{ttt} - \Delta x^2\big[2u_x u_{xx} + uu_{xxx} + \frac{1}{4}u_{xxxxx}\big]. \tag{21}$$

Furthermore, we have the linear stability condition $dt \leq \Delta x^3/4 \Rightarrow \Delta t^2 \sim O(\Delta x^6)$.

Results for other common schemes follow similarly. The Goda scheme is given by

$$\frac{1}{\Delta t}\left(u_n^{m+1} - n_n^m\right) + \frac{1}{\Delta x}\left[u_{n+1}^{m+1}\left(u_n^m + u_{n+1}^m\right) - u_{n-1}^{m+1}\left(u_n^m + u_{n-1}^m\right)\right] + \frac{1}{2\Delta x^3}\left(u_{n+2}^{m+1} - 2u_{n+1}^{m+1} + 2u_{n-1}^{m+1} - u_{n-2}^{m+1}\right) = 0. \tag{22}$$

Denoting

$$K[u] = u_t + 6uu_x + u_{xxx}, \tag{23}$$

this results in the perturbed KdV equation:

$$K[u] = -\Delta t\big[4uu_{xt} + 2u_x u_t + u_{xxxt} + \frac{1}{2}u_{tt}\big] - \Delta x^2\big[uu_{xxx} + 2u_x u_{xx} + \frac{1}{4}u_{xxxxx}\big]. \tag{24}$$

This scheme is unconditionally stable for $\Delta t << \Delta x$.

Taha introduced a scheme based on Kruskal's implicit scheme for $u_t + u_{xxx} = 0$. Taha provided a discretization for the uu_x term. It is given as

$$\frac{u_n^{m+1} - u_n^m}{\Delta t} + \frac{u_{n+2}^{m+1} - 3u_{n+1}^{m+1} + 3u_n^{m+1} - u_{n-1}^{m+1}}{2\Delta x^3} + \frac{u_{n+1}^m - 3u_n^m + 3u_{n-1}^m - u_{n-2}^m}{2\Delta x^3} + 3\Big[\frac{\theta}{4\Delta x}\left[(u_{n+1}^{m+1})^2 - (u_{n-1}^{m+1})^2 + (u_{n+1}^m)^2 - (u_{n-1}^m)^2\right] + \frac{1-\theta}{2\Delta x}\left[u_n^{m+1}(u_{n+1}^{m+1} - u_{n-1}^{m+1}) + u_n^m(u_{n+1}^m - u_{n-1}^m)\right]\Big] = 0. \tag{25}$$

Here θ is an adjustable parameter, which Taha had claimed to have obtained the best results for $\theta = 2/3$. The perturbed equation is

$$K[u] = -\Delta x\Delta t\big[\frac{1}{4}u_{xxxxt}\big] - \Delta x^2\big[\frac{1}{4}u_{xxxxx} + 3\theta u_x u_{xx} + uu_{xxx}\big] - \Delta t^2\big[\frac{1}{6}u_{ttt} + \frac{1}{4}u_{xxxtt} + \frac{3}{2}u_x u_{tt} + 3u_{xt}u_t\big]. \tag{26}$$

This scheme is also unconditionally stable for $\Delta t \sim O(\Delta x)$

The highlight of the Taha-Ablowitz paper is their scheme, which is based on the inverse scattering transform[1,2]. It is

$$\begin{aligned}\frac{u_n^{m+1}-u_n^m}{\Delta t} =& \frac{1}{2\Delta x^3}\left[u_{n-1}^{m+1}-3u_n^{m+1}+3u_{n+1}^{m+1}-u_{n+2}^{m+1}+u_{n-2}^{m}-3u_{n-1}^{m}+3u_n^m-u_{n+1}^m\right]\\ &-\frac{3}{2\Delta x}\left[(u_n^m)^2-(u_n^{m+1})^2\right]-\frac{1}{2\Delta x}\left[u_{n+1}^{m+1}(u_n^{m+1}+u_{n+1}^{m+1}+u_{n+2}^{m+1})\right.\\ &\left.-u_{n-1}^m(u_n^m+u_{n-1}^m+u_{n-2}^m)\right].\end{aligned} \tag{27}$$

This leads to

$$\begin{aligned}K[u]=&-\Delta t^2\left[\frac{3}{2}uu_{xtt}+\frac{3}{2}u_xu_{tt}+\frac{1}{4}u_{xxxtt}+3u_tu_{xt}+\frac{1}{6}u_{ttt}\right]\\ &-\Delta x^2\left[2uu_{xxx}+4u_xu_{xx}+\frac{1}{4}u_{xxxxx}\right]\\ &-\Delta x\Delta t\left[2uu_{xxt}+3u_xu_{xt}+2u_{xx}u_t+\frac{1}{4}u_{xxxxt}\right].\end{aligned} \tag{28}$$

This scheme is unconditionally stable for $\Delta t \sim O(\Delta x)$.

4.2 Perturbation Results

We are now ready to apply the results of perturbation analysis for the various perturbed KdV equations found in the last section. Using equations (13)-(16), we have the following:

Zabusky-Kruskal

$$\eta_t = 0 \qquad x_{1t} = -\frac{4}{5}\eta^4\Delta x^2+\frac{128}{9}\eta^8\Delta t^2 \tag{29}$$

$$x_c = (4\eta^2-\frac{4}{5}\eta^4\Delta x^2+\frac{128}{9}\eta^8\Delta t^2)t. \tag{30}$$

Goda

$$\eta_t = -\frac{64}{15}\eta^7\Delta t \qquad x_{1t} = -\frac{64}{15}\eta^5\Delta t-\frac{4}{5}\eta^4\Delta x^2 \tag{31}$$

$$u_{shelf} \sim -\frac{64}{15}\eta^5\Delta t \tag{32}$$

In this case there is damping and a shelf forms behind the soliton, which is given in (32). Integrating equation (31), we have

$$\eta = \eta_0[1+6\eta_0^6 At]^{-1/6}. \tag{33}$$

This gives the amplitude as

$$A = \frac{16}{15}\Delta t. \tag{34}$$

From these results, the soliton position is

$$x_c = \frac{1}{A}\eta^{-4} - \eta^{-1} - \frac{2\Delta x^2}{5A}\eta^{-2} - \frac{1}{A}\eta_0^{-4} + \eta_0^{-1} + \frac{2\Delta x^2}{5A}\eta_0^{-2}. \tag{35}$$

Kruskal-Taha

$$\eta_t = 0 \qquad x_{1t} = -[\frac{68}{45} - \frac{16}{15}\theta]\eta^4\Delta x^2 - \frac{16}{3}\eta^6\Delta x\Delta t - \frac{64}{9}\eta^8\Delta t^2. \tag{36}$$

$$x_c = (4\eta^2 - [\frac{68}{45} - \frac{16}{15}\theta]\eta^4\Delta x^2 - \frac{16}{3}\eta^6\Delta x\Delta t - \frac{64}{9}\eta^8\Delta t^2)t. \tag{37}$$

Taha-Ablowitz

$$\eta_t = 0 \qquad x_{1t} = -\frac{64}{9}\eta^8\Delta t^2 - \frac{44}{15}\eta^4\Delta x^2 + \frac{464}{45}\eta^6\Delta x\Delta t. \tag{38}$$

$$x_c = (4\eta^2 - \frac{64}{9}\eta^8\Delta t^2 - \frac{44}{15}\eta^4\Delta x^2 + \frac{464}{45}\eta^6\Delta x\Delta t)t. \tag{39}$$

5. NLS Schemes

5.1 Truncation Error

In a similar fashion we can study finite difference schemes for the NLS equation. In this case the perturbed equation takes the general form

$$iq_t - q_{xx} - 2|q|^2q = \mathrm{E}[q] \tag{40}$$

As before, we will turn to Taha and Ablowitz[12] for several schemes and produce the results for the perturbation analysis, where the pertrubation terms result from the local truncation error. Three schemes are:

Classic Explicit Scheme:

$$i\frac{q_n^{m+1} - q_n^{m-1}}{2\Delta t} = \frac{q_{n+1}^m - 2q_n^m + q_{n-1}^m}{\Delta x^2} + 2(q^2)_n^m q_n^m. \tag{41}$$

The local truncation error is

$$iq_t - q_{xx} - 2|q|^2q = \frac{\Delta x^2}{12}q_{xxxx} - \frac{i\Delta t^2}{6}q_{ttt}, \tag{42}$$

and the linear stability condition is $\Delta t \leq \Delta x^2/4$.

The remaining schemes can be found in the literature[12]. Their truncation error are

Implicit-Explicit Scheme:

$$E[q] = \frac{\Delta t}{2}[-(q|q|^2)_t] + \Delta x^2[(q|q|^2)_{xx} + \frac{1}{12}q_{xxxx}] + \Delta t^2[\frac{1}{4}q_{xxtt} - \frac{i}{6}q_{ttt}] \tag{43}$$

Local IST (Ladik) Scheme:

$$\begin{aligned} E[q] =& \frac{-\Delta t}{2}\frac{d}{dt}[iq_t - q_{xx} - 2|q|^2 q] \\ &+ \Delta x^2[\frac{1}{12}q_{xxxx} + |q|^2 q_{xx}] \\ &+ \Delta x \Delta t[\frac{1}{2}|q|^2 q_{xt} - \frac{1}{2}qq_x q_t^* - q^* q_x q_t] \\ &+ \frac{\Delta t^2}{4}[\frac{i}{3}q_{ttt} - 3q_t(|q|^2)_t] \end{aligned} \tag{44}$$

5.2 Perturbation Results

Empolying the perturbation results in equations (18), we have for the above NLS schemes:

Classic Explicit

$$\begin{aligned} \eta_t = 0, \qquad \xi_t = 0 \\ x_{1t} = -\frac{8}{3}\Delta x^2 \xi(\eta^2 + \xi^2) + \frac{32}{3}\Delta t^2[-2\eta^2\xi^3 + 3\eta^4\xi + 3\xi^5] \\ \beta_{1t} = \frac{4}{9}\Delta x^2[-6\eta^2\xi^2 - 7\eta^4 + 9\xi^4] + \frac{32}{3}\Delta t^2[13\eta^2\xi^4 - 15\eta^4\xi^2 - \eta^6 - 5\xi^6] \end{aligned} \tag{45}$$

Implicit-Explicit

$$\eta_t = \frac{64}{3}\Delta t \eta^3(\eta^2 - \xi^2)$$

$$\xi_t = -\frac{128}{5}\eta^4 \xi \Delta t$$

$$x_{1t} = \frac{8}{3}\Delta x^2 \xi(5\eta^2 - \xi^2) + \frac{16}{3}\Delta t^2[-10\eta^2\xi^3 + 9\eta^4\xi - 3\xi^5]$$

$$\beta_{1t} = \frac{4}{9}\Delta x^2[-42\eta^2\xi^2 + 77\eta^4 + 9\xi^4] + \frac{16}{3}\Delta t^2[5\eta^2\xi^4 - 37\eta^4\xi^2 + 5\xi^6 - 5\eta^6] \quad (46)$$

Local IST (Ladik)

$$\eta_t = 0, \qquad \xi_t = 0$$

$$x_{1t} = \frac{8}{3}\Delta x^2 \xi(\eta^2 - \xi^2) + \frac{32}{3}\Delta x \Delta t \eta^2 \xi^2 + \frac{16}{3}\Delta t^2 \xi[-4\eta^2\xi^2 + 3\eta^4 - 3\xi^4]$$

$$\begin{aligned}\beta_{1t} &= \frac{4}{3}\Delta x^2[2\eta^2\xi^2 + 7\eta^4 + 3\xi^4] \\ &+ \frac{16}{9}\Delta x \Delta t \eta^2 \xi(12\xi^2 - 14\eta^2) \\ &+ \frac{16}{3}\Delta t^2[-\eta^2\xi^4 + 3\eta^4\xi^2 + \eta^6 + 5\xi^6]\end{aligned} \quad (47)$$

6. Discussion

We have presented the results of a study of the effects of several finite difference schemes for the KdV and NLS equations. In each case we retained the leading order terms from Taylor expansions in the finite difference schemes to obtain the approximate perturbed evolution equations, which we are actually solving with these schemes. Using the mesh widths, Δx and Δt as the small parameters in the equations, we can determine the effects of the local truncation errors on the soliton's shape and speed.

We have compared the theoretical results found in this paper with actual numerical compuations; however, due to a lack of space, we will report these results in a future publication. Generally speaking, the simulations support the analysis given here within the validity of the perturbation analysis. Furthermore, such an analysis can lead to new schemes by adding the discretization of the truncation error to give a more accurate scheme. Such a modification has been reported in an earlier paper for the Zabusky-Kruskal scheme[5].

7. Acknowledgements

Many thanks go to Dr. C.J. Knickerbocker and Mr. M. Hays for many discussions and the computer simulations, which will be reported in joint work at a later time.

8. References

1. M. J. Ablowitz and H. Segur, *Solitons and the Inverse Scattering Transform* (SIAM, Philadelphia, 1981).
2. C. S. Gardner, C. S., J. M. Greene, M. D. Kruskal, and R. M. Miura, *Phys. Rev. Lett.* **19** (1967) 1095.
3. A. Hasegawa and F. Tappert, *Appl. Phys. Lett.* **23** (1973) 142.
4. R. L. Herman,*J. Phys. A* **23** (1990) 2327.
5. R.L. Herman and C. J. Knickerbocker, *J. Comp. Phys.* **104** (1993) 50.
6. V. I. Karpman and E. M. Maslov, *Sov. Phys. JETP* **48** (1978) 252.
7. D. J. Kaup, *SIAM J. Appl. Math.* **31** (1976) 121.
8. D. J. Kaup and A. C. Newell, *Proc. Roy. Soc. Lond.* **A361** (1978) 413.
9. J. W. Miles, *J. Fluid Mech.* **106** (1981) 131.
10. A. C. Newell, in *Solitons,* edited by R. K. Bullough and P. J. Caudrey, (Springer Verlag, New York, 1980), p. 177.
11. T.R. Taha and M. J. Ablowitz, *J. Comp. Phys.* **55** (1984) 203.
12. T.R. Taha and M. J. Ablowitz, *J. Comp. Phys.* **55** (1984) 231.
13. A. C. Vliegenthart, *J. Eng. Math.* **5** (1971) 131.
14. H. Washimi and T. Taniuti, *Phys. Rev. Lett.* **17** (1966) 996.
15. N. J. Zabusky and M.D. Kruskal, *Phys. Rev. Lett.* **15** (1965) 240.

SOLITON PHENOMENA IN PERIODIC SYSTEMS

TATESHI IIZUKA
Department of Physics, Faculty of Science,
Ehime University,
Matsuyama 790-77, Japan.

Abstract

Nonlinear waves in a one-dimensional periodic system are investigated. If the system is linearized, it has Bloch type solutions. It is shown that envelopes of the Bloch wave obey the nonlinear Schrödinger equation, the result of which suggests soliton propagations in the periodic system. Our method is applicable to many nonlinear periodic systems.

1. Introduction and Main Results

Recently A.Campa et al.[1] have studied nonlinear diatonic lattices. Using a reductive perturbative technique, they derived the nonlinear Schrödinger (NLS) equation describing the modulations of the optical or acoustic mode. Formerly, the same method has been used in case of the quartic nonlinearity[2]. As expected from the theory, stable solitary wave is observed numerically[1,2]. The results exhibit soliton phenomena in nonlinear diatonic lattices. The aim of this letter is to extend the technique into a wide class of nonlinear periodic systems and to present a new concept of soliton phenomena[3].

It is well known that the NLS equation is completely integrable[4] and has soliton solutions. In general, the NLS equation represents modulations of monochromatic waves in one-dimensional nonlinear dispersive systems, such as anharmonic lattices, optical fiber, water wave and plasma. In derivations of the NLS equation from *homogeneous* systems, we usually employ the reductive perturbation method (RPM).

In the present paper we shall extend the reductive perturbation technique to nonlinear periodic systems, in which the carrier wave is not the monochromatic but the Bloch wave. Here, the 'Bloch wave' is referred to as a solution of a linear periodic system which has the same property as that of the Bloch wave function in the quantum mechanics. In actual, many linearized periodic systems such as periodic lattices and water waves over a periodically uneven bottom have Bloch type solutions.

The next section is devoted to investigations of linear waves in a periodic system. This analysis is useful in constructing the RPM in nonlinear periodic systems. In §2, an extended RPM will be applied to the nonlinear periodic Klein-Gordon equation (3.2a). In conclusion, the envelope of the Bloch wave is shown to obey the NLS equation (see Eq. (3.13))

$$i\frac{\partial U}{\partial \tau} - \frac{\partial^2 U}{\partial \xi^2} + bU^2\overline{U} = 0. \tag{1.1}$$

Assuming a symmetry condition (4.1) we will show that constant b is real in §3, Thus, solitons are expected to propagate in the system.

Our method is probably applicable to many nonlinear periodic systems and if $b \in \Re$ is satisfied, ***soliton phenomena in the periodic systems are observed.*** This picture provides an extension of the soliton concept into periodic systems.

2. Bloch Waves in Linear Systems

In this section we shall give rather general method for a linearized wave $\Psi(x,t)$ in a one-dimensional periodically inhomogeneous system. The variables x and t stand for space and time coordinates respectively. Assume that there exist special solutions $\Psi = Y(x,\omega)\exp(-i\omega t)$, where ω is a real constant. Here, function $Y(x,\omega)$ is assumed to satisfy the following condition

$$Y(x+L,\omega) = \mathrm{e}^{ikL}Y(x,\omega), \tag{2.1}$$

where L is the period of the system and real constant k is connected with ω by a dispersion relation $\omega = \omega(k)$. Since the Bloch wave function of the quantum mechanics is defined by (2.1), we shall term the solution $Y(x,\omega)\exp(-i\omega t)$ the 'Bloch wave'. Exchange the sign of k corresponds to taking complex conjugation of the Bloch wave and then relations $\omega(-k) = -\omega(k)$ and $Y(x,-k) = Y^*(x,k)$ hold. We have denoted the complex conjugation by $*$. From now on, we term the constants k and ω wavenumber and angular frequency, respectively. In actual, the above assumption (2.1) is realized in many periodic lossless wave systems such as water waves and periodic lattices[3]. The dispersion relations exhibit band structures.

Since the system is linearized, general solution of $\Psi(x,t)$ is expressed by an integral form

$$\Psi(x,t) = \int_{-\infty}^{+\infty} \rho(\omega)Y(x,\omega)\mathrm{e}^{-i\omega t}\mathrm{d}\omega = 2\int_0^{\infty} \rho(\omega)\mathrm{Re}\Big(Y(x,\omega)\mathrm{e}^{-i\omega t}\Big)\mathrm{d}\omega, \tag{2.2}$$

that is regarded as linear combination of the Bloch waves. Density $\rho(\omega)$ has been set real without loss of the generality. Moreover, $\rho(\omega)$ is an even function owing to the reality of $\Psi(x,t)$. Thus, the second euqation in (2.2) holds, where Re indicates the real part.

From now on we assume that $\rho(\omega)$ is distributed nearby angular frequencies $\omega = \pm\omega_0$ in a band. Let us denote the width of its distribution by a real smallness parameter ε. Introducing Ω and $\tilde{\rho}(\Omega)$ as

$$\omega = \omega_0 + \varepsilon\Omega, \tag{2.3a}$$
$$\rho(\omega) = \varepsilon^{-1}\tilde{\rho}(\Omega) \qquad (\text{for} \quad \omega > 0), \tag{2.3b}$$

we can rewrite (2.2) into

$$\Psi(x,t) = 2\int \tilde{\rho}(\Omega)\mathrm{Re}\Big(Y(x,\omega_0+\varepsilon\Omega)\mathrm{e}^{-i(\omega_0+\varepsilon\Omega)t}\Big)\mathrm{d}\Omega, \tag{2.4}$$

where the integral range is virtually confined in a interval($\sim O(1)$) around $\Omega = 0$.

Replacing ω_0 in (2.4) by ω, we shall calculate a perturbation expansion of (2.4) with respect to ε. Dividing $Y(x,\omega)$ into $(Y(x,\omega)\exp(-ikx))\cdot\exp(+ikx)$ we have

$$\begin{aligned}
Y(x,\omega+\varepsilon\Omega) =& \Big\{Y(x,\omega)\mathrm{e}^{-ikx} + \varepsilon\Omega\frac{\partial}{\partial\omega}\Big(Y(x,\omega)\mathrm{e}^{-ikx}\Big) + \frac{\varepsilon^2\Omega^2}{2}\frac{\partial^2}{\partial\omega^2}\Big(Y(x,\omega)\mathrm{e}^{-ikx}\Big)\Big\} \\
&\times \exp i\Big\{kx + \varepsilon\Omega\frac{\mathrm{d}k}{\mathrm{d}\omega}x + \frac{\varepsilon^2\Omega^2}{2}\frac{\mathrm{d}^2k}{\mathrm{d}\omega^2}x\Big\} + O(\varepsilon^3) \\
=& \Big(Y(x,\omega) + \varepsilon\Omega V(x,\omega) + \varepsilon^2\Omega^2 W(x,\omega)\Big)\exp i\Big\{\varepsilon\Omega\frac{\mathrm{d}k}{\mathrm{d}\omega}x + \frac{\varepsilon^2\Omega^2}{2}\frac{\mathrm{d}^2k}{\mathrm{d}\omega^2}x\Big\} \\
&+ O(\varepsilon^3),
\end{aligned} \tag{2.5}$$

with

$$V(x,\omega) = \mathrm{e}^{ikx}\frac{\partial}{\partial\omega}\Big(Y(x,\omega)\mathrm{e}^{-ikx}\Big), \tag{2.6a}$$
$$W(x,\omega) = \mathrm{e}^{ikx}\frac{1}{2}\frac{\partial^2}{\partial\omega^2}\Big(Y(x,\omega)\mathrm{e}^{-ikx}\Big). \tag{2.6b}$$

It is easily shown that $V(x,\omega)$ and $W(x,\omega)$ have the same property as that of the Bloch solution (2.1)

$$V(x+L,\omega) = \mathrm{e}^{ikL}V(x,\omega), \tag{2.7a}$$
$$W(x+L,\omega) = \mathrm{e}^{ikL}W(x,\omega). \tag{2.7b}$$

Substitution (2.5) into the integral form (2.4) gives

$$\begin{aligned}
\Psi(x,t) =& Y(x,\omega)\mathrm{e}^{-i\omega t}\Psi_+^{(1)} + Y^*(x,\omega)\mathrm{e}^{+i\omega t}\Psi_-^{(1)} \\
&+ \varepsilon V(x,\omega)\mathrm{e}^{-i\omega t}\Psi_+^{(2)} + \varepsilon V^*(x,\omega)\mathrm{e}^{+i\omega t}\Psi_-^{(2)} \\
&+ \varepsilon^2 W(x,\omega)\mathrm{e}^{-i\omega t}\Psi_+^{(3)} + \varepsilon^2 W^*(x,\omega)\mathrm{e}^{+i\omega t}\Psi_-^{(3)} + O(\varepsilon^3),
\end{aligned} \tag{2.8}$$

where $\Psi_\pm^{(j)} (j = 1,2,3)$ have been defined by

$$\Psi_\pm^{(j)} = \int d\Omega \tilde{\rho}(\Omega)\Omega^{j-1} \exp(\pm i)\Big\{\Omega\varepsilon(\frac{dk}{d\omega}x - t) + \Omega^2 \frac{\varepsilon^2}{2}\frac{d^2k}{d\omega^2}x\Big\}. \tag{2.9}$$

Here, we should note that $\Psi_\pm^{(j)}(x,t)$ are slowly varying functions of x and t. Moreover, neglecting $O(\varepsilon)$ in (2.8), we see that $\Psi_+^{(1)}$ and $\Psi_-^{(1)}$ represent the envelopes of the Bloch waves $Y(x,\omega)\exp(-i\omega t)$ and $Y^*(x,\omega)\exp(+i\omega t)$, respectively.

Let us introduce new independent variables ξ and τ that describe the slow evolution of $\Psi_\pm^{(j)}(x,t)$;

$$\xi = \varepsilon(\frac{dk}{d\omega}x - t), \tag{2.10a}$$

$$\tau = \frac{\varepsilon^2}{2}\frac{d^2k}{d\omega^2}x. \tag{2.10b}$$

With these variables, (2.9) is rewritten as

$$\Psi_\pm^{(j)}(\xi,\tau) = \int d\Omega \tilde{\rho}(\Omega)\Omega^{j-1} \exp(\pm i)\Big\{\Omega\xi + \Omega^2\tau\Big\}. \tag{2.11}$$

from which it is straightforward to obtain the free Schrödinger equations for $\Psi_\pm^{(1)}(\xi,\tau)$

$$i\frac{\partial \Psi_\pm^{(1)}}{\partial \tau} = \pm \frac{\partial^2 \Psi_\pm^{(1)}}{\partial \xi^2}. \tag{2.12}$$

The expansion (2.8) is expressed by the envelope functions $\Psi_\pm^{(1)}(\xi,\tau)$ as

$$\begin{aligned}\Psi(x,t) =& Y(x,\omega)e^{-i\omega t}\Psi_+^{(1)}(\xi,\tau) + Y^*(x,\omega)e^{+i\omega t}\Psi_-^{(1)}(\xi,\tau)\\ &+\varepsilon V(x,\omega)e^{-i\omega t}(-i)\frac{\partial \Psi_+^{(1)}(\xi,\tau)}{\partial \xi} + \varepsilon V^*(x,\omega)e^{+i\omega t} i\frac{\partial \Psi_-^{(1)}(\xi,\tau)}{\partial \xi}\\ &+\varepsilon^2 W(x,\omega)e^{-i\omega t}(-1)\frac{\partial^2 \Psi_+^{(1)}(\xi,\tau)}{\partial \xi^2} + \varepsilon^2 W^*(x,\omega)e^{+i\omega t}(-1)\frac{\partial^2 \Psi_-^{(1)}(\xi,\tau)}{\partial \xi^2}\\ &+O(\varepsilon^3).\end{aligned} \tag{2.13}$$

It is notable that up to $O(\varepsilon^2)$, the above expansion and the conditions (2.7) are invariant under the following gauge transformation;

$$\tau \rightarrow \tau + \varepsilon^2\chi(x,\omega), \tag{2.14a}$$

$$W(x,\omega) \rightarrow W(x,\omega) - i\chi(x,\omega)Y(x,\omega), \tag{2.14b}$$

$$W^*(x,\omega) \rightarrow W^*(x,\omega) + i\chi(x,\omega)Y^*(x,\omega), \tag{2.14c}$$

where the auxiliary function $\chi(x,\omega)$ is arbitrary which satisfies

$$\chi(x+L,\omega) = \chi(x,\omega). \tag{2.15}$$

Let us summerize the results of the present section. We have considered linear waves $\Psi(x,t)$ in a periodic system, the solution of which is a linear combination of the Bloch waves (2.2). Provided that the spctrum $\rho(\omega > 0)$ is distributed only in a small interval ($\sim O(\varepsilon)$) around an angular frequency ω, $\Psi(x,t)$ is calculated into the perturbation expansion (2.13). Up to $O(\varepsilon)$, $\Psi_{\pm}^{(1)}$ defined in (2.9) are regarded as slowly changing envelopes of the Bloch waves. With the independent variables ξ and τ in (2.10), the envelopes $\Psi_{\pm}^{(1)}$ obey the free Schrödinger equation (2.12). The result is analogous to that of the homogeneous systems where the carrier wave is the monochromatic wave.

The expansion (2.13) and the relations (2.7) are invariant under the gauge transfromation (2.14). As long as (2.15) is satisfied, $\chi(x,\omega)$ can be chosen an arbitrary complex function. However, when the balance between the dispersive and the nonlinear effects are taken, explicit form of $\chi(x,\omega)$ should be determined. In another words, the gauge is fixed. We will see this process using a dynamic model in the next section.

3. The Nonlinear Schrödinger Equation

In this section we investigate nonlinear waves in a one-dimensional periodic system. The waves are assumed to have the following two properties;
1)The amplitude is small but finite such that the nonlinear effects are sufficiently weak but not negligible.
2)The waves are slow modulations of the Bloch wave due to the dispersion and the nonlinearities. The wave length of the Bloch wave $l_B \equiv 1/k$ is of the same order as the period of the system L.
Owing to *2)*, the following relation holds

$$L \sim l_B = k^{-1} \ll l_E, \tag{3.1}$$

where l_E is the characteristic wave length the envelope. The aim of the present section is derivation of nonlinear evolution equations for the envelope of the Bloch wave. For this end we shall refer to the perturbation expansion introduced in the previous section.

For example we consider a periodic nonlinear Klein-Gordon equation for $u(x,t)$;

$$\frac{\partial^2 u}{\partial t^2} - \frac{\partial^2 u}{\partial x^2} + m(x)u + \kappa u^3 = 0, \tag{3.2a}$$

$$m(x+L) = m(x). \tag{3.2b}$$

If (3.2a) is linearized, substitution $u = Y(x,\omega)e^{-i\omega t}$ gives

$$-\omega^2 Y - \frac{\partial^2 Y}{\partial x^2} + m(x)Y = 0. \tag{3.3}$$

This is the stationary Schrödinger equation in the periodic potential $m(x)$. Therefore, there exist Bloch solutions[5] which satisfy (2.1).

Remarking that the amplitude of $u(x,t)$ is small($\sim O(\varepsilon)$), we apply alomst the same perturbation expansion as (2.13) to $u(x,t)$;

$$\begin{aligned} u(x,t) =& \varepsilon Y(x,\omega)\mathrm{e}^{-i\omega t}U(\xi,\tau) + \varepsilon Y^*(x,\omega)\mathrm{e}^{+i\omega t}\overline{U}(\xi,\tau) \\ &+\varepsilon^2 V(x,\omega)\mathrm{e}^{-i\omega t}(-i)\frac{\partial U(\xi,\tau)}{\partial \xi} + \varepsilon^2 V^*(x,\omega)\mathrm{e}^{+i\omega t} i\frac{\partial \overline{U}(\xi,\tau)}{\partial \xi} \\ &+\varepsilon^3 W(x,\omega)\mathrm{e}^{-i\omega t}(-1)\frac{\partial^2 U(\xi,\tau)}{\partial \xi^2} + \varepsilon^3 \overline{W}(x,\omega)\mathrm{e}^{+i\omega t}(-1)\frac{\partial^2 \overline{U}(\xi,\tau)}{\partial \xi^2} \\ &+\varepsilon^3 X(x,\omega)\mathrm{e}^{-3i\omega t}\Big(U(\xi,\tau)\Big)^3 + \varepsilon^3 X^*(x,\omega)\mathrm{e}^{+3i\omega t}\Big(\overline{U}(\xi,\tau)\Big)^3 \\ &+O(\varepsilon^4), \end{aligned} \tag{3.4}$$

where $Y(x,\omega)\exp(-i\omega t)$ is the Bloch wave. Terms in the 4th row have been added to absorb the nonlinear effect. Referring (2.10), (2.6) and the gauge transformation (2.14), we have defined ξ, τ, $V(x,\omega)$, $W(x,\omega)$ and $\overline{W}(x,\omega)$ by

$$\xi = \varepsilon(\frac{\mathrm{d}k}{\mathrm{d}\omega}x - t), \tag{3.5a}$$

$$\tau = \varepsilon^2(\frac{1}{2}\frac{\mathrm{d}^2 k}{\mathrm{d}\omega^2}x + \chi(x,\omega)). \tag{3.5b}$$

$$V(x,\omega) = \mathrm{e}^{ikx}\frac{\partial}{\partial\omega}\Big(Y(x,\omega)\mathrm{e}^{-ikx}\Big), \tag{3.5c}$$

$$W(x,\omega) = \mathrm{e}^{+ikx}\frac{1}{2}\frac{\partial^2}{\partial\omega^2}\Big(Y(x,\omega)\mathrm{e}^{-ikx}\Big) - i\chi(x,\omega)Y(x,\omega), \tag{3.5d}$$

$$\overline{W}(x,\omega) = \mathrm{e}^{-ikx}\frac{1}{2}\frac{\partial^2}{\partial\omega^2}\Big(Y^*(x,\omega)\mathrm{e}^{+ikx}\Big) + i\chi(x,\omega)Y^*(x,\omega), \tag{3.5e}$$

where the auxiliary function $\chi(x,\omega)$ satisfies (2.15) and will be fixed later. Dependent variable $U(\xi,\tau)$ is unknown and $\overline{U}(\xi,\tau)$ is given by

$$\overline{U}(\xi,\tau) = \Big(U(\xi,\tau^*)\Big)^*. \tag{3.6}$$

Up to $O(\varepsilon)$, $U(\xi,\tau)$ and $\overline{U}(\xi,\tau)$ are regarded as the envelopes of the Bloch waves $Y(x,\omega)\cdot\exp(-i\omega t)$ and $Y^*(x,\omega)\exp(+i\omega t)$, respectively. Let us substitute (3.4) into the nonlinear Klein -Gordon equation (3.2a) and compare the coefficients of $\varepsilon^n\exp(-il\omega t)$ ($n = 1,2,3,\cdots$, $l = 0,\pm1,\pm2,\cdots$).

At $n = 1, l = \pm1$ and at $n = 2, l = \pm1$ we confirm the Schrödinger equation (3.3) and its differentiation with respect to ω, respectively. Equation which gives $X(x,\omega)$ is obtained at $n = 3, l = \pm3$;

$$-9\omega^2 X(x,\omega) - \frac{\partial^2 X(x,\omega)}{\partial x^2} + m(x)X(x,\omega) + \kappa Y^3(x,\omega) = 0. \tag{3.7}$$

At $n=3, l=1$ we obtain

$$iP(x,\omega)\Big(i\frac{\partial U}{\partial \tau}-\frac{\partial^2 U}{\partial \xi^2}\Big)+3\kappa|Y(x,\omega)|^2 Y(x,\omega)U^2\overline{U}=0, \tag{3.8}$$

$$P(x,\omega)=\frac{1}{Y(x,\omega)}\frac{\partial}{\partial x}\Big\{Y^2(x,\omega)\Big(\frac{\partial \chi(x,\omega)}{\partial x}+\frac{1}{2}\frac{\mathrm{d}^2 k}{\mathrm{d}\omega^2}\Big)\Big\}. \tag{3.9}$$

We can remove variable x from the Eq.(3.8) as follows. Since $\chi(x,\omega)$ has not been fixed yet, we may add a gauge condition;

$$bP(x,\omega)=-3i\kappa|Y(x,\omega)|^2 Y(x,\omega), \tag{3.10}$$

where b is a constant. Owing to the periodicity of $\chi(x,\omega)$ (2.15), Eqs. (3.9) and (3.10) give explicit form of b and $\chi(x,\omega)$

$$b=-3\Big(L\frac{\mathrm{d}^2 k}{\mathrm{d}\omega^2}\Big)^{-1}\frac{\kappa \mathrm{e}^{-ikL}}{\sin kL}\int_{d'}^{L+d'}\mathrm{d}x' Y^{-2}(x')\int_{x'}^{x'+L}\mathrm{d}x''|Y(x'')|^2 Y^2(x''), \tag{3.11}$$

$$\chi(x,\omega)=-\frac{1}{2}\frac{\mathrm{d}^2 k}{\mathrm{d}\omega^2}x-\frac{3\kappa \mathrm{e}^{-ikL}}{2b\sin kL}\int_0^x \mathrm{d}x' Y^{-2}(x')\int_{x'}^{x'+L}\mathrm{d}x''|Y(x'')|^2 Y^2(x''), \tag{3.12}$$

where d' is an arbitrary real number which does not change the value of b. In deriving (3.12), we have set $\chi(0,\omega)=0$ without loss of the generality. Hence, explicit forms of τ, $W(x,\omega)$ and $\overline{W}(x,\omega)$ are determined through (3.5b), (3.5d) and (3.5e), respectively. At last (3.8) is reduced to the NLS equation

$$i\frac{\partial U}{\partial \tau}-\frac{\partial^2 U}{\partial \xi^2}+bU^2\overline{U}=0. \tag{3.13}$$

4. Solitons in Periodic Systems

We do not know whether the coefficient b given by (3.11) is real or not. However, if $m(x)$ is symmetric function, b is shown to be real in the following way.

Let us assume

$$m(-x+d)=m(x+d)\quad (0\le d\le L), \tag{4.1}$$

that is $m(x)$ is symmetric with respect to $x=d$. Provided that $Y(x,\omega)$ is a solution of (3.3), $\tilde{Y}(x,\omega)\equiv Y(-x+2d,\omega)$ is also its solution. Then, Eq.(2.1) gives

$$\tilde{Y}(x+L,\omega)=Y(-x+2d-L,\omega)=\mathrm{e}^{-ikL}Y(-x+2d,\omega)=\mathrm{e}^{i(-k)L}\tilde{Y}(x,\omega), \tag{4.2}$$

which is equivalent to

$$Y(-x+2d,\omega)=Y(x,-\omega)=Y^*(x,\omega). \tag{4.3}$$

Using (4.3) and setting $d' = d$, we take the complex conjugation of (3.11);

$$
\begin{aligned}
b^* &= -3\Big(L\frac{\mathrm{d}^2k}{\mathrm{d}\omega^2}\Big)^{-1}\frac{\kappa e^{+ikL}}{\sin kL}\int_d^{L+d}\mathrm{d}x' Y^{-2}(-x'+2d)\int_{x'}^{x'+L}\mathrm{d}x''|Y(x'')|^2Y^2(-x''+2d).\\
&= -3\Big(L\frac{\mathrm{d}^2k}{\mathrm{d}\omega^2}\Big)^{-1}\frac{\kappa e^{-ikL}}{\sin kL}\int_d^{L+d}\mathrm{d}\tilde{x}Y^{-2}(\tilde{x})\int_{\bar{x}}^{\tilde{x}+L}\mathrm{d}\bar{x}|Y(\bar{x})|^2Y^2(\bar{x})\\
&= b,
\end{aligned}
\tag{4.4}
$$

with

$$\bar{x} = -x'' + 2d + 2L \tag{4.5a}$$

$$\tilde{x} = -x' + 2d + L \tag{4.5b}$$

Thus, the coefficient b is real. It is well known that the NLS equation (3.13) for real b has soliton solutions. Even if $\tau = \tau(x)$ is complex, we analytically continue the soliton solution $U(\xi, \tau)$ into complex τ plane;

$$U = \sqrt{\frac{-2}{b}}A\,\mathrm{sech}\Big(A\xi(x,t) + 2BA\tau(x)\Big)\exp\Big\{iA\xi(x,t) - i(A^2 - B^2)\tau(x)\Big\}, \tag{4.6}$$

where we have assumed $b < 0$ and A, B are real parameters. This implies that ***solitons are observed as envelopes of the Bloch wave.*** Since both $\partial U/\partial\xi$ and $\partial^2 U/\partial\xi^2$ are solitary waves in x-space, $u(x,t)$ given by the expansion (3.4) is also a solitary wave. In case of $b > 0$, dark solitons are observed.

Moreover, the initial value problem of the NLS equation has been solved analytically[4] This mathematical result can be applied to our system in the following way. At first let us set envelope fuction $U(-\infty < t < +\infty)$ at the origin $x = 0$. Remarking (3.5a) and (3.5b) we see

$$U(x = 0, t) = U(\xi = -\varepsilon t, \tau = 0). \tag{4.7}$$

The second step is to construct $U(\xi, \tau \in \Re)$ from $U(\xi, 0)$ by means of the inverse scattering method of the NLS equation. However, we must remark that 'time' $\tau = \tau(x)$ possibly be complex. Analytic continuation of $U(\xi,\tau)$ into the complex τ plane gives the form of $U(x,t)$. Similarly we can calculate $\partial U/\partial\xi$, $\partial^2 U/\partial\xi^2$, $\overline{U}$, $\partial\overline{U}/\partial\xi$ and $\partial^2\overline{U}/\partial\xi^2$ as functions of x and t. At last, using the perturbation expansion (3.4) we obtain $u(x,t)$ up to $O(\varepsilon^3)$.

5. Summery and Discussions

In this letter we have extended the reductive perturbation method(RPM) to a nonlinear periodic system. There, the Bloch wave is introduced as the carrier wave. Using the method we have shown that the envelope is governed by the nonlinear Schrödinger(NLS) equation (3.13). The procedure of driving the NLS equation from (3.2a) is summerized as follows.

*1)*Solve the linearized equation (3.3) and obtain the Bloch type solution $Y(x,\omega)$. The dispersion relation $\omega = \omega(k)$ is derived simultaneously.
*2)*Introduce the auxiliary function $\chi(x,\omega)$ which satisfies (2.15). Set variables ξ, τ, $V(x,\omega)$, $W(x,\omega)$ and $\overline{W}(x,\omega)$ as (3.5).
*3)*Expand $u(x,t)$ as (3.4), substitute it into the nonlinear equation (3.2a) and compare the coefficients of $\varepsilon^n \exp(-il\omega t)$ ($n = 1,2,3,\cdots$, $l = 0,\pm1,\pm2,\cdots$). At (n,l)=(3,1) the following type of equation is obtained.

$$P(x)\left(i\frac{\partial U}{\partial \tau} - \frac{\partial^2 U}{\partial \xi^2}\right) + Q(x)U^2\overline{U} = 0 \tag{5.1}$$

4) Assume that $P(x)$ is proportional to $Q(x)$ ($Q(x) = bP(x)$, b is a constant). This assumption fixes the gauge $\chi(x,\omega)$ and b is also obtained. Thus, explicit form of τ $W(x,\omega)$ and $\overline{W}(x,\omega)$, are determined.
5) At last NLS equation for the envelope of the Bloch wave is derived;

$$i\frac{\partial U}{\partial \tau} - \frac{\partial^2 U}{\partial \xi^2} + bU^2\overline{U} = 0, \tag{5.2}$$

The above procedure can be applied also to anharmonic periodic lattices[3]. Hence, remarking that the RPMs are common to many homogeneous systems, we believe that the extended RPM is useful for a large class of nonlinear periodic systems.

Provided that b in (5.2) is real, soliton propagations are observed. *This suggests us a new concept of the soliton phenomena because the carrier wave is not the monochromatic wave but the Bloch wave.* In our case, symmetry of $m(x)$ gives a sufficient condition for $b \in \Re$. In case of the anharmonic periodic lattices, we have shown that b is real if the nonlinearities of the springs are quartic[3]. For other cases, under what physical conditions $b \in \Re$ holds? We do not have a general method to answer this question.

The Schrödinger equaiton (3.3) is generally difficult to solve, though exact form of $Y(x,\omega)$ are obtained for some spacial $m(x)$. Calculation in these cases remains an our future work. We must also confirm our result by numerical experiments.

Throughout this letter it has been assumed that the angular freqency ω is finite. Similarly to the homogeneous systems, we expect that the KdV type evolution equation is derived in the limiting case $\omega, k \to 0$. In actual, KdV equation was obtained from the continuum limit of the acoustic mode in the anharmonic diatonic lattices[6]. Thus, it is interesting to extend our method to zero-wavenumber limit.

Our method is applicable only to the periodic system in which the modulation wave is the Bloch wave. For more general cases, it seems that threre is no such procedure.

6. References

1. A. Campa, A. Giansanti, A. Tenenbaum, D. Levi and O. Ragnisco, *Phys. Rev.* B **48** (1993) 10168.

2. S. Pnevmatikos, N. Flytzanis, ans M. Remoissenet, *Phys.Rev.* B **33** (1986) 2308.

3. T. Iizuka,*J.Phys.Soc.Jpn* **63** (1994) no.12 in press.

4. V.E. Zakharov and A.B. Shabat, *Sov. Phys. JETP* **34** (1972) 62.

5. S.P. Novikov, S.V. Manakov, L.P. Pitaevskii, and V.E. Zaknarov, *Theory of Solitons. The Inverse Scattering Method* (Plenum New York, 1984).

6. N. Yajima and J. Satsuma, Prog. Theor. Phys. **62** (1979) 370.

ONE CENTER EXPANSION IN QUANTUM MECHANICS OF MOLECULES

Rimantas J. RAKAUSKAS, Juozas ŠULSKUS
Dept. of Molecular Dynamics, Vilnius University,
Saulėtekio 9, 2054 Vilnius, Lithuania

ABSTRACT

Using Fourier transform and plane wave expansion in spherical harmonics, the analytical expression of the Slater type atomic wavefunction (STF), Cartesian Gaussian (CGF) and harmonic oscilator functions (HOF) transformed to another center were found.

1. Introduction

The Cartesian Gaussian-type functions were first proposed and introduced into quantum chemistry by Boys[1] and linear combinations of them have been extensively used in the last years as basis set orbitals in atomic and molecular calculations. This is due to the fact that all multicenter integrals needed can be evaluated in closed form[2,3]. But these functions do not form an orthonormal set. For this reason Wulfman[4] proposed the use of the orthogonal complete set of harmonic oscillator wave functions as basis set. Three-dimensional harmonic oscillator functions play the important role in nuclear shell calculations and the mathematical techniques for dealing with HOF have been developed by nuclear physicists.

The mathematical formalism of the Talmi transformation[5] makes it possible to evaluate analytically the molecular multicenter integrals over HOF in closed form[6]. If spherical polar coordinates are used, the molecular integrals over HOF can be represented in terms of, again, HOF, depending only on the distance of the centers and some known quantities as Moshinsky brackets[7], Clebsh-Gordan coefficients and the confluent hypergeometric function[8]. Because the HOF form a complete orthonormal set, HOF can be used as a basis set in LCAO calculations. Moreover, other types of orbitals can be expanded in terms of HOF and molecular integrals over these orbitals can be obtained analytically from integrals over HOF. But with current supercomputers the vast majority of molecular quantum- mechanical calculations are carried out with CGF basis sets by using packages POLYATOM/2, GAUSSIAN92, HONDO7[9], GAMESS.

The usage of CGF in determination of adiabatic potential energy surfaces (PES) for molecular dynamics purposes have problems, mainly due to the immensely great number of CGF to be used, because CGF, and also HOF, do not resemble physically meaningful one-electron wave functions for the one-nucleus one-electron Coulomb problem, like exponentially decaying hydrogenic or Slater-type functions.

The usage of STF leads to a correct representation of the behaviour of wave functions at very short or very long distances from the nucleus, but difficulties arise during the evaluation of multicenter integrals with STF for polyatomic molecules. The contraction of mixed basis sets, using CGF, HOF and STF, allows a reduction of the number of basis functions used for solving problems dealing with electron-molecule scattering, molecular dynamics, large amplitude molecular vibrations, excited electronic states, photoionization of molecules.

2. Transformation formulas

Analytical transformation of Slater type AO localised in center with radius vector $\vec{R}_b$ to the center $\vec{R}_a$ (Cartesian coordinate systems of centers a and b are parallel) have the form[10]:

$$r_b^{n-1} Y_{l,m}\left(\Omega_{\vec{r}_b}\right)\exp(-\xi r_b) = \sum_{l',m',l'',m''} \sqrt{\frac{(2l+1)(2l''+1)}{4\pi(2l'+1)}} \times$$

$$\times C(l,m;l'',m'';l',m') R_{nll'l''}^{(\mathrm{STF})}(\xi, r_a, R_{ab}) Y_{l'',m''}\left(\Omega_{\vec{R}_{ab}}\right) Y_{l',m'}\left(\Omega_{\vec{r}_a}\right), \tag{1}$$

were $C(l,m;l'',m'';l',m') = \begin{bmatrix} l & l'' & l' \\ 0 & 0 & 0 \end{bmatrix} \begin{bmatrix} l & l'' & l' \\ m & m'' & m' \end{bmatrix}$ - multiplication of Clebsh-Gordan coefficients. $R^{(\mathrm{STF})}{}_{nll'l''}(\xi, r_a, R_{ab})$ - radial part of transformation. It depends on quantum numbers n, l, l', l'', parameter ξ and AO localisation vector $\vec{R}_{ab}$:

$$R^{(\mathrm{STF})}{}_{nll'l''}(\xi, r_a, R_{ab}) = (-1)^l \pi^{\frac{3}{2}} \xi^{-(n+1)} (\xi r_a)^{-(l'+1)} (\xi R_{ab})^{-(l''+1)} \times$$

$$\times 2^{-(n+l+l'+l'')} \frac{\Gamma\left(\frac{n-l}{2}+1\right)\Gamma\left(\frac{n-l-1}{2}+1\right)\Gamma(n+l+2)}{\Gamma(n+1)} \times$$

$$\times \left\{ \exp\left[-\xi\left|r_a - R_{ab}\right|\right] \sum_{\alpha=0}^{n+l''} \sum_{\beta=0}^{n+l'} A_{\alpha\beta}(nll'l'')(\mp \xi r_a)^{\beta} (\pm \xi R_{ab})^{\alpha} - \right.$$

$$-\exp\left[-\xi\left|r_a+R_{ab}\right|\right]\sum_{\alpha=0}^{n+l''}\sum_{\beta=0}^{n+l'}A_{\alpha\beta}(nll'l'')(\xi r_a)^{\beta}(\xi R_{ab})^{\alpha}\Bigg\}, \tag{2}$$

were upper signs in $\{\mp\}$ belongs to $R^{(\mathrm{STF})}{}_{nll'l''}(\xi,r_a,R_{ab})$ with $r_a \le R_{ab}$ and lower - for the case $r_a > R_{ab}$. $A_{\alpha\beta}(\ldots)$- elements of the matrix $\|A\|$ with dimension $(n+l''+1)\times \times(n+l'+1)$:

$$A_{\alpha\beta}(nll'l'')=$$

$$=\sum_{k=0}^{\left[\frac{n-1}{2}\right]}\sum_{q=0}^{l'}\sum_{v=0}^{l''}\sum_{k=0}^{s}\frac{(-1)^{(s+k)}2^{(q+v-k)}\Gamma(2l'-q+1)\Gamma(2l''-v+1)}{\Gamma\left(\frac{n-l}{2}-k+1\right)\Gamma\left(\frac{n-l-1}{2}-k+1\right)}\times$$

$$\times\frac{\Gamma(n+k+1)\Gamma(l-l'-l''+2k+q+v+1)}{\Gamma\left(l+k+\frac{3}{2}\right)\Gamma(k+1)\Gamma(q+1)\Gamma(v+1)\Gamma(k+1)}\times$$

$$\times\frac{1}{\Gamma(l'-q+1)\Gamma(l''-v+1)\Gamma(q-k+1)\Gamma(z+1)\Gamma(n-s-z+1)}\times$$

$$\times\frac{1}{\Gamma(l-l'-l''+2k+q+v-s+k+1)}, \tag{3}$$

were $s=q+v+n-\alpha-\beta$, $z=\alpha-v$ with conditions $0\le s\le n$, $0\le z\le n-s$.

The transformation to another center of HOF

$$\varphi^{\mathrm{HOF}}(\vec{r}_A)=N^{\mathrm{HOF}}(\alpha;n,l,m)\times$$

$$\left|\vec{r}-\vec{R}_A\right|^{l}L_n^{l+\frac{1}{2}}\left(2\alpha r^2\right)Y_{l,m}\left(\Omega_{\vec{r}_A}\right)\exp\left(-\alpha\left|\vec{r}-\vec{R}_A\right|^2\right),$$

$$N^{\mathrm{HOF}}(\alpha;n,l,m)=(2\alpha)^{\frac{l}{2}+\frac{5}{4}}\left[\frac{n!}{\alpha\Gamma\left(n+l+\frac{3}{2}\right)}\right]^{\frac{1}{2}}. \tag{4}$$

is of analogous form with STF, Eq. (1).

Radial part for the case of HOF $R^{(\mathrm{HOF})}_{nll'l''}(\xi, r_a, R_{ab})$ takes the form:

$$R^{(\mathrm{HOF})}_{nll'l''}(\xi, r_a, R_{ab}) = (-1)^{(n+l+l'')} i^{(l+l'+l'')} (2\pi)^{\frac{3}{2}} (2\alpha)^{-\left(l+\frac{3}{2}\right)} \alpha^{\frac{1}{2}(l+l'+l''+3)} \times$$

$$\times r_a^{\left(l'+\frac{1}{2}\right)} R_{ab}^{\left(l''+\frac{1}{2}\right)} \sum_{m=0}^{n} \frac{(-1)^m}{2^m m!} \binom{n+l+\frac{1}{2}}{n-m} \times$$

$$\times \sum_{s=0}^{\infty} (-1)^s \frac{\Gamma\left(s+m+\frac{1}{2}(l+l'+l''+3)\right)}{s!\,\Gamma\left(s+l'+\frac{3}{2}\right)} {}_2F_1\left(-s, -\left(s+l'+\frac{1}{2}\right), l''+\frac{3}{2}, \frac{R_{ab}^2}{r_a^2}\right) \alpha^s r_a^{2s} . \tag{5}$$

The analytical expansion of SGF in the case of linear molecules takes a form:

$$\left|\vec{r}-\vec{R}_A\right|^{2n_A+l_A} Y_{l_A,m_A}\left(\Omega_{\vec{r}_A}\right) \exp\left(-\alpha\left|\vec{r}-\vec{R}_A\right|^2\right) = (2l_A+1)!!\,(2l_A+1)^{1/2} \left(\frac{z_A}{R_A}\right)^{L+l_A} \times$$

$$\times R_A^{2n_A} \exp\left(-\alpha\left(r^2+R_A^2\right)\right) R_A^{l_A} \sum_{L=|m_A|}^{\infty} Y_{L,m_A}(\Omega_{\vec{r}}) \frac{1}{(2L+1)^{1/2}} V_{L,m_A}^{(l_A,m_A)}(r), \tag{6}$$

$$V_{L,m_A}^{(l_A,m_A)}(r) = \left\{ \sum_{l=0}^{n_A} \left[(2l+1) \sum_{t=l}^{2n_A-l} T_{l,t}^{(2n_A)} \left(\frac{r}{R_A}\right)^t \right] \times \right.$$

$$\times \sum_{l_2=0}^{l_A} \frac{(2l_2+1)}{(2l_A-2l_2+1)!\,(2l_2+1)!!} \left(\frac{r}{R_A}\right)^{l_A-l_2} C(l_A, m_A; l_2, 0; l_A-l_2, m_A) \times$$

$$\left. \times \sum_{l_e} (2l_e+1) i_{l_e}(2\alpha R_A r) \left[\sum_{L'} C(l,0; l_A-l_2, m_A; L', m_A) C(L', m_A; l_e, 0; L, m_A) \right] \right\}, \tag{7}$$

were $T_{l,t}^{(2n_A)} = (-1)^l \frac{(2n_A+1)}{(t-l)!!(t+l+1)!!(2n_A-t-l)!!(2n_A-t+l+1)!!}$, $i_{l_e}(2\alpha R_A r)$ - modified Bessel function,

The program for determination of the radial part of the expansion, Eq. (7), for any l_A, m_A and L was written using symbolic computation system REDUCE[11]. In table 1 we present the analytical expressions of $V_{L,m_A}^{(l_A,m_A)}(r)$ for some SGF's. The radial part of the expansion $V_{L,m_A}^{(l_A,m_A)}(r)$ is a sum of modified Bessel functions multiplied by the polynomials of $\left(\frac{r}{R_A}\right)$, were r is the distance between the origin of coordinate system and electron and R_A is the distance from the origin of coordinate system and center A. The methods of evaluation of Bessel functions are well developed.

Table 1. The radial parts $V_{L,m_A}^{(l_A,m_A)}(r)$ of the expansion Eq. (11)

Bessel function, Eq. (7)	Polynomial at Bessel function	Bessel function, Eq. (7)	Polynomial at Bessel function
$n_A=0,\; l_A=1,\; m_A=0$		$n_A=0,\; l_A=2,\; m_A=0$	
i_{L-1}	$\frac{1}{3}L\left(\frac{r}{R_A}\right)$	i_{L-2}	$\frac{L(L-1)}{10(2L-1)}\left(\frac{r}{R_A}\right)^2$
i_L	$\frac{1}{3}(2L+1)$	i_{L-1}	$\frac{2}{15}L\left(\frac{r}{R_A}\right)$
i_{L+1}	$\frac{1}{3}(L+1)\left(\frac{r}{R_A}\right)$	i_L	$\frac{L(L+1)(2L+1)}{(2L+3)(2L-1)}\left(\frac{r}{R_A}\right)^2 + \frac{1}{15}(2L+1)$
$n_A=1,\; l_A=0,\; m_A=0$		i_{L+1}	$\frac{2}{15}(L+1)\left(\frac{r}{R_A}\right)$
i_{L-1}	$-2L\left(\frac{r}{R_A}\right)$	i_{L+2}	$\frac{(L+1)(L+2)}{10(2L+3)}\left(\frac{r}{R_A}\right)^2$
i_L	$(2L+1)\left(\frac{r}{R_A}\right) + (2L+1)$		
i_{L+1}	$-2(L+1)\left(\frac{r}{R_A}\right)$		

3. Conclusions

The use of both the analytical and the numerical approach for the transformation of Slater-type, spherical Gaussian and harmonic oscillator functions allows the effective determination of molecular multicenter integrals, one center expansions of molecular wave functions and molecule-electron interaction potential with necessary accuracy, enables to use vector processors effectively.

The transformation formulas (Eq. 5) can be applied for Cartesian Gaussian functions transformation after preliminary transformation of CGF to HOF basis. The algorithms developed for multicenter integrals with STF[10] can be applied for HOF and the evaluation of multicenter integrals of mixed basis sets (STF and CGF) may be performed.

4. References

1. S. P. Boys, *Proc. Roy. Soc.* **A200** (1950) 542.
2. L. Shavitt, *Methods in Computational Physics* **2** (1963) 1.
3. V. R. Saunders, *Methods in Computational Molecular Physics* (1983).
4. Wulfman, *J. Chem. Phys.* **33** (1960) 1567.
5. I. Talmi, A. de-Shavit, *Nuclear Shell Theory.* (Academic,NY, 1963).
6. V. Vanagas, *Teor. i eksper. khem.* **3** (1967) 24 (in Russian).
7. M. Moshinsky, *Nucl. Phys.* **13** (1959) 104.
8. D. K. Maratis, *J. Chem. Phys.* **71** (1979) 917.
9. M. Dupuis, J. D. Watts, H. O. Villar, G. J. B. Hurst, *Comp. Phys. Commun.* **52** (1989) 415.
10. B. S. Aleksandrov, A. B. Bolotin, N. P. Pošiūnaitė, R. J. Rakauskas, V. K. Shugurov, *Multicenter integrals* (Vilnius, Vilnius University, 1981), p. 220, (in Russian).
11. G. Rayna, *REDUCE software for algebraic computation* (New York, Springer-Verlag, 1987), p. 329.

CONFORMATIONAL EFFECTS IN ACETYL METHOXYPHENOLS

Rimantas ŠADŽIUS, Juozas ŠULSKUS, Rimantas J. RAKAUSKAS
Dept. of Molecular Dynamics, Vilnius University, Saulėtekio 9
Vilnius, 2054 Lithuania

ABSTRACT

Using the semiempirical AM1 method, conformational analysis has been performed of 4-acetyl-2-methoxyphenol (**I**) and 5-acetyl-2-methoxyphenol (**II**), common formula $(CH_3CO)(CH_3O)C_6H_3OH$, which are simplified structures of corresponding (3-dimethylamino propionyl) methoxyphenols demonstrating different activity *in vivo*. Potential barriers of acetyl rotation and molecular dipole moments have also been estimated *ab initio* within several basis sets, demonstrating qualitative agreement between methods. A two-dimensional potential surface of simultaneous internal rotation of $-OH$ and $-OCH_3$ groups in 2-methoxyphenol (**III**) has been studied indicating a complicated character of the collective rotation of the groups. Planar arrangement of acetyl with respect to the benzene ring is the most stable conformation of both **I** and **II**, although two possible orientations of acetyl differ by 0.6 kcal/mol. Moreover, **I** is additionally stabilized by about 0.3 kcal/mol as compared with **II**.

1. Introduction

Aryl ketones are known as biologically active compounds with antiinflammatory features. Their activity has been shown to depend essentially on their conformation and electronic charge distribution. Planar arrangement of the vicinal carboxy group with respect to the aromatic skeleton seems to be one of significant properties determining the above activity[1].

We have performed semiempirical conformational analysis of 4-acetyl-2-methoxyphenol (**I**) and 5-acetyl-2-methoxyphenol (**II**), common formula $(CH_3CO)(CH_3O)C_6H_3OH$, which are simplified structures of corresponding (3-dimethylamino propionyl) methoxy phenols (see Figs. 1 and 2) demonstrating different activity *in vivo* despite being close isomers. Several significant points on the potential surface were verified using *ab initio* calculations.

2. Methods

The semiempirical AM1 method[2] was chosen for potential surface calculations since it proved to reproduce more reliably the internal rotation barriers when conjugated π-systems were involved[3]. The MOPAC program package was used[4].

Ab initio calculations of several geometrical configurations of the investigated

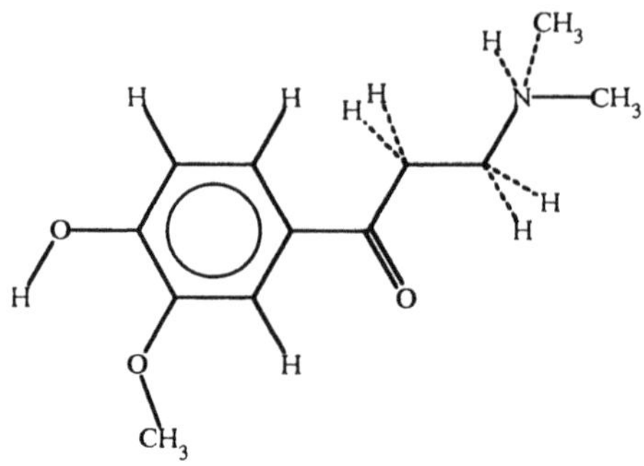

Fig. 1. An example of a biologically active antiinflammatory agent (3-dimethylamino propionyl) methoxy phenol. Bold lines depict the bonds for the most stable conformation that lie virtually in plane.

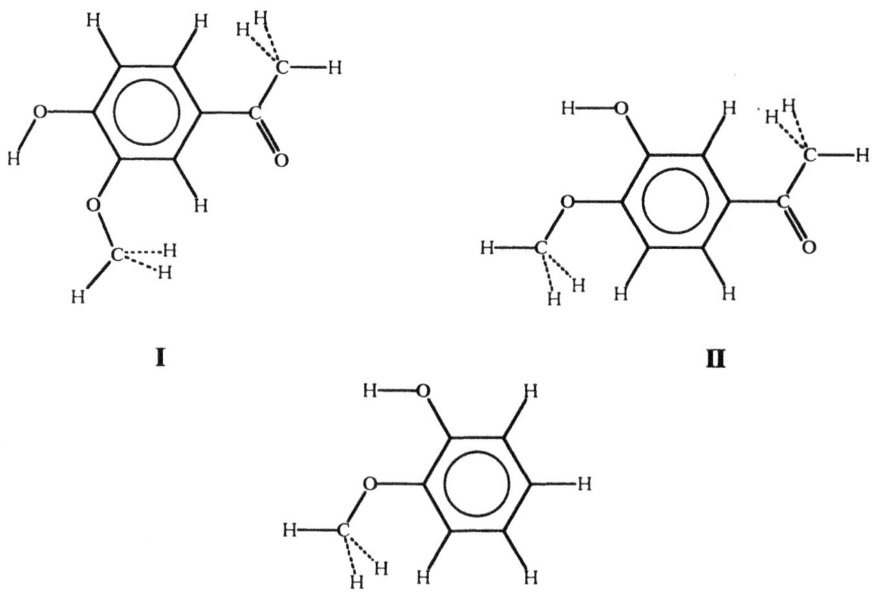

I

II

III

Fig. 2. The structure of (**I**) 4-acetyl-2-methoxy phenol, (**II**) 5-acetyl-2-methoxy phenol, and (**III**) 2-methoxy phenol. Solid lines denote the bonds lying virtually in the plane of the benzene ring. The conformations shown are the most stable ones corresponding to -OH and -OCH_3 torsion angles of (180°; 180°), i.e. hydroxyl looking in and methoxyl out.

molecules were carried out by the Hartree–Fock method using the program package HONDO7 [5], with shared–exponent Gaussian basis sets STO–3G, –4G [1*s*1*p*], and –4G [2*s*2*p*]). The latter two basis sets were used with *ab initio* effective core potentials (ECP) for atoms C and O [6].

3. Potential surface of internal rotation

A two-dimensional semiempirical potential surface was investigated describing simultaneous internal rotation of the adjacent –OH and $–OCH_3$ groups around the C(aryl)–O bonds in 2-methoxyphenol (**III**) (see Fig. 3). Corresponding dihedral angles were chosen as active coordinates. Zero values of the coordinates corresponded to a planar conformation (–OH out, $–OCH_3$ in, cf. Fig. 2), which was expected to be a global minimum for sterical reasons. The rest of the skeleton was maintained planar, though bond lengths and valence angles were fully optimized for each point.

Table 1 gives characterization of the stationary points on the surface.

Table 1. Characteristics of stationary points on the internal rotation potential surface of methoxyl $–OCH_3$ and hydroxyl –OH in **III**. Minima (L), maxima (H), and saddle points (S)

Stationary point	Energy, kcal/mol	With respect to global minimum	–OH rotation angle, deg	$–OCH_3$ rotation angle, deg
L1	-58.65	0.00	180	180
L2	-56.12	2.53	0	23
S1	-56.07	2.58	0	0
L3	-55.17	3.48	0	180
S2	-54.60	4.05	-5	120
S3	-54.46	4.19	-90	45
S4	-54.10	4.55	-70	175
H1	-53.25	5.40	-75	115
S5	-53.00	5.65	110	55
H2	-52.45	6.20	80	100
H3	-50.80	7.85	180	0

Note: positions and energies of the maxima and saddle points – an estimate.

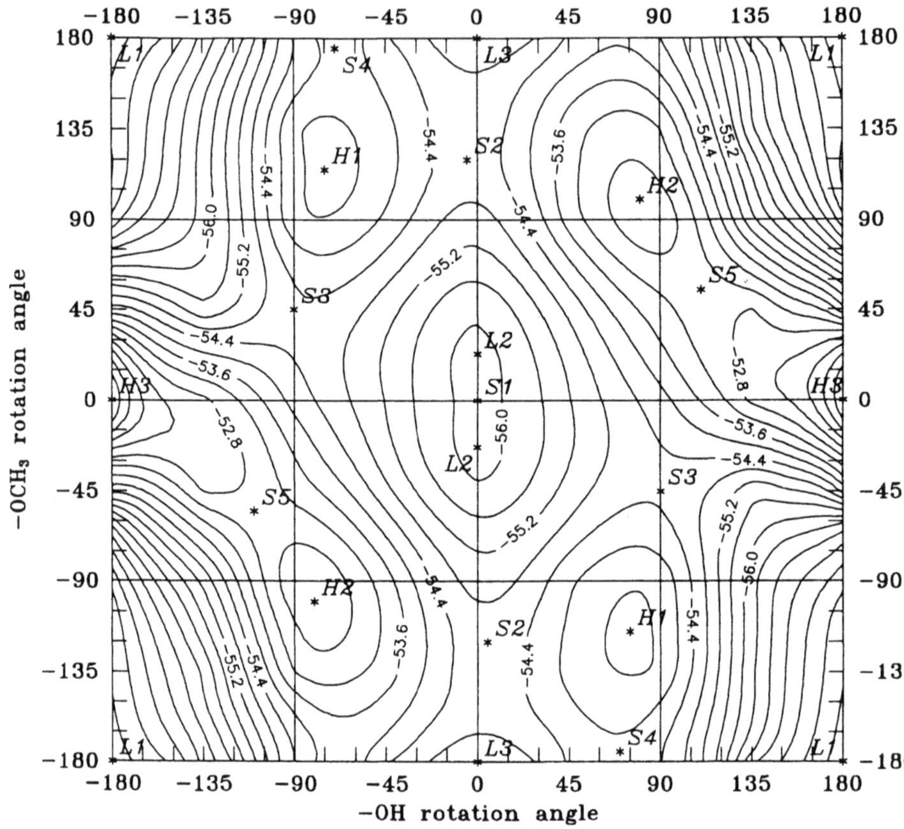

Fig. 3. The internal rotation potential surface of methoxyl $-OCH_3$ and hydroxyl -OH in **III**. Minima (*L*), maxima (*H*), and saddle points (*S*) are listed in Table 1.

Surprisingly, the global minimum of the potential is achieved when both groups are in plane with phenyl, but –OH looking to the inside and $-OCH_3$ to the outside. In the reverse arrangement (–OH out, $-OCH_3$ in), $-OCH_3$ moves in a shallow groove with two minima, shifted symmetrically by a dihedral ±23° from the planar position, with a small barrier (of about 0.05 kcal/mol) between them. Another shallow minimum corresponds to the out-out conformation, but it is separated by rather low barriers from the other minima. The global maximum corresponds to the planar crowded in-in conformation.

Some inequivalence between respective quadrants of the plot in Fig. 3 is caused by surface approximation errors.

4. *Ab initio* electronic structure results

Planar and perpendicular arrangement of the carboxyl group with respect to the benzene ring in **I** and **II** were investigated *ab initio*. No geometry optimization was performed. The calculations predict that the planar configuration is the most stable one for both molecules. The rotational barrier of carboxyl group determined using different basis sets is presented in Table 2. It usually increased with the extension of the basis set due to lowering of the total energy of the planar conformation.

The dipole moment of the conformation with the carboxyl perpendicular to the benzene ring (see Table 2) was comparatively stable with respect to the basis extension. Meanwhile, for the planar conformation the dipole moment was less by an order and therefore its direction was very sensitive to the choice of the basis set. For instance, in STO–3G and ECP [1*s*1*p*] bases, the directions were almost opposite.

Table 2. The *ab initio* rotational barrier of the carboxyl group in **I** and dipole moments for planar and perpendicular arrangement of this group obtained in various basis sets

Basis set	Rotational barrier, kcal/mol	Dipole moment μ (in D) for conformations:	
		perpendicular	planar
STO–3G	1.40	2.5084	0.3334
ECP [1s1p]	3.95	3.9593	0.2776
ECP [2s2p]	5.40	4.1601	0.4054

5. Discussion

The position of acetyl determines the overall distribution of bond lengths and atomic charges in the benzene ring, as well as particular energies of stationary points on the above potential surface. Planar arrangement of acetyl with respect to the benzene ring is the most stable conformation, although two possible orientations of carbonyl differ by 0.6 kcal/mol. Moreover, **I** is additionally stabilized by about 0.3 kcal/mol as compared with **II**. The results are qualitatively supported by *ab initio* calculations.

The dipole moment of the molecules (see Table 3) is most sensitive to internal rotation and differs significantly for the two isomers, **I** and **II**. The lowest value is obtained for the most stable conformation of the most stable isomer **I**. Interestingly, the antiinflamatory agent corresponding to that structure (Fig. 1) shows a greater biological activity.

However, from the dynamical point of view, the quantum-mechanical problem of two coupled rotators should be solved within the two-dimensional potential obtained. Only then one would have an answer to the question about stable rigid conformers, if any, or the observed molecular geometry averaged this or that way

Table 3. Internal rotation barriers (in kcal/mol) for the $-COCH_3$ group with respect to the benzene ring and the dipole moment (μ, in D) of separate conformations

Torsion angle	I					II		
	Abs.	*Rel.*	STO-3G	μ	μ (STO–3G)	*Abs.*	*Rel.*	μ, D
0°	-97.52	0.00	0.000	0.70	0.3334	-97.08	0.00	3.02
90°	-95.20	2.32	1.397	3.12	2.5084	-95.07	2.01	4.23
180°	-96.92	0.60	0.497	4.24		-96.50	0.58	5.40

Notes:

The reference for the torsion angle of the $-COCH_3$ group (first column) is the planar conformation shown in Fig. 2.

The energies and dipole moments were calculated by the semiempirical AM1 method with full geometry optimization.

Abs. denotes an absolute value of the heat of formation (in kcal/mol), and *Rel.* is the value relative to the global minimum.

The "STO–3G" columns give *ab initio* estimates for **I** within the STO–3G basis set at frozen geometry.

over nuclear motion. Despite a great difference in moments of inertia of the two rotating groups (–OH and $-OCH_3$), coupling between them is in this case too strong to permit ordinary variable separation or normal-mode approximation. Note that cross-section profiles of the potential surface corresponding to different constant values of one rotation angle differ even qualitatively from each other, e.g. by the number and types of extrema. Therefore, quite complex collective motion of the two functional groups should be supposed within the molecule.

6. References

1. P. Gaidelis, E. Udrėnaitė, L. Gumbaragytė, D. Kopilevich, D. Ramanauskas, and V. Daukšas, in *The II Baltic Meeting on Pharmacology and Clinical Pharmacology (Abstr.)* (1990, Tallinn), p. 37.
2. M.J.S. Dewar, E.G. Zoebish, E.F. Healy, and J.J.P. Stewart, *J. Am. Chem. Soc.* **107** (1985) 3902.
3. G. Buemi, F. Zuccarello, and A. Raudino, *J. Mol. Struct. (Theochem)* **164** (1988) 379.
4. J.J.P. Stewart, *MOPAC – A General Molecular Orbital Package. Ver. 3.10* (F.J. Seiler Res. Lab., U.S. Air Force Acad., Col. Spgs, CO, 1983).
5. M.Dupuis, J.D.Watts, H .O.Villar, and G.J.B.Hurst, *Comp. Phys. Commun.* **52** (1989) 415.
6. W.J.Steven, H.Basch, and M.Krauss, *J. Chem. Phys.* **82** (1985) 284.

Perturbation Theory as Dynamical Theory

V.I.Yukalov and E.P.Yukalova

Department of Mathematics and Statistics
Queen's University, Kingston
Ontario K7L 3N6, Canada

Abstract

A new method is presented for calculating the spectra of Hamiltonians in quantum mechanics. This method permits to treat perturbation theory from the point of view of dynamical theory. In this method a sequence of perturbative approximations is put into one - to - one correspondence with a trajectory of a dynamical system with discrete time, called the approximation cascade. Approximate fixed points of the latter define the self - similar approximations for an effective limit of a perturbative sequence. The stability of the method is controlled by calculating the mapping multipliers and Lyapunov exponents. As an illustration, the self - similar approximations for the spectra of several Hamiltonians are constructed.

1 Introduction

Perturbation theory is the main tool for treating realistic problems in quantum mechanics, statistical physics and field theory. However, practically always perturbation theory yields divergent series [1]. There are several numerical techniques helping to improve the behaviour of perturbation series. But quite often these techniques do not provide a remedy for treating divergences. Thus, there are many examples when perturbation series are not Padé summable or Borel summable [2], when the Hill - determinant method does not work [3], and when even rather complicated nonlinear transformations, like the Levin transformation, are not able to get out of divergences [4]. The momentum - method iterative approach, based on recurrence relations between different - order terms, is also divergent [5]. Another difficulty in dealing with the resummation techniques mentioned is the necessity to know a number of terms of perturbation series, usually about ten of them or more. These techniques are useless when one has in hands just a few first terms, which is the standard situation for the majority of realistic problems. Finally, the numerical techniques do not allow, or at least are not convenient for, an analytical investigation of results, especially in the case of several parameters or variables.

In order to obtain sufficiently simple analytical formulae providing good accuracy even being based on a couple of perturbative terms, a new approach has been recently advanced, called the method of self - similar approximations [6,7]. The foundation of this method has been based on the renormalization group theory [6-8]. As far as a renormalization group can be considered as a dynamical system, there should exist direct analogies between this method and dynamical theory. Moreover, the language of dynamical theory [9-11] can be not only very appropriate for finding such analogies, but also for providing firm grounds for some resummation tricks heuristically used in perturbation theory [12-14].

In the present paper we proffer the reformulation of the method of self - similar approximations to the language of dynamical theory. We exemplify the approach by calculating the energy levels of an anharmonic oscillator. A very attractive feature of our method, as we show, is that its higher - order approximations are practically as simple as the first one.

2 Scheme of method

Suppose we need to find a function $f(g)$ of a variable $g \in \mathbf{R}$. We are speaking about one function of one variable just for simplity. The generalization to the case of several functions of several variables is straightforward. Assume that the sought function satisfies an equation that cannot be solved exactly, and the use of the standard perturbation theory yields divergent series. The application of our approach to such problems consists of the following steps.

(1) *Renormalization by governing functions.*

Let us incorporate into perturbation theory a trial parameter z (or a set of trial parameters), the perturbative approximations $F_k(g,z)$, where $k \in \mathbf{Z}^+ \equiv \{0,1,2,\ldots\}$, becoming dependent on the latter. Then, in each $F_k(g,z)$ we substitute $z_k(g)$ instead of z so that the sequence $\{f_k(g)\}_{k=0}^{\infty}$ of terms

$$f_k(g) \equiv F_k(g, z_k(g)) \tag{1}$$

would be convergent,

$$\lim_{k\to\infty} f_k(g) = f(g).$$

Because of their role of governing the convergence, the functions $z_k(g)$ have been called the governing functions. This idea of renormalizing perturbation theory by introducing governing functions defined by some self - consistent conditions has been advanced first if ref. [15] and used for describing anharmonic crystals in several papers, e.g. in [16-18]. Later the renormalized perturbation theory has been developed by many authors and has been applied to various systems, acquiring the names of modified perturbation theory, optimized perturbation theory, variational perturbation theory, ε - expansion or δ - expansion [19-27]. The convergence of the renormalized perturbation theory for some models has also been demonstrated [20,28,29]. This theory, although being quite successful for many applications suffers two main deficiences. First, its accuracy is not, sometimes, sufficiently high. Second, the choice of governing functions in this theory is always heuristic. These two deficiences can be overcome in the method of self - similar approximations [6,7] for which the renormalized perturbation theory is only the first step.

(2) *Construction of approximation cascade.*

Define the coupling functions $g_k(f)$ by the equation

$$F_0(g, z_k(g)) = f; \qquad g = g_k(f) \tag{2}$$

and the inverse coupling function $g_k^{-1}(g)$ as

$$g_k^{-1}(g) = F_0(g, z_k(g)), \tag{3}$$

so that

$$g_k^{-1}(g_k(f)) = f, \qquad g_k(g_k^{-1}(g)) = g. \tag{4}$$

Introduce the notation

$$y_k(f) \equiv f_k(g_k(f)). \tag{5}$$

The set $\{z_k(g)\}$ of the governing functions can be called the government. Let us require for the government to satisfy the property of self - similarity

$$y_{k+p}(f) = y_k(y_p(f)). \tag{6}$$

The endomorphizm (5), according to (6), has the semi - group attributes: $y_k \cdot y_p = y_{k+p}$, $y_0 = 1$. Therefore, the semigroup $Y \equiv \{y_k\}$ forms a dynamical system with discrete time, which we call the approximation cascade. The trajectory $T(f) \equiv \{y_k(f)\}_{k=0}^{\infty}$ of this cascade is, by construction, bijective to the sequence $S(g) \equiv \{f_k(g)\}_{k=0}^{\infty}$ of perturbation approximants (1). Each point $y_k(f)$ of the cascade trajectory is in one - to -one correspondence with a term $f_k(g)$ of the perturbative sequence. The relation inverse to (5) reads

$$f_k(g) = y_k(g_k^{-1}(g)). \tag{7}$$

In this way, the attractive fixed point

$$y^*(f) = \lim_{k\to\infty} y_k(f) \tag{8}$$

of the approximation cascade corresponds to the sought function $f(g)$.

(3) *Embedding of cascade into flow.*

To deal with dynamical systems with continuous time is easier than with those with discrete time. Therefore, let us embed the approximation cascade into a flow. For a variable $t \in \mathbf{R}^+ \equiv [0, \infty)$ we define a function $y(t, f)$ satisfying the self - similar relation

$$y(t + t', f) = y(t, y(t', f)) \tag{9}$$

and the embedding property

$$y(k, f) \equiv y_k(f) \qquad (k \in \mathbf{Z}^+). \tag{10}$$

The semigroup $\bar{Y} \equiv \{y(t, \cdot)\}$ of endomorphisms $y(t, \cdot)$ is what we call the approximation flow. By this construction, the approximation cascade Y is embedded into the approximation flow $\bar{Y}$, whose trajectory $\bar{T}(f) \equiv \{y(t, f)\}$ contains the whole cascade trajectory:

$$Y \subset \bar{Y}, \qquad T(f) \subset \bar{T}(f).$$

Differentiating (9), we can write for the approximation flow the evolution equation

$$\frac{d}{df} y(t, f) = v(y(t, f)), \tag{11}$$

in which $v(y)$ is the flow velocity. Integrating (11) between $t = k$ and $t = k + 1$ we obtain the evolution integral

$$\int_{y_k(f)}^{y_{k+1}(f)} \frac{dy}{v(y)} = 1. \tag{12}$$

(4) *Definition of quasifixed points.*

To intergrate (12), we have to write down the cascade velocity, which can be done by the Euler discretization of (11). This yields

$$v_k(f) \equiv F_{k+1}(g_k, z_k) - F_k(g_k, z_k) + (z_{k+1} - z_k)\frac{\partial}{\partial z_k} F_k(g, z_k), \tag{13}$$

where $g_k = g_k(f)$ and $z_k = z_k(g_k(f))$. A fixed point would correspond to $v_k(f) = 0$, which would give us only a relation between F_{k+1}, F_k, z_{k+1} and z_k . To define these functions explicitly, we assume that not the total velocity is zero but only its part, namely that the following quasifixed - point condition holds:

$$(z_{k+1} - z_k)\frac{\partial}{\partial z_k} F_k(g_k, z_k) = 0. \tag{14}$$

Then the cascade velocity at the quasifixed point $y_k^*(f)$ is $v_k^*(y_k^*)$, where

$$v_k^*(f) = F_{k+1}(g_k, z_k) - F_k(g_k, z_k). \tag{15}$$

The quasifixed point itself can be defined from the evolution integral

$$\int_{y_k(f)}^{y_k^*(f)} \frac{dy}{v_k^*(y)} = 1 \qquad (k \geq 1). \tag{16}$$

The zero - order quasifixed point is, by definition, $y_0^*(f) \equiv y_1(f)$. Using (7), we can easily convert (16) into

$$\int_{f_k(g)}^{f_k^*(g)} \frac{df}{v_k^*(f)} = 1 \qquad (k \geq 1) \tag{17}$$

with the assumption that $f_0^*(g) \equiv f_1(g)$. The quantities $f_k^*(g)$ are just what is called [6,7] the self - similar approximations for the sought function $f(g)$.

(5) *Analysis of stability conditions.*

The method of self - similar approximations gives us a unique opportunity, absent in perturbation theory, to check its stability at each step [7]. To this end, we define the mapping multipliers $\mu_k(f)$ and their images $M_k(g)$ by the expressions

$$\mu_k(f) \equiv \lim_{x \to y_k^*(f)} \left| \frac{\partial}{\partial x} y_k(x) \right| ,$$

$$M_k(g) \equiv \mu_k(g_k^{-1}(g)). \tag{18}$$

Also, we introduce the local Lyapunov exponents

$$\lambda_k(f) \equiv \lim_{x \to y_k^*(f)} \frac{\partial}{\partial x} v_k^*(x),$$

$$\Lambda_k(g) \equiv \lambda_k(g_k^{-1}(g)), \tag{19}$$

which are analogous to the local Lyapunov exponents used in dynamical theory [30-33]. We shall say that the method is locally stable if

$$\mu_k(f) < 1, \qquad \lambda_k(f) < 0,$$

$$M_k(g) < 1, \qquad \Lambda_k(g) < 0. \tag{20}$$

Thus, the scheme is completed.

3 Illustration of method

To exemplify our approach, we shall calculate the spectrum of a quantum - mechanical problem. It is worth mentioning that some ideas of dynamical theory [34,35] and renormalization - group theory [36,37] have been exploited for calculating quantum - mechanical energies, although in a way very different of ours.

As an example, it is instructive to take the touchestone model of an anharmonic oscillator with the Hamiltonian

$$H = -\frac{1}{2}\frac{d^2}{dx^2} + \frac{1}{2}x^2 + gx^4. \tag{21}$$

The trial parameter z can be incorporated into the initial Hamiltonian

$$H_0 = -\frac{1}{2}\frac{d^2}{dx^2} + \frac{1}{2}z^2x^2. \tag{22}$$

Following the scheme of the previous section we find for the spectrum of (21) the sequence of self - similar approximations

$$e_k^*(n,g) = \left(n + \frac{1}{2}\right) f_k^*(n,g), \tag{23}$$

where

$$f_k^*(n,g) = \sqrt{1 + u_k^*(n,g)} \tag{24}$$

and u_k^* satisfies the equation

$$u_k^* = u_k \exp\left\{\sum_{p=0}^{k-1} \frac{C_k^p}{k-p}\left[\frac{1}{(u_k^*)^{k-p}} - \frac{1}{u_k^{k-p}}\right] + 2A_n\right\}, \tag{25}$$

in which

$$u_k = u_k(n,g) = f_k^2(n,g) - 1,$$

$$f_k(n,g) = \frac{e_k(n,g)}{n+1/2},$$

and $A_k = A_k(n)$. Because of the lack of space we will not write down A_k explicitly, but notice that

$$A_k \approx -\frac{1}{2}(0.1)^k, \qquad (k \to \infty).$$

To check the accuracy of the self - similar approximations (23), we have calculated the errors

$$\varepsilon_k^*(n,g) \equiv \left[\frac{e_k^*(n,g)}{e(n,g)} - 1\right] \cdot 100\% \tag{26}$$

as compared to the numerical results [38] given by a numerical solution of the Schrödinger equation. If we omit the levels with $n = 2,3$ and 4, then $\varepsilon_0^* \leq 2\%$, $\varepsilon_1^* \leq 0.4\%$, $\varepsilon_2^* \leq 0.2\%$ and $\varepsilon_3^* \leq 0.08\%$. An occasional jump of the error at some few levels is, probably, due to a local instability of the approximation cascade.

To analise the stability, we need to find the mapping multipliers (18) and Lyapunov exponents (19). For all mapping multipliers we have

$$M_k(n,g) \leq 1 \qquad (n \in \mathbf{Z}^+,\ g \in \mathbf{R}^+), \tag{27}$$

so that the stability condition (20) holds true. For the Lyapunov exponents (19) we get

$$\lambda_k(f) = \lim_{\xi \to y_k^*(f)} A_k \left(1 + \frac{2k+1}{\xi^2}\right)\left(1 - \frac{1}{\xi^2}\right)^k \tag{28}$$

and

$$\Lambda_k(n,g) = A_k(n)\left[1 + \frac{2k+1}{1+u_k^*(n,g)}\right]\left[1 - \frac{1}{1+u_k^*(n,g)}\right]^k . \tag{29}$$

As far as $u_k^*(n,g) > 0$ for all $g > 0$ and any k and n, the sign of (29) is defined by that of A_k,

$$sign\Lambda_k(n,g) = signA_k(n). \tag{30}$$

For $A_k(n)$ we have the following inequalities:

$$-\frac{1}{48} \leq A_1(n) \leq \frac{1}{144},$$

$$-0.002496 \leq A_2(n) \leq 0.003001,$$

$$-0.000525 \leq A_3(n) \leq -0.000439.$$

These inequalities show that $\Lambda_k(n,g)$ for $k = 1,2$ become positive for some energy levels. The positiveness of the local Lyapunov exponents signifies the occurence of local chaos. However, this happens only at two first steps, at the third step the motion stabilizes since $\Lambda_3(n,g) < 0$ for all n and g. To make an estimate for Λ_k as k increases, let us use the notation

$$\sigma \equiv \lim_{k\to\infty}\left(1 - \frac{1}{1+u_k^*}\right), \qquad 0 \leq \sigma \leq 1,$$

whose limits are defined by the inequality $0 \leq u_k^* < \infty$. Then (29) yields

$$\Lambda_k \simeq 2A_k(1-\sigma)\sigma^k \to 0 \qquad (k \to \infty)$$

with $A_k \approx -\frac{1}{2}(0.1)^k$.

In conclusion, let us summarize the main results. We have shown that the language of dynamical theory provides an elegant formulation for the method of self - similar approximations. Moreover, this formulation is not solely aesthetically attractive but gives a prescription for calculating higher - order terms of self - similar approximations and allows to check the stability of the method. For the spectrum of an anharmonic oscillator we have calculated the self - similar approximations up to third order. As we have discussed earlier [39,40] the precision of our method is better than that of other known analytical methods involving the same number of perturbative terms.

References

[1] B.Simon, Bull. Am. Math. Soc. **24**, 303 (1991).

[2] A.Bray, T.McCarthy, M.Moore, J.Reger, and A.Young, Phys. Rev. B **36**, 2212 (1987).

[3] M.Tater, J.Phys. A **20**, 2438 (1987).

[4] E.Weniger, J.Cizek and F.Vinette, J. Math. Phys. **34**, 571 (1993).

[5] F.M.Fernandez and J.F.Ogilvie, Phys. Lett. A **178**, 11 (1993).

[6] V.I.Yukalov, J. Math. Phys. **32**, 1235 (1991).

[7] V.I.Yukalov, J.Math. Phys. **33**, 3994 (1992).

[8] V.I.Yukalov, Proc. Lebedev Phys. Inst. **188**, 297 (1991).

[9] P.R.Halmos, Lectures on Ergodic Theory (Chelsea, New York, 1956).

[10] J.Eckmann and D.Ruelle, Rev. Mod. Phys. **57**, 617 (1985).

[11] D.Ruelle, Chaotic Evolution and Strange Attractors (Cambridge University, Cambridge, 1989).

[12] V.I.Yukalov and E.P.Yukalova, Int. J. Mod. Phys. B **7**, 2367 (1993).

[13] V.I.Yukalov and E.P.Yukalova, Nuovo Cimento B **108**, 1017 (1993).

[14] V.I.Yukalov and E.P.Yukalova, in: Nonlinear Evolution Equations and Dynamical Systems, eds. V.Makhankov, I.Puzynin and O.Pashaev (World Scientific, Singapore, 1993), p.18.

[15] V.I.Yukalov, Mosc. Univ. Phys. Bull. **31**, 10 (1976).

[16] V.I.Yukalov, Theor. Math. Phys. **28**, 652 (1976).

[17] V.I.Yukalov, Physica A **89**, 363 (1977).

[18] V.I.Yukalov, Ann. Physik **36**, 31 (1979).

[19] W.E.Caswell, Ann. Phys. **123**, 153 (1979).

[20] I.Halliday and P.Suranyi, Phys. Rev. D **21**, 1529 (1980).

[21] P.M.Stevenson, Phys. Rev. D **23**, 2916 (1981).

[22] P.M.Stevenson, Phys. Rev. D **24**, 1622 (1981).

[23] J.Killingbeck, J. Phys. A **14**, 1005 (1981).

[24] J.O.Akeyo and H.F.Jones, Imperial College Preprint **TP-91-92-27** (London, 1992).

[25] J.O.Akeyo and H.F.Jones, Imperial College Preprint **TP-92-93-14** (London, 1993).

[26] J.H.Field, Ann. Phys. **226**, 209 (1993).

[27] A.Okopinska, Ann. Phys. **228**, 19 (1993).

[28] I.Buckley, A.Duncan and H.F.Jones, Phys. Rev. D **47**, 2554 (1993).

[29] A.Duncan and H.F.Jones, Phys. Rev. D **47**, 2560 (1993).

[30] H.Fujisaka, Prog. Theor. Phys. **70**, 1264 (1983).

[31] H.Shibata, Physica A **196**, 93 (1993).

[32] H.Shibata and R.Ishizaki, Physica A **197**, 130 (1993).

[33] B.Eckhardt and D.Yao, Physica D **65**, 100 (1993).

[34] R.L.Hall, J. Phys. A **19**, 2079 (1986).

[35] R.L.Hall, J. Math. Phys. **28**, 457 (1987).

[36] E.Papp, Phys. Lett. A **178**, 231 (1993).

[37] E.Papp, Phys. Rev. A **48**, 4091 (1993).

[38] F.T.Hioe, D. MacMillen and E.W.Montroll, Phys. Rep. **43**, 305 (1978).

[39] E.P.Yukalova and V.I.Yukalov, J. Phys. A **26**, 2011 (1993).

[40] E.P.Yukalova and V.I.Yukalov, Phys. Lett. A **175**, 27 (1993).

Time-Advance Algorithms based on Hamilton's Principle

H. RALPH LEWIS and PETER J. KOSTELEC
Dartmouth College
Department of Physics and Astronomy
Hanover, New Hampshire 03755-3528, USA

Abstract

Hamilton's principle is applied to derive a class of numerical algorithms for systems of ordinary differential equations when the equations are derivable from a Lagrangian. This is an important extension into the time domain of an earlier use of Hamilton's principle to derive algorithms for the spatial operators in Maxwell's equations. In that work, given a set of expansion functions for spatial dependences, the Vlasov-Maxwell equations were replaced by a system of ordinary differential equations in time; but the question of solving the ordinary differential equations was not addressed. Advantageous properties of the new time-advance algorithms have been identified analytically and by numerical comparison with other methods, such as Runge-Kutta and symplectic algorithms. This approach to time advance can be extended to include partial differential equations and the Vlasov-Maxwell equations. An important objective of the research is to develop improved numerical methods for application to equations that arise in connection with plasma problems. One interesting issue that can be studied is whether a simulation completely based on Hamilton's principle can be used to obtain a convergent computation of average properties, such as the electric energy, even when the underlying particle motion is characterized by sensitive dependence on initial conditions.

1. Introduction

In the description of plasmas and other systems, differential and integro-differential equations frequently arise which are derivable from a Lagrangian. Examples are the equations of motion of a charged particle in a time-dependent electromagnetic field; the Maxwell equations; the Vlasov-Maxwell system of equations for a collisionless plasma; wave equations, including a large class of nonlinear wave equations. Other examples are various models of competition or conflict, such as the time-dependent Lotka-Volterra equations that describe a predator-prey system and also find application in areas such as mode coupling in plasma physics.[1] Thus, development of algorithms that provide reliable numerical solution of Lagrangian systems of equations is of considerable practical importance. In addition to requirements of accuracy, numerical stability, and speed, there is the need to capture the important qualitative features of the solutions, for example, the phase-space topology of solutions of particle equations of motion. In a many-body problem, there is the need to determine average quantities accurately, even though most details may be unimportant. For example, in the one-dimensional Vlasov-Poisson system, a quantity of particular interest is the energy of the electric field, which is an average quantity determined by the detailed motion of the many plasma particles. Usually, numerical algorithms are based directly on approximations of the governing equations. However, in the case of Lagrangian systems of equations (or their Hamiltonian equivalents), Hamilton's variational principle can be used as the basis for deriving numerical algorithms. This has the advantage that the numerical solutions are then based on a global principle that uniquely determines the evolution of the system. It is in contrast to attempting to mimic or preserve some particular properties of the system such as, for example, conservation of the Hamiltonian for an autonomous system, or preservation of the symplectic nature of the mapping from initial data to final data.

Hamilton's principle was introduced for deriving approximation schemes for the Vlasov-Maxwell system in 1968.[2] Stated briefly, given a set of expansion functions for representing the spatial dependence of the electromagnetic potentials, and given a set of expansion functions for representing the dependence of the particle variables on the phase-space of their initial conditions, a system of ordinary differential equations in time was derived whose solution determines an approximation to the solution of the exact Vlasov-Maxwell system. Specialization to a collection of point particles gave particle-in-cell simulation schemes. For example, with the Vlasov-Poisson system in two cartesian spatial dimensions, if the expansion functions for the scalar potential are linear splines ("tent functions"), then this application of Hamilton's principle leads directly to the standard particle-in-cell (PIC) scheme with "area weighting."

It is well known that time-advance algorithms are very relevant to accurate plasma simulation. Indeed, Lewis, Barnes and Melendez[3] raised the question of whether a new type of time-advance algorithm could fundamentally improve the quality of plasma simulations. The original application of Hamilton's principle to plasma simulation did not address this question; no time-advance algorithms were derived. Nevertheless, there was some belief that practical value would accrue to having time-advance algorithms based on Hamilton's principle. For example, problems with numerical stability might be avoided, or it might be possible to reduce the effects of so-called numerical collisions. Until recently, there has been little progress toward obtaining such algorithms. Godfrey derived an algorithm to which he referred in an unpublished conference proceedings.[4] Although he found the algorithm advantageous regarding accuracy and stability, he also found it too elaborate and costly for practical use. In a later unpublished conference proceedings,[5] he referred to a practicable approximate implementation of the algorithm that was sufficiently accurate for a two-dimensional PIC plasma simulation code called CCUBE. Details of the algorithms were not presented in these proceedings and the work appears not to have been pursued to examine the merits and disadvantages of time-advance algorithms based on Hamilton's principle. Eastwood has described a "virtual-particle" space-time algorithm for electromagnetic plasma simulation[6] whose derivation relies on Hamilton's principle. However, his derivation of the time-advance algorithm is inadequate because the conditions that render Hamilton's principle equivalent to specifying a particular solution of Hamilton's equations are not addressed. The work presented in this paper represents an initial attempt to determine the properties of time-advance algorithms based on Hamilton's principle and to examine their merits and disadvantages in comparison to other approaches. The scope of the present work is limited to ordinary differential equations.

The spirit of an initial-value problem is that of extrapolation: Initial data are to be *projected* to a later time according to the action of governing equations. However, the spirit of Hamilton's principle is that of interpolation: The solution *between* an initial time and a final time is to be determined from data at both the initial and the final times according to the action of governing equations. The key to applying Hamilton's principle to derive time-advance algorithms is realizing how to bring the extrapolatory nature of the problem into consonance with the interpolatory nature of the principle. There are two steps in using Hamilton's principle to derive a time-advance algorithm. First, a class of functions with undetermined parameters is chosen for approximating the solution of the equations over a specific time interval. For example, the approximating functions could be expansions in basis functions that are polynomials in time. This step, in one form or another, is part of any derivation of a time-advance algorithm. Second, Hamilton's principle is applied ***exactly*** within the class of approximating functions to obtain the conditions that the parameters must satisfy. Those conditions involve the unknown value of the solution at the final time. However, the value of the solution at the final time can be expressed in terms of the initial data and the undetermined parameters. Therefore, the conditions on the undetermined parameters can be expressed completely in terms of the initial data. For a given class of functions used to represent approximate solutions, Hamilton's principle specifies a *unique* algorithm.

Using Hamilton's principle in a restricted function space (the space of approximating functions) is different than approximating the Lagrange or Hamilton equations of motion. If the function space were not restricted, then using Hamilton's principle would be equivalent to solving the exact equations of motion. However, if the function space be restricted to a set of approximating functions (for example, polynomials in time of some degree), then one may expect the approximation achieved with Hamilton's principle to be different than that obtained by applying some particular approximate method to the equations of motion directly. In fact, the algorithms based on Hamilton's principle are equivalent to a Galerkin approximation of the differential equations. Hamilton's principle determines a particular weight function for the Galerkin method and constrains the function space in which the Galerkin's method is applied. The issue of which algorithms are best is complex, and it can only be resolved by detailed numerical and analytical comparison and contrast of various methods with one another.

The following progress toward developing and understanding algorithms based on Hamilton's principle for systems of ordinary differential equations has been achieved:

- A general formulation has been developed for initial-value problems.

- Algorithms for general Lagrangians have been defined in terms of polynomial approximating functions.

- The behavior of the algorithms for small values of the timestep has been determined. It has been shown that the algorithms have a *unique* solution in the limit of zero timestep. As a result, they can be implemented in a numerically stable manner and they can be made explicit to any order in the timestep.

- A general theorem has been demonstrated, which is an analog of the relation $\frac{dH}{dt} = \frac{\partial H}{\partial t}$ that holds for all Hamiltonian systems. Finding this theorem is part of our general program of discovering the properties of algorithms based on Hamilton's principle. If Hamilton's principle can be the starting point for deriving algorithms that are optimal in a useful sense, then one would suppose that there would be analogs for those algorithms of every general property of exact Hamiltonian systems.

• Analytical and numerical comparisons of algorithms based on Hamilton's principle have been carried out with some other algorithms. Among the algorithms used for comparison were symplectic algorithms. Symplectic algorithms mirror Hamilton's equations in that the mapping of generalized coordinates and their conjugate momenta from one time to a timestep later is canonical, as is the case for the exact solution of Hamilton's equations. That is an appealing property, although the question of whether it is the best property to strive for is open. After all, the main objective is to approximate a *particular solution* of the equations; achieving an approximation that represents a canonical mapping as precisely as possible may or may not be valuable in the context of a particular problem. The comparisons of symplectic algorithms with algorithms based on Hamilton's principle indicated that the properties of the two kinds of algorithms are similar, but that the algorithms based on Hamilton's principle often approximate exact solutions better. It might have been supposed that algorithms based on Hamilton's principle would also be symplectic; that is generally not the case. However, even when the algorithm based on Hamilton's principle *was* symplectic, its overall performance, in comparison to that of a symplectic algorithm of the same order that was not derived from Hamilton's principle, was generally better at approximating the exact solution.

In Sec. 2 we describe the formulation of the use of Hamilton's principle for deriving algorithms for Lagrangian systems of ordinary differential equations. Some numerical results and comparisons with other methods are given in Sec. 3.

2. The Method

a) *theoretical background*

Consider a system of ordinary differential equations which can be derived from a Lagrangian associated with N degrees of freedom. Denote the Lagrangian by $L(q,\dot{q},t)$, where the arguments q and $\dot{q}$, respectively, are sets of generalized coordinates and generalized velocities $q^{(i)}$ and $\dot{q}^{(i)}$, $1 \le i \le N$. The generalized coordinates satisfy the Lagrange equations

$$\frac{d}{dt}L_{\dot{q}^{(i)}}(q,\dot{q},t) = L_{q^{(i)}}(q,\dot{q},t) \qquad (1 \le i \le N)\ , \tag{2.1}$$

which comprise a system of N 2nd-order equations in time. [The subscript notation is used to denote partial differentiation with respect to the arguments of $L(q,\dot{q},t)$].

A particular solution of Lagrange's equations can be specified in many ways. As an example, for an *initial-value problem,* the solution is required to satisfy $2N$ independent conditions at the initial time. Typical initial data that uniquely specify a particular solution are

$$q^{(i)}(t_1) = q_1^{(i)}, \qquad \dot{q}^{(i)}(t_1) = \dot{q}_1^{(i)} . \tag{2.2}$$

The particular solution specified by the data given in (2.2) could, for example, also be specified by the values of that particular solution at two distinct times, say t_1 and t_2,

$$q^{(i)}(t_1) = q_1^{(i)}, \qquad q^{(i)}(t_2) = q_2^{(i)} . \tag{2.3}$$

There is a one-to-one correspondence between the values $(q_1, \dot{q}_1)$ and the values (q_1, q_2) which plays an important role in our application of Hamilton's principle. This can be expressed most clearly by explicitly indicating the dependence of the solution on either $(q_1, \dot{q}_1)$ or (q_1, q_2):

$$q^{(i)}(t) \;\rightarrow\; f^{(i)}(t; q_1,\dot{q}_1) \qquad \text{or} \qquad q^{(i)}(t) \;\rightarrow\; g^{(i)}(t; q_1,q_2) . \tag{2.4}$$

Then, $q_2^{(i)}$ is expressed in terms of $(q_1, \dot{q}_1)$ by

$$q_2^{(i)} = f^{(i)}(t_2; q_1,\dot{q}_1) ; \tag{2.5a}$$

and $\dot{q}_1^{(i)}$ is expressed in terms of (q_1, q_2) by

$$\dot{q}_1^{(i)} = \frac{d}{dt} g^{(i)}(t; q_1,q_2)\bigg|_{t=t_1} . \tag{2.5b}$$

Lagrange's equations (2.1), supplemented by data that specify a unique solution [for example, (2.2) or (2.3)], are necessary and sufficient for the validity of *Hamilton's principle.* This principle concerns the so-called *action integral,* $\int_{t_1}^{t_2} L[q(t),\dot{q}(t),t]dt$, where here $q(t)$ denotes an arbitrary function which assumes values (q_1, q_2) at times (t_1,t_2). As discussed above, these values specify a unique solution of Lagrange's equations. Let the symbol δ denote functional variations of $q(t)$. Because $q(t)$ assumes values (q_1, q_2) at times (t_1, t_2), the variations satisfy

$$\delta q^{(i)}(t_1) = \delta q^{(i)}(t_2) = 0 \ . \tag{2.6}$$

Hamilton's principle states that

$$0 = \delta \int_{t_1}^{t_2} L[q(t), \dot{q}(t), t] dt \tag{2.7}$$

will be satisfied *if and only if* $q(t)$ is the particular solution of Lagrange's equations that is uniquely specified by (q_1, q_2). Requiring the variational condition (2.7) is equivalent to specifying a particular solution of Lagrange's equations.

b) *the approximation technique*

We employ two steps in order to derive a numerical algorithm from Hamilton's principle.

Step One: Choose a restricted function space for approximating $q(t)$, where the elements of the function space satisfy the constraints (2.3).

Step Two: Apply Hamilton's principle ***exactly*** within the restriced function space for $q(t)$. In conjunction with Step One, this use of Hamilton's principle defines a unique numerical algorithm.

The **only difference** between this derivation of a numerical algorithm and the derivation of the exact Lagrange equations is that the functional variations satisfying constraints (2.6) are carried out within a restricted function space in the case of the numerical algorithm and within an unrestricted function space in the case of the exact differential equations.

Denote the approximating function for $q^{(i)}(t)$ by $Q^{(i)}(t, \alpha; q_1, q_2)$, where $Q^{(i)}(t, \alpha; q_1, q_2)$ is a *prescribed* function of t, q_1, q_2, and a set of $M_q^{(i)}$ independent parameters $\alpha_j^{(i)}$. For all values of the parameters $\alpha_j^{(i)}$, in analogy to (2.3), we require that the approximating functions $Q^{(i)}(t, \alpha; q_1, q_2)$ satisfy the same constraints as the functions $q^{(i)}(t)$:

$$Q^{(i)}(t_1,\alpha;q_1,q_2)=q_1^{(i)}\,, \qquad Q^{(i)}(t_2,\alpha;q_1,q_2)=q_2^{(i)} \tag{2.8}$$

for all values of the parameters $\alpha_j^{(i)}$. The parameters $\alpha_j^{(i)}$ are to be varied when applying Hamilton's principle; the parameters (q_1, q_2) are not to be varied. In analogy to (2.6), the variations of $Q^{(i)}(t,\alpha;q_1,q_2)$ satisfy

$$\delta Q^{(i)}(t_1,\alpha;q_1,q_2)=\sum_{j=2}^{M_q^{(i)}+1} Q^{(i)}_{\alpha_j^{(i)}}(t_1,\alpha;q_1,q_2)\delta\alpha_j^{(i)}=0\,, \tag{2.9a}$$

$$\delta Q^{(i)}(t_2,\alpha;q_1,q_2)=\sum_{j=2}^{M_q^{(i)}+1} Q^{(i)}_{\alpha_j^{(i)}}(t_2,\alpha;q_1,q_2)\delta\alpha_j^{(i)}=0\,. \tag{2.9b}$$

The parameters $\alpha_j^{(i)}$ are to be determined by exact satisfaction of Hamilton's principle within the restricted function space represented by the functions $Q^{(i)}(t,\alpha;q_1,q_2)$. That is,

$$0=\delta\int_{t_1}^{t_2} L(Q,\dot{Q},t)dt$$

$$=\int_{t_1}^{t_2}\sum_{i=1}^{N}\left\{L_{q^{(i)}}(Q,\dot{Q},t)\sum_{j=2}^{M_q^{(i)}+1} Q^{(i)}_{\alpha_j^{(i)}}(t,\alpha;q_1,q_2)\delta\alpha_j^{(i)}+L_{\dot{q}^{(i)}}(Q,\dot{Q},t)\sum_{j=2}^{M_q^{(i)}+1}\dot{Q}^{(i)}_{\alpha_j^{(i)}}(t,\alpha;q_1,q_2)\delta\alpha_j^{(i)}\right\}dt$$

$$=\int_{t_1}^{t_2}\sum_{i=1}^{N}\left\{L_{q^{(i)}}(Q,\dot{Q},t)-\left[\frac{d}{dt}L_{\dot{q}^{(i)}}(Q,\dot{Q},t)\right]\right\}\sum_{j=2}^{M_q^{(i)}+1} Q^{(i)}_{\alpha_j^{(i)}}(t,\alpha;q_1,q_2)\delta\alpha_j^{(i)}dt\,. \tag{2.10}$$

Because the parameters $\alpha_j^{(i)}$ are independent, (2.10) is equivalent to the system of equations

$$0=\int_{t_1}^{t_2}\left\{L_{q^{(i)}}(Q,\dot{Q},t)-\left[\frac{d}{dt}L_{\dot{q}^{(i)}}(Q,\dot{Q},t)\right]\right\}Q^{(i)}_{\alpha_j^{(i)}}(t,\alpha;q_1,q_2)dt\,,$$

$$\text{for}\quad 1\le i\le N\,,\quad 2\le j\le M_q^{(i)}+1\,. \tag{2.11}$$

The functions $Q^{(i)}(t,\alpha;q_1,q_2)$ have been introduced as approximating functions. However, it should be noted that (2.11) applies to the exact solution of Lagrange's equations if, for each i, the functions $Q^{(i)}(t,\alpha;q_1,q_2)$ are complete in the space of functions whose elements satisfy (2.3). In that case, the system of equations (2.11) implies that Lagrange's equations are satisfied exactly on an arbitrary open interval (t_1,t_2) .

Eq. (2.11) is a fundamental equation. However, as written, it determines the parameters α in terms of the quantities (q_1, q_2). Although the quantities q_2 are not specified in advance for an initial-value problem, we need the parameters α in terms of the quantities that *are* specified in advance. In order to achieve this, we modify (2.11) by replacing the quantities q_2 with their expressions in terms of the parameters α and the quantities that are specified in advance. So modified, (2.11) will depend only on the parameters α and the quantities that are specified in advance; the solution will give α in terms of the quantities that are specified in advance.

With initial data q_1 and $\dot{q}_1$, the quantities q_2 are expressed in terms of the parameters α and the initial data by solving the analogue of (2.5b) in the restricted function space:

$$\dot{q}_1^{(i)} = \dot{Q}^{(i)}(t_1,\alpha;q_1,q_2)\,, \qquad 1 \le i \le N\,, \tag{2.12}$$

where

$$\dot{Q}^{(i)}(t,\alpha;q_1,q_2) \equiv \frac{\partial}{\partial t} Q^{(i)}(t,\alpha;q_1,q_2)\,. \tag{2.13}$$

c) <u>*expansion in basis functions*</u>

At this point the condition (2.11) is completely general. The times t_1 and t_2 are arbitrary and the approximating functions $Q^{(i)}(t,\alpha;q_1,q_2)$ are arbitrary. For particular problems, various sophisticated choices would undoubtedly be useful. However, for the development of general-purpose algorithms, we take $Q^{(i)}(t,\alpha;q_1,q_2)$ to be a polynomial in time t:

$$Q^{(i)}(t,\alpha;q_1,q_2) = q_1^{(i)}(1-\tau) + q_2^{(i)}\tau + \sum_{j=2}^{M_q^{(i)}+1} \alpha_j^{(i)} \chi_j(\tau)\,, \tag{2.14}$$

where

$$\tau \equiv \frac{(t-t_1)}{\Delta}\,, \quad \Delta \equiv (t_2 - t_1)\,, \tag{2.15}$$

$$\chi_j(0) = \chi_j(1) = 0\,, \tag{2.16}$$

$$\int_0^1 \chi_i(\tau)\chi_j(\tau)\, d\tau = \delta_{ij}\,. \tag{2.17}$$

Then (2.11) for Lagrange's equations becomes

$$0 = \int_0^1 \left\{ L_{q^{(i)}}(Q,\dot{Q},t) - \left[\frac{d}{dt} L_{\dot{q}^{(i)}}(Q,\dot{Q},t)\right]\right\} \chi_j(\tau) d\tau\,,$$

$$\text{for} \quad 1 \le i \le N\,, \quad 2 \le j \le M_q^{(i)} + 1\,. \tag{2.18}$$

Eq. (2.18) represents a Galerkin approximation of (2.1) in the linear vector space defined by the functions $\chi_j(\tau)$, which vanish at the endpoints $\tau = 0$ and $\tau = 1$.

3. Numerical results

For purposes of numerical comparison, we discuss results for eight algorithms: second-order Runge-Kutta (RK2) and fourth-order Runge-Kutta (RK4);[7] algorithms based on Hamilton's principle, approximating with quadratic, cubic and quartic functions (denoted by HAMQ, HAMC and HAMQuar, respectively); and second-, third- and fourth-order symplectic integration algorithms developed by Candy and Rozmus[8] (denoted CR2, CR3 and CR4, respectively). The Candy-Rozmus algorithms of second order is not unique, whereas their fourth-order algorithm is unique. The algorithms that we denote by CR2, CR3 and CR4 are those specified in Table II of Ref. 8. All computations were performed on a NeXTstation, using Mathematica© and C code.

We consider the motion of a particle with unit mass in a one-dimensional, time-independent potential, so that the Lagrangian is

$$L(q,\dot{q}) = \frac{\dot{q}^2}{2} - V(q)\,. \tag{3.1}$$

The energy,

$$E(q,\dot{q}) = \frac{\dot{q}^2}{2} + V(q)\,, \tag{3.2}$$

is conserved by the exact differential equations. The examples are for the harmonic oscillator potential, the potential for a pendulum, and for a double-well potential:

$$V(q) = \begin{cases} \dfrac{q^2}{2}\,, & \text{(harmonic oscillator)} \\ -\cos(q)\,, & \text{(pendulum)} \\ \left(q^2 + \dfrac{1/16}{1/16+q^2} - 1\right). & \text{(double - well)} \end{cases} \tag{3.3}$$

The double-well potential, which was discussed by Friedman and Auerbach[9] and by Auerbach and Friedman,[10] is graphed in Fig. 1. The horizontal line in the graph corresponds to the energy for the initial conditions $(q_0,\dot{q}_0) = (1,-0.09446)$, which were used in the computations with the double-well potential.

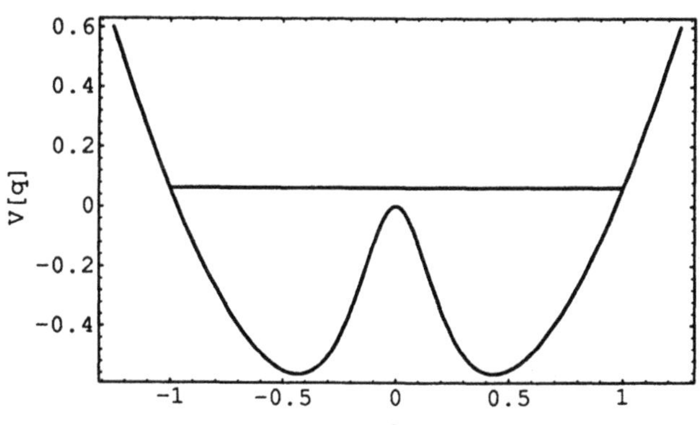

Figure 1: Double-well potential and energy corresponding to $(q_0,\dot{q}_0) = (1,-0.09446)$.

The timestep that was used for all of the computations that are displayed is $\Delta = 1/10$.

Qualitatively and semi-quantitatively the comparison among the various methods is the same for the nonlinear problems as for the harmonic oscillator. Thus, the major points of comparison among the methods can be exhibited with the harmonic oscillator, for which all of the algorithms can be solved analytically and completely understood. It has been shown that the second-, third- and fourth-order algorithms based on Hamilton's principle applied to the harmonic oscillator are in fact symplectic. However, when applied to the nonlinear problems, the algorithms are not symplectic.

We first consider the energy as a function of time for the harmonic oscillator as computed by the various algorithms at the discrete times $t_n = n\Delta$. The initial conditions are $(q_0, \dot{q}_0) = (1,0)$. In Fig. 2, the energy is shown for HAMQ, CR2 and RK2; in Fig. 3 the energy is shown for HAMQuar and CR4. The correct energy is $1/2$.

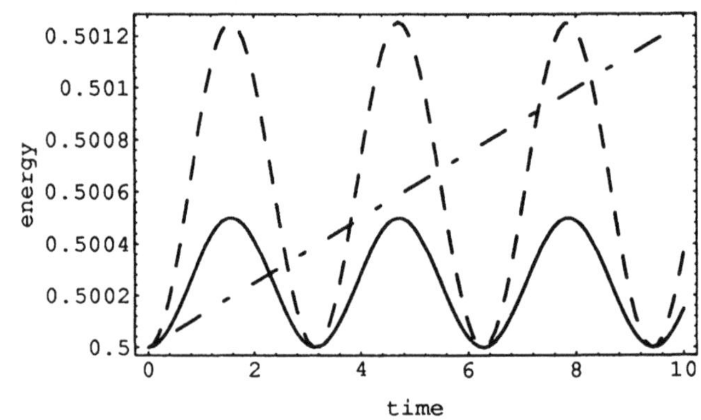

Figure 2: Energy for harmonic oscillator. HAMQ (solid), CR2 (dash-dash), RK2 (dash-dot). $(q_0, \dot{q}_0) = (1,0)$, $\Delta = 0.1$.

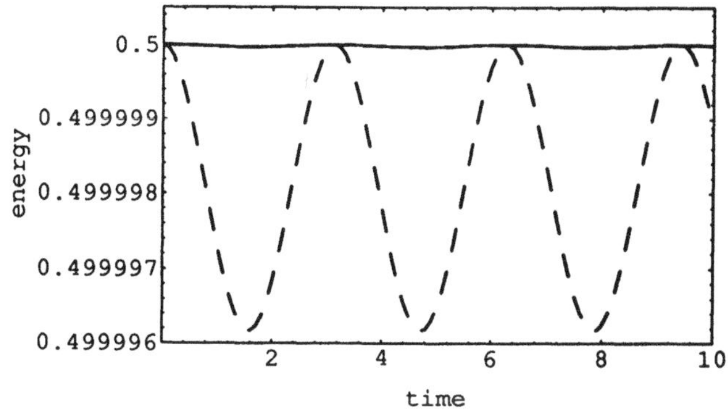

Figure 3: Energy for harmonic oscillator. HAMQuar (solid), CR4 (dash). $(q_0, \dot{q}_0) = (1,0)$, $\Delta = 0.1$.

We see in Figs. 2-3 that, except for RK2, the energy oscillates in time. The RK2 energy increases monotonically, as can be demonstrated analytically. As is evident in Figs. 2 and 3, for $\Delta = 1/10$, the algorithms based on Hamilton's principle approximate the true energy more accurately than the Candy-Rozmus algorithms of corresponding order. However, CR3 performs more accurately than HAMC for that value of Δ; the relative performance for other values of Δ is examined below in connection with Fig. 5.

For the harmonic oscillator there is a very important qualitative difference between the behavior of the Runge-Kutta algorithms on the one hand and the algorithms based on Hamilton's principle and the Candy-Rozmus algorithms on the other. RK2 and RK4 applied to the harmonic oscillator are *never periodic;* in fact, the ratio of the energy at one time to the ratio one timestep earlier is a constant that depends only on Δ. This means that the topology of the phase space with Runge-Kutta is *always incorrect.* In contrast, for the algorithms based on Hamilton's principle and the Candy-Rozmus algorithms, the solutions are *periodic for sufficiently small values of the timestep.* As a result, with those algorithms, *for sufficiently small values of the timestep,* the topology of the phase space is *always correct.*

The algorithms based on Hamilton's principle differ from the Candy-Rozmus algorithms with respect to the *maximum* value of the timestep for which the harmonic oscillator solution is periodic (so that the topology of the phase space is correct). The maximum value of Δ for an algorithm based on Hamilton's principle is *larger* than the maximum value for the Candy-Rozmus algorithm of the same order. For example, for HAMQ the maximum value of Δ is $\sqrt{10}$, whereas the maximum value for CR2 is 2. (Recall that the period of our harmonic oscillator is $2\pi = 6.28\ldots$.) The phase-space topology is correct over a considerably larger range of stepsize with HAMQ than with CR2.

In order to explore in more detail the comparison between the algorithms based on Hamilton's principle and the Candy-Rozmus algorithms, it is instructive to examine the ratio of the amplitudes of oscillation of the energy for the algorithms of corresponding order. The amplitude R for each algorithm can be calculated analytically and their ratios are functions of Δ. In Figs. 4 and 5 we plot the ratios $(R_{\mathrm{HAMQ}})/(R_{\mathrm{CR2}})$ and $(R_{\mathrm{HAMC}})/(R_{\mathrm{CR3}})$ as functions of stepsize.

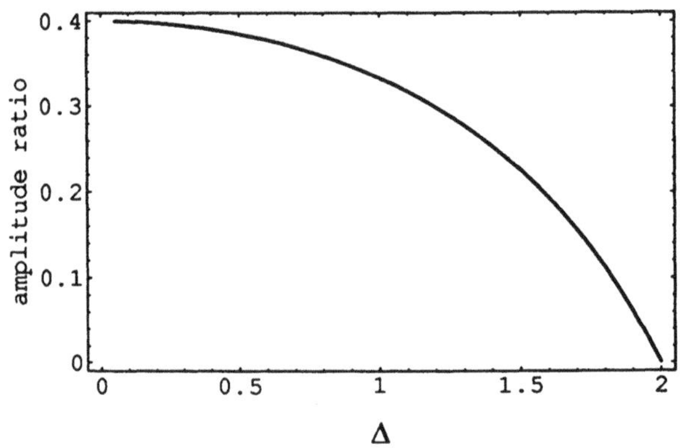

Figure 4: Harmonic oscillator. $(R_{\mathrm{HAMQ}})/(R_{\mathrm{CR2}})$ versus stepsize.

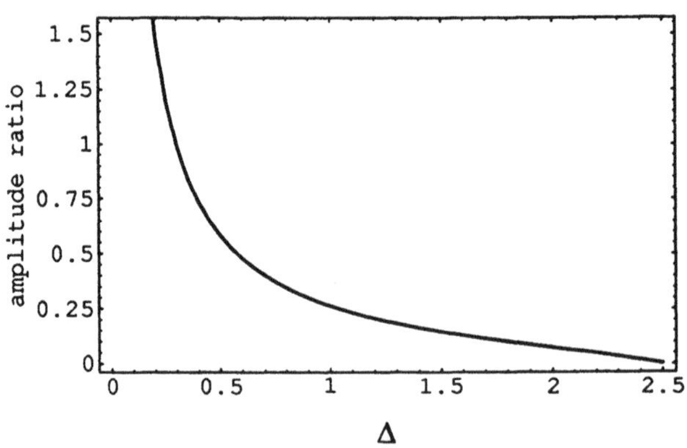

Figure 5: Harmonic oscillator. $(R_{\mathrm{HAMC}})/(R_{\mathrm{CR3}})$ versus stepsize.

In Fig. 5 we see that $(R_{\mathrm{HAMC}})/(R_{\mathrm{CR3}})$ diverges as $\Delta \to 0$. Nevertheless, the amplitude R for each method approaches 0 rapidly, so that the variation of the energy actually becomes very small as $\Delta \to 0$. For $\Delta \geq 0.3$, the ratio $(R_{\mathrm{HAMC}})/(R_{\mathrm{CR3}})$ is less than unity, so that $R_{\mathrm{HAMC}} < R_{\mathrm{CR3}}$ for that range of step sizes. The third-order algorithm based on Hamilton's principle performs *better* in regard to variation of the energy than CR3 for *large* stepsizes. The second-order algorithm based on Hamilton's principle performs better in regard to variation of the energy than CR2 for all stepsizes.

Acknowledgments. We thank Simon G. Shepherd for discussions. This work was supported by the United States Air Force Office of Scientific Research.

References

1. R. K. Colegrave and F. M. F. El-Sabaa, *Arch. Mech.* **38** (1986) 403.
2. H. Ralph Lewis, *Methods in Computational Physics*, **9** (1970) 307-338.
3. H. R. Lewis, D. C. Barnes, and K. J. Melendez, *J. Comput. Phys.* **69** (1987) 267.
4. B. B. Godfrey and L. E. Thode, "Galerkin Difference Schemes for Plasma Simulation Codes," *Proceedings of the Seventh Conference on Numerical Simulation of Plasmas (Courant Institute, New York University, June 2-4, 1975)*, pp. 87-91.
5. B. B. Godfrey, "Application of Galerkin's Method to Particle-in-Cell Plasma Simulation," *Proceedings of the Eighth Conference on Numerical Simulation of Plasmas (Del Monte Hyatt House, Monterey, California, June 28-30, 1978)*, paper PE-3.
6. J. W. Eastwood,*Computer Physics Communications* **64** (1991) 252.
7. See, for example, G. H. Golub and J. M. Ortega, *Scientific Computing and Differential Equations* (Academic Press, Boston, 1992).
8. J. Candy and W. Rozmus, *J. Comput. Phys.* **92** (1991) 230.
9. A. Friedman and S. P. Auerbach, *J. Comput. Phys.* **93** (1991) 171.
10. S. P. Auerbach and A. Friedman, *J. Comput. Phys.* **93** (1991) 189.

Generalization of Giacomini's Results Concerning Invariants for One-Dimensional Time-Dependent Potentials

S. BOUQUET
Commissariat a l'Energie Atomique (CEA)
Centre d'Etudes de Vaujours-Moronvilliers
Courtry, France

and

H. R. LEWIS
Dartmouth College
Department of Physics and Astronomy
Hanover, New Hampshire 03755-3528, USA

Abstract

Giacomini has developed a theoretical framework that defines exact invariants for certain one-dimensional time-dependent potentials and he solved the relevant equations to determine a class of such potentials. We present additional solutions of the equations, thereby expanding the class of potentials of the Giacomini type. In addition, we discuss possibilities of generalizing Giacomini's method by using the formulation of Lewis, Leach, Bouquet, and Feix.

1. Introduction

Given a non-autonomous one-degree-of-freedom Hamiltonian $H(q,p,t)$ associated with a one-dimensional potential $V(q,t)$,

$$H(q,p,t) = \frac{1}{2}p^2 + V(q,t) , \tag{1}$$

we seek an exact invariant $I(q,p,t)$, which is defined by

$$\frac{dI}{dt} = I_t(q,p,t) + [I,H]$$

$$= I_t(q,p,t) + I_q H_p - I_p H_q \ . \qquad (2)$$

Giacomini has developed a theoretical framework that defines exact invariants for certain one-dimensional time-dependent potentials and he solved the relevant equations to determine a class of such potentials.[1] Giacomini's result begins with the ansatz that the invariant is expressible as a function of two arguments, $V(q,t)$ and p:

$$I(q,p,t) \to I[V(q,t),p] = I(v,p) \qquad \{v = V(q,t)\} \qquad (3)$$

Then (2) can be written as

$$\left[\frac{V_t(q,t)}{V_q(q,t)} + p\right] I_v[V(q,t),p] - I_p[V(q,t),p] = 0 \ . \qquad (4)$$

In order for (4) to have a solution, $V(q,t)$ must satisfy

$$V_t(q,t) = f[V] V_q(q,t) \ , \qquad (5)$$

where f is an arbitrary function. The implicit solution of (5) is

$$\Phi[V] - t - \frac{q}{f[V]} = 0 \ , \qquad (6)$$

where also Φ is an arbitrary function. As a result, the equation for $I(v,p)$ is

$$[f(v) + p] I_v(v,p) - I_p(v,p) = 0 \ , \qquad (7)$$

whose characteristic equation is

$$\frac{dv}{dp} + f(v) + p = 0 \ . \qquad (8)$$

Giacomini analytically treated the case

$$f(v) = a_1 v^2 + a_2 v + a_3 \ , \qquad (9)$$

where (a_1, a_2, a_3) are arbitrary constants, in terms of Airy functions.

2. New Solutions

We can obtain additional solutions of Giacomini's relations (7) and (8) by defining a function $X(v)$ as

$$X(v) = \frac{1}{p + f(v)}, \tag{10}$$

where p is determined as a function of v by (8). By differentiating (10) with respect to v and using (8) we find that $X(v)$ satisfies the equation

$$\frac{dX}{dv} = X^3 - \frac{df}{dv} X^2 . \tag{11}$$

Eq. (11) can be transformed into a separable form by means of a transformation from $X(v)$ to $U(v)$ defined by

$$U(v) = \frac{[X(v) - X_2(v)]}{[X_1(v)]^{1/3}}, \tag{12}$$

where

$$X_1(v) = -\frac{1}{3}\frac{d^2 f}{dv^2} - \frac{2}{27}\left(\frac{df}{dv}\right)^3, \qquad X_2(v) = \frac{1}{3}\frac{df}{dv} . \tag{13}$$

Eq. (11) then has the form

$$\frac{dU}{dv} = X_1^{2/3}\left[1 - \left(3X_1^{-2/3}X_2^2 + \frac{1}{3}X_1^{-5/3}\frac{dX_1}{dv}\right)U + U^3\right]. \tag{14}$$

In order that (14) be separable, the coefficient of U must be a constant, which we denote by K. This gives the following condition on $f(v)$:

$$\left[9\frac{d^3 f}{dv^3} + 15\frac{d^2 f}{dv^2}\left(\frac{df}{dv}\right)^2 + 2\left(\frac{df}{dv}\right)^5\right]^3 = \left(\frac{K}{3}\right)^3\left[9\frac{d^2 f}{dv^2} + 2\left(\frac{df}{dv}\right)^3\right]^5 \tag{15}$$

We have found three values of K for which (15) can be solved.

2.1 *Case 1*: $K = \frac{3}{2^{2/3}}$

This choice of K leads to

$$f(v) = a_2 v + a_3 , \tag{16}$$

which is a case analyzed by Giacomini.[1]

2.2 *Case 2*: $K = 3\frac{A^2 - 6}{\left[2A\left(A^2 - 9\right)\right]^{2/3}} ,$ A = constant $\neq \pm 3$

In this case, we find

$$f(v) = A\sqrt{v} . \tag{17}$$

The invariant is given by

$$I\left[V(q,t),p\right] = \ln\left(2V + p^2 + A\sqrt{V}p\right) - A\int^{\sqrt{V}/p} \frac{dy}{\left(2y^2 + Ay + 1\right)} \tag{18}$$

and, from (6), the potential is determined from

$$\Phi[V] - t - \frac{q}{A\sqrt{V}} = 0 . \tag{19}$$

2.3 *Case 3*: $K = 0$

With this choice, (15) reduces to

$$9\frac{d^3 f}{dv^3} + 15\frac{d^2 f}{dv^2}\left(\frac{df}{dv}\right)^2 + 2\left(\frac{df}{dv}\right)^5 = 0 . \tag{20}$$

This equation can be integrated completely in terms of 3 arbitrary constants. Once that is done, the allowed potential/invariant pairs can be specified implicitly.

3. A Potential Avenue for Extending Giacomini's Result

What underlies Giacomini's result is the ansatz (3). So far, attempts to find potential/invariant pairs by generalizing that ansatz have failed. Perhaps the implications of a generalized ansatz could be analyzed by using recent results of Lewis, Leach, Bouquet and Feix.[2] In this section we put Giacomini's result in the context of the results of Lewis, Leach, Bouquet and Feix.

3.1 *The Result of Lewis, Leach, Bouquet and Feix*

All Hamiltonians $H(q,p,t)$ for which a specified function $\tilde{P}(q,p,t)$ is an invariant are given by

$$H(q,p,t) = K(t) - F_t[\tilde{P}(q,p,t),t] - \int^q G_t[q',\tilde{P}(q,p,t),t]\,dq' \ , \tag{21}$$

where $G(q,P,t)$ is the inverse of $\tilde{P}(q,p,t)$,

$$p = G(q,P,t) \quad \leftrightarrow \quad P = \tilde{P}(q,p,t) \ , \tag{22}$$

and where $F(P,t)$ and $K(t)$ are arbitrary functions. For a given $H(q,p,t)$, the function $F(P,t)$ can be determined and a canonically conjugate second invariant can be constructed.

Eq. (21) is related to the Hamilton-Jacobi equation through a generating function $S(q,P,t)$. The generating function is defined by

$$S(q,P,t) = \int^q G(q',P,t)\,dq' + F(P,t) - \int^t K(t')\,dt' \ , \tag{23}$$

so that $G(q,P,t)$ can be expressed as

$$G(q,P,t) = S_q(q,P,t) \ . \tag{24}$$

Then (21) can be written in terms of (q,P,t) as

$$H\left[q,S_q(q,P,t),t\right] + S_t(q,P,t) = 0 \ , \tag{25}$$

which is the Hamilton-Jacobi equation.

3.2 *Giacomini's Result in Terms of* (q,P,t)

The ansatz in terms of (q,P,t) that is analogous to (3) is to demand that $G(q,P,t)$ be expressible as a function of the two arguments $V(q,t)$ and P:

$$G(q,P,t) = \overline{G}[V(q,t),P] \, . \tag{26}$$

This ansatz also leads to Giacomini's result, now expressed in terms of (q,P,t). A natural generalization of (26) would be to allow the two arguments to be other functions of (q,P,t). That is,

$$G(q,P,t) = \overline{G}(x,y) = \overline{G}\left[f(q,P,t),h(q,P,t)\right] , \tag{27}$$

where

$$x = f(q,P,t) \, , \quad y = h(q,P,t) \, . \tag{28}$$

Perhaps new potential/invariant pairs can be found by means of (27) with some choices of $f(q,P,t)$ and $h(q,P,t)$.

4. References

1. H. Giacomini, *J. Phys. AL, Math. Gen.* **23** (1990) 865.

2. H. R. Lewis, P. G. L. Leach, S. Bouquet and M. R. Feix, *J. Math. Phys.* **33** (1992) 591.

LIST OF PARTICIPANTS

Last Name	First Name	Country	Affiliation	EMail
Ablowitz	Mark	United States	Univ. of Colorado	markjab@newton.colorado.edu
Baker	Sarah	England	Univ. of Leeds	amt5sb@leeds.ac.uk
Baryakhtar	Igor	Ukraine	National Academy of Sciences, Ukraine	igbar%azbuka.kharkov.uaorelsy.ussr.eu.net
Berman	Gennady	United States	Los Alamos	gpb@goshawk.lanl.gov
Beyer	William	United States	Los Alamos	beyer@lanl.gov
Biondini	Gino	Italy	Univ. of Perugia	biondini@vaxpg.pg.infn.it
Bishop	Alan	United States	Los Alamos	arb@dove.lanl.gov
Blumenfeld	Raphael	United States	Los Alamos	rafi@goshawk.lanl.gov
Bourtsev	Serguei	United States	Clarkson Univ.	burtsevs@sun.mcs.clarkson.edu
Brand	Helmut	Germany	Univ. of Bayreuth	hrb@cnls.lanl.gov or btp511@btp3x6.phy.uni-bayreuth.de
Bubelev	Engeliy	Russia	Joint Inst. for Nuclear Research	bubelev@main1.jinr.dubna.su
Cai	David	United States	Los Alamos	cai@viking.lanl.gov
Calogero	Francesco	Italy	Univ. Roma	calogero@sci.uniroma1.it
Camassa	Roberto	United States	Los Alamos	roberto@goshawk.lanl.gov
Carroll	Robert	United States	Univ. of Illinois	rcarroll@uiuc.edu
Chakravarty	Sarbarish	United States	Univ. of Colorado	chuck@newton.colorado.edu
Channell	Paul	United States	Los Alamos	pchannell@lanl.gov
Cohen	Donald	United States	Los Alamos	cohen@goshawk.lanl.gov
Cruz-Pacheco	Gustavo	United States	Univ. of Arizona	gustavo@goshawk.lanl.gov gustavo@math.arizona.edu or
De Lillo	Silvana	Italy	Univ. Di Perugia	delillo@perugia.infn.it
Deissler	Robert	United States	NASA Lewis Research Center	fsrjd@icomp01.lerc.nasa.gov
Dix	Daniel	United States	Univ. of South Carolina	dix@math.scarolina.edu
Doering	Charles	United States	Los Alamos	doering@cnls.lanl.gov
Donets	Evgueni	Russia	Joint Inst. for Nuclear Research	donets@lhe26.jinr.dubna.su
Doolen	Gary	United States	Los Alamos	gdd@lanl.gov
Fordy	Allan	England	Univ. of Leeds	allan@amsta.leeds.ac.uk
Gabitov	Ildar	Russia	Landau Institute	ildar@cnls.lanl.gov
Gerstl	Sig	United States	Los Alamos	sig@lanl.gov
Gilson	Claire	United Kingdom	Glasgow Univ.	
Goldstein	Raymond	United States	Princeton Univ.	gold@davinci.princeton.edu
Gomez	Camilo	United States	Los Alamos	cgomez@lanl.gov
Habib	Salman	United State	Los Alamos	habib@eagle.lanl.gov
Hanagandi	Vijaykumar	United States	Los Alamos	vmh@killdeer.lanl.gov
Herbst	Ben	United States	Univ. of Colorado	herbst@newton.colorado.edu
Heredero	Rafael	Spain	Univ. Complutense de Madrid	rhh@fis.ucm.es
Herman	Russell	United States	Univ. of North Carolina	herman@seq.cms.uncwil.edu
Hietarinta	Jarmo	Finland	Univ. of Turku	hietarin@utu.fi
Holm	Darryl	United States	Los Alamos	dholm@lanl.gov
Hyman	Mac	United States	Los Alamos	mac@cnls.lanl.gov
Iizuka	Takeshi	Japan	Ehime Univ.	iizuka@sci.ehime-u.ac.jp
Kajiwara	Kenji	Japan	Doshisha Univ.	kkajiwar@dualc.doshisha.ac.jp
Kimura	Yoshifumi	United States	Univ. of Colorado	kimura@oak.mmm.ucar.edu
Kodama	Yuji	United States	Ohio State Univ.	kodama@math.ohio-state.edu
Kovalev	Alexander	Ukraine	Academy of Sciences of Ukraine	phys29@ilt.kharkov.ua
Kupershmidt	Boris	United States	Univ. of Tennesse	bkupersh@sparc2000.utsi.edu
Lambert	Franklin	Belgium	Brussels' Free Univ.	
Levermore	Charles	United States	Inst. for Advanced Study	lvrmr@cnls.lanl.gov lvrmr@math.arizona.edu
Lewis	H. Ralph	United States	Dartmouth College	ralph.lewis@dartmouth.edu
Lu	Liang	United States	Univ. of Maryland	
Makhankov	Aleksei	Russia	Joint Inst. for Nuclear Research	mvg@lcta16.jinr.dubna.su
Makhankov	Vladimir	United States	Los Alamos	mvg@cnls.lanl.gov
Manas	Manuel	Spain	Univ. Complutense	fite108@sis.ucm.es
Margolin	Len	United States	Los Alamos	len@lanl.gov
Mel'nikov	Viktor	Russia	Joint Inst. for Nuclear Research	melnikov@theor.jinrc.dubna.su
Miller	Peter	United States	Univ. of Arizona	millerpd@math.arizona.edu
Mineev	Mark	United States	Los Alamos	mineev@goshawk.lanl.gov
Mokhov	Oleg	Russia	Steklov Institute	mokhov@class.mian.su
Nakao	Takenobu	Japan	Univ. of Tokyo	nakao@monet.phys.s.u-tokyo.ac.jp
Nimmo	Jonathan	Scotland	Univ. of Glasgow	john@maths.gla.ac.uk
Nucci	Maria Clara	Italy	Univ. Di Perugia	clara@ipguniv.unipg.it
Patil	Rajendra	United States	Los Alamos	rbp@demos.lanl.gov

Last Name	First Name	Country	Affiliation	EMail
Rakauskas	Rimantas	Lithuania	Vilnius Univ.	rimantas.rakauskas@ff.vu.lt
Robinson	James	United Kingdom	Cambridge Univ.	jcr14@amtp.cam.ac.uk
Rodin	Yury	United States	Wayne State Univ.	rodin@math.wayne.edu
Rosenau	Philip	Israel	Tel-Aviv University	philip@cnls.lanl.gov, rosenau@math.tau.ac.il
Sabatier	Pierre	France	Univ. Montpellier II	sabatier@lpm.univ-montp2.fr or rcp@lpm.univ-montp2.fr
Sanyuk	Valerii	Russia	People's Friendship Univ.	sanyuk@udn.msk.su
Saveliev	Mikhail	Russia	Inst. for High Energy Physics	saveliev@m10.ihep.su
Saxena	Avadh	United States	Los Alamos	abs@anyon.lanl.gov
Schober	Constance	United States	Univ. of Colorado, Boulder	schober@newton.colorado.edu
Sobehart	Jorge	United States	Los Alamos	sobehart@goshawk.lanl.gov
Stark	Donald	United States	Los Alamos	dstark@cnls.lanl.gov
Sung	Li-Yeng	United States	Univ. of South Carolina	sung@math.scarolina.edu
Tang	Xianzhu	United States	Columbia Univ.	tang@physics.wm.edu
Taranenko	Nina	United States	Los Alamos	nlt@lanl.gov
Taranenko	Yuri	United States	Los Alamos	yuri@sstcx1.lanl.gov
Yajima	Tetsu	Japan	Univ. of Tokyo	yajimat@mmm.t.u-tokyo.ac.jp
Yukalov	Vyacheslav	Canada	Queen's Univ.	yukalovv@jeff-lab.queensu.ca
Zagrodzinski	Jerzy	Poland	Polish Academy of Sciences	zagro@beta1.ifpan.edu.pl
Zhang	Jin	United States	Caltech	jzhang@cco.caltech.edu or jzhang@cnls.lanl.gov

www.ingramcontent.com/pod-product-compliance
Ingram Content Group UK Ltd.
Pitfield, Milton Keynes, MK11 3LW, UK
UKHW020401150926
8624IPUK00004B/287

9 789810 222192